ESO ASTROPHYSICS SYMPOSIA
European Southern Observatory

Series Editor: Bruno Leibundgut

N.C. Santos L. Pasquini A.C.M. Correia
M. Romaniello (Eds.)

Precision Spectroscopy in Astrophysics

Proceedings of the ESO/Lisbon/Aveiro Conference held in Aveiro, Portugal, 11-15 September 2006

Springer

Volume Editors

Nuno C. Santos
Centro de Astrofisica da
Univ. do Porto
Rua das Estrelas
4150-762 Porto
Portugal

Alexandre C. M. Correia
Universidade Aveiro
Depto. Fisica
Campus Universitário de Santiago
3810 Aveiro
Portugal

Luca Pasquini
European Southern Observatory
Karl-Schwarschild-Str. 2
85748 Garching
Germany

Martino Romaniello
European Southern Observatory
Karl-Schwarschild-Str. 2
85748 Garching
Germany

Series Editor

Bruno Leibundgut
European Southern Observatory
Karl-Schwarzschild-Str. 2
85748 Garching
Germany

ISBN 978-3-642-44283-4   Springer Berlin Heidelberg New York

Springer is a part of Springer Science+Business Media
springer.com

Softcover re-print of the Hardcover 1st edition 2008

Cover design: WMXDesign, Heidelberg
Typesetting: by the authors
Production: Integra Software Services Pvt. Ltd., Puducherry, India

Printed on acid-free paper   55/3180/Integra   5 4 3 2 1 0

Preface

In the last decade we have witnessed impressive advancements in the accuracy of Doppler shift measurements in astronomy, as well as of high-precision spectroscopy in general. Even if the main (and most recognized) driver for this development is the search for exo-planets, extremely interesting applications include the analysis of QSO absorption lines to determine the variability of physical constants and the analysis of the isotopic ratio of interesting species, such as ^{6}Li/^{7}Li in metal poor stars or D/H in QSO absorption lines. Another field in strong expansion is the determination of stellar oscillations through radial velocity measurements, a technique that is providing interesting results, sometimes in apparent contradiction with photometric determinations.

The use of high precision/resolution spectroscopy is intimately connected to the ability to collect a large number of photons. Indeed, the random measurement uncertainty depends on the inverse of the signal-to-noise ratio. Therefore, high accuracy usually requires a large photon collecting capability. Not only do the scientific domains using this technique benefit tremendously from the use of 8-meter class telescopes, but they will also fully exploit the tremendous gain provided by future Extremely Large Telescopes (ELTs), as clearly shown by the preliminary studies of high resolution spectrographs for ELTs. And even if most applications so far have been at optical wavelengths, IR high-resolution spectroscopy should soon approach the same accuracy regime.

With this motivation in mind, we proposed to gather together scientists to discuss topics related to various aspects of high precision spectroscopy.

In a collaboration between ESO, the Center for Astronomy and Astrophysics of the University of Lisbon (CAAUL), and the University of Aveiro, the conference "Precision Spectroscopy in Astrophysics" was thus organized. During the week from 11 to 15 September 2006, about 100 scientists gathered in the pleasant town of Aveiro, near the northern Portuguese Atlantic coast. Between excellent talks and posters, Portuguese cuisine, and some boat-trips and barbecues, this conference gave the opportunity to discuss the different topics mentioned above in a relaxed but fruitful atmosphere.

For this great success we would like to deeply thank all the participants who made it possible. This conference was done for you and by you. Some persons and institutions were, however, of fundamental importance in organizing the process. First of all, the SOC members, whose suggestions produced the final conference program. The LOC made a wonderful work, and we would like to thank in particular Susana Fernandes, Eugenia Carvalho, and Britt Sjöberg, for their invaluable dedication and help, also in the most complicated moments.

Finally, we would like to thank the three institutions that made this event possible, ESO, the University of Lisbon, and in particular the University of Aveiro for having provided all the necessary infrastructures for the conference venue. These proceedings could not see the light without the patient and careful work of Pam Bristow.

Garching, Lisbon, Aveiro
July 2007

Nuno C. Santos
Luca Pasquini
Alexandre C.M. Correia
Martino Romaniello

Local Organizing Committee:

Nuno C. Santos (co-chair, CAAUL), Alexandre C.M. Correia (co-chair, University of Aveiro), Xavier Bonfils (CAAUL), Sergio Sousa (CAAUL/CAUP), Susana Fernandes (University of Aveiro), Eugénia Carvalho (CAAUL), Britt Sjöberg (ESO)

Scientific Organizing Committee:

Beatriz Barbuy (Brazil), Jacqueline Bergeron (France), Dainis Dravins (Sweden), Artie Hatzes (Germany), Garik Israelian (Spain), David Lambert (USA), Michel Mayor (Switzerland), Paolo Molaro (Italy), Mario J. Monteiro (Portugal), Luca Pasquini (ESO, co-chair), Max Pettini (UK), Martino Romaniello (ESO, co-chair), Nuno C. Santos (Portugal, co-chair)

Contents

Part V Asteroseismology/Oscillations

Part VI Planets

Part VIII Posters

The conference participants

List of Participants

Aerts, Conny
Katholieke Universiteit Leuven
conny@ster.kuleuven.be

Aldenius, Maria
Lund Observatory
maria@astro.lu.se

Ammler, Matthias
AIU Jena
ammler@astro.uni-jena.de

Aret, Anna
Tartu Observatory
aret@aai.ee

Barbuy, Beatriz
University of Sao Paulo
barbuy@astro.iag.usp.br

Benz, Willy
University of Bern
wbenz@space.unibe.ch

Biazzo, Katia
ESO Garching; INAF - Catania
kbiazzo@oact.inaf.it,
kbiazzo@eso.org

Bigot, Lionel
Observatoire de la Côte d'Azur
lbigot@obs-nice.fr

Bonfils, Xavier
CAAUL
xavier.bonfils@oal.ul.pt

Brandão, Isa
FCUP
isabrandao@tugamail.com

Catanzaro, Giovanni
INAF - Catania Observatory
gca@oact.inaf.it

Chand, Hum
IUCAA Pune
hcverma@iucaa.ernet.in

Cochran, Anita
University of Texas at Austin
anita@barolo.as.utexas.edu

Cochran, William
McDonald Observatory
wdc@astro.as.utexas.edu

Correia, Alexandre C.M.
University of Aveiro
acorreia@fis.ua.pt

Couetdic, Jocelyn
IMCCE Paris
couetdic@imcce.fr

Courtin, Régis
LESIA / CNRS - Observatoire de Paris
Regis.Courtin@obspm.fr

D'Odorico, Sandro
ESO Garching
sdodoric@eso.org

Da Silva, Licio
Observatorio Nacional - Brazil
licio@on.br

Dessauges-Zavadsky, Miroslava
Geneva Observatory
miroslava.dessauges@obs.unige.ch

Dravins, Dainis
Lund Observatory
dainis@astro.lu.se

Ecuvillon, Alexandra
Instituto de Astrofisica de Canarias
aecuvill@iac.es

Ellison, Sara
University of Victoria
sarae@uvic.ca

Fabbian, Damian
Mt. Stromlo Observatory
damian@mso.anu.edu.au

Fechner, Cora
Hamburger Sternwarte
cfechner@hs.uni-hamburg.de

Ferlet, Roger
Institut d'Astrophysique de Paris/CNRS/UPMC
ferlet@iap.fr

Ferreira-Rodrigues, Ana Mónica
Universidade Rio de Janeiro
anamonica.rodrigues@gmail.com

Figueira, Pedro
Observatoire de Genève
pedro.figueira@obs.unige.ch

Foellmi, Cédric
ESO + LAOG
cfoellmi@eso.org

Forveille, Thierry
Observatoire de Genève
Thierry.Forveille@obs.ujf-grenoble.fr

Fox, Andrew
Institut d'Astrophysique de Paris
fox@iap.fr

Funayama, Hitoshi
Kobe University
funayama@kobe-u.ac.jp

Furesz, Gabor
CfA Cambridge
gfuresz@cfa.harvard.edu

Garcia Perez, Ana
University of Hertfordshire
aegp@astro.uu.se,
agperez@herts.ac.uk

Grundahl, Frank
University of Aarhus
fgj@phys.au.dk

Hartmann, Michael
Thüringer Landessternwarte Tautenburg
michael@tls-tautenburg.de

Hatzes, Artie
Thüringer Landessternwarte Tautenburg
artie@tls-tautenburg.de

Hubrig, Swetlana
ESO Chile
shubrig@eso.org

James, Gaël
ESO Chile
gjames@eso.org

Jenkins, Edward
Princeton University
ebj@astro.princeton.edu

Johansson, Sveneric
Lund Observatory
Johansson@astro.lu.se

Jones, Hugh
University of Hertfordshire
hraj@star.herts.ac.uk

Jurkovic, Monika
University of Szeged
mojur@titan.physx.u-szeged.hu

Käufl, Hans Ulrich
ESO Garching
hukaufl@eso.org

Kanekar, Nissim
NRAO Socorro
nkanekar@aoc.nrao.edu

Kerber, Florian
ESO Garching
fkerber@eso.org

Letokhov, Vladilen
Inst. Spectroscopy Moscow / Lund University
letokhov@isan.troitsk.ru

Levshakov, Sergei
Physico-Technical Institute, St. Petersburg
lev@astro.ioffe.rssi.ru

Lopez, Sebastian
U. de Chile
slopez@das.uchile.cl

Lovis, Christophe
Observatoire de Genève
christophe.lovis@obs.unige.ch

Ludwig, Hans
Observatoire de Paris-Meudon
Hans.Ludwig@obspm.fr

Margheim, Steven
Gemini Observatory
smargheim@gemini.edu

Martinez-Fiorenzano, Aldo
INAF - Fundacion Galileo Galilei
fiorenzano@tng.iac.es

Martins, Carlos
CFP (U. Porto) & DAMTP (U. Cambridge)
c.j.a.p.martins@damtp.cam.ac.uk,
cmartins@fc.up.pt

Mayor, Michel
Observatoire de Genève
michel.mayor@obs.unige.ch

Melo, Claudio
ESO Chile
cmelo@eso.org

Mkrtichian, David
Sejong University
davidm@sejong.ac.kr,
davidmkrt@gmail.com

Monteiro, Mario
CAUP
mjm@astro.up.pt

Moreira Morais, Maria Helena
Universidade de Coimbra
hmorais@mat.uc.pt

Morel, Thierry
Katholieke Universiteit Leuven
thierry@ster.kuleuven.be

Muirhead, Philip
Cornell University
muirhead@astro.cornell.edu

Murphy, Michael
IoA Cambridge
mim@ast.cam.ac.uk

Nguyen, Duy
University of Toronto
nguyen@astro.utoronto.ca

Niedzielski, Andrzej
Nicolaus Copernicus University
aniedzi@astri.uni.torun.pl

Nieva, Maria Fernanda
Dr. Remeis Sternwarte Bamberg
nieva@sternwarte.uni-erlangen.de

Nissen, Poul Erik
University of Aarhus
pen@phys.au.dk

Pasquini, Luca
ESO Garching
lpasquin@eso.org

Paulson, Diane
NASA GSFC
diane.b.paulson@gsfc.nasa.gov

Pepe, Francesco
Observatoire de Genève
Francesco.Pepe@obs.unige.ch

Pereira, Tiago
Mount Stromlo Observatory
tiago@mso.anu.edu.au

Petitjean, Patrick
Institut d'Astrophysique de Paris
petitjean@iap.fr

Piskunov, Nikolai
Uppsala University
piskunov@astro.uu.se

Prochaska, Jason
UCO/Lick Observatory
xavier@ucolick.org

Prochter, Gabriel
UCO/Lick Observatory
prochter@ucolick.org

Quirrenbach, Andreas
Landessternwarte Heidelberg
A.Quirrenbach@lsw.uni-heidelberg.de

Reiners, Ansgar
Hamburger Sternwarte
areiners@hs.uni-hamburg.de

Rodler, Florian
MPIA Heidelberg
rodler@mpia.de

Roederer, Ian
University of Texas at Austin
iur@astro.as.utexas.edu

Romaniello, Martino
ESO Garching
mromanie@eso.org

Santos, Nuno
CAAUL
nuno@oal.ul.pt

Setiawan, Johny
MPIA Heidelberg
setiawan@mpia.de

Sousa, Sérgio
CAUP & CAAUL
sousasag@astro.up.pt

Sziladi, Katalin
University of Szeged
szkati@titan.physx.u-szeged.hu

Thompson, Rodger
Steward Observatory
rthompson@as.arizona.edu

Toyota, Eri
Kobe University
toyota@kobe-u.ac.jp

Uttenthaler, Stefan
ESO Garching
suttenth@eso.org

Vinko, Jozsef
University of Szeged
vinko@physx.u-szeged.hu

Weise, Patrick
MPIA Heidelberg
weise@mpia.de

Zucker, Shay
Tel Aviv University
shayz@post.tau.ac.il

Zych, Berkeley
IoA Cambridge
bjz@ast.cam.ac.uk

Part I

Abundances and Isotopes

Lithium Isotopic Abundances in Stars

Poul Erik Nissen[1] and Martin Asplund[2]

[1] Department of Physics and Astronomy, University of Aarhus, DK-8000 Aarhus C, Denmark pen@phys.au.dk
[2] Research School of Astronomy and Astrophysics, Australian National University, Cotter Road, Weston ACT 2611, Australia martin@mso.anu.edu.au

Summary. The Li isotope ratio, ^{6}Li/^{7}Li, in stars can be determined from the isotopic shift in the Li I 670.8 nm resonance line. Because of the small effect this however requires truly precision spectroscopy: spectral resolving power $R \geq 10^5$ and $S/N \geq 500$. In this review we discuss the method and what one can learn from Li isotopic abundances in terms of Big-Bang nucleosynthesis, cosmic ray production of Li, stellar structure, and planet formation. Some instrumental problems and the need for new instrumentation are briefly discussed.

1 Determination of the Li Isotope Ratio

In determining Li isotopic abundances we have the advantage that wavelengths and gf-values of the components of the Li I 670.8 nm resonance line are known with superior accuracy [17]. The isotopic shift of the ^{6}Li doublet relative to the ^{7}Li doublet is +0.160 Å corresponding to 7.1 km s^{-1}. Hence, one might think that a measurement of the center-of-gravity wavelength of the Li line would be a simple and straightforward method to determine the ^{6}Li/^{7}Li ratio in stars. The accuracy is, however, limited by possible errors in the laboratory wavelengths of the reference lines needed to correct for the radial velocity shift of the star. Furthermore, differences in convective blueshifts of the Li line and the reference lines should be taken into account, and for the more metal-rich stars the Li line is affected by weak blends as discussed in Sect. 3. In practice, this method is therefore prone to systematic errors.

An alternative method to obtain ^{6}Li/^{7}Li consists of a model atmosphere synthesis of the observed profile of the Li line, since a mixture of ^{6}Li and ^{7}Li results in a wider profile than in the case of one isotope alone. It requires that the stellar line broadening function is determined from other spectral lines. In plane-parallel (1D) models, the broadening is due to a combination of rotation and macroturbulence, and the question arises if one can use the same macroturbulence parameter for the Li line and the reference lines. In 3D hydrodynamical models, line broadening due to atmospheric velocity fields is already accounted for and the only parameter to be determined from comparison lines is the projected stellar rotation velocity, $v_{\rm rot} \sin i$. Furthermore, line asymmetries due to convective motions are modelled in 3D calculations, whereas one has to assume a symmetric line broadening profile in the 1D synthesis.

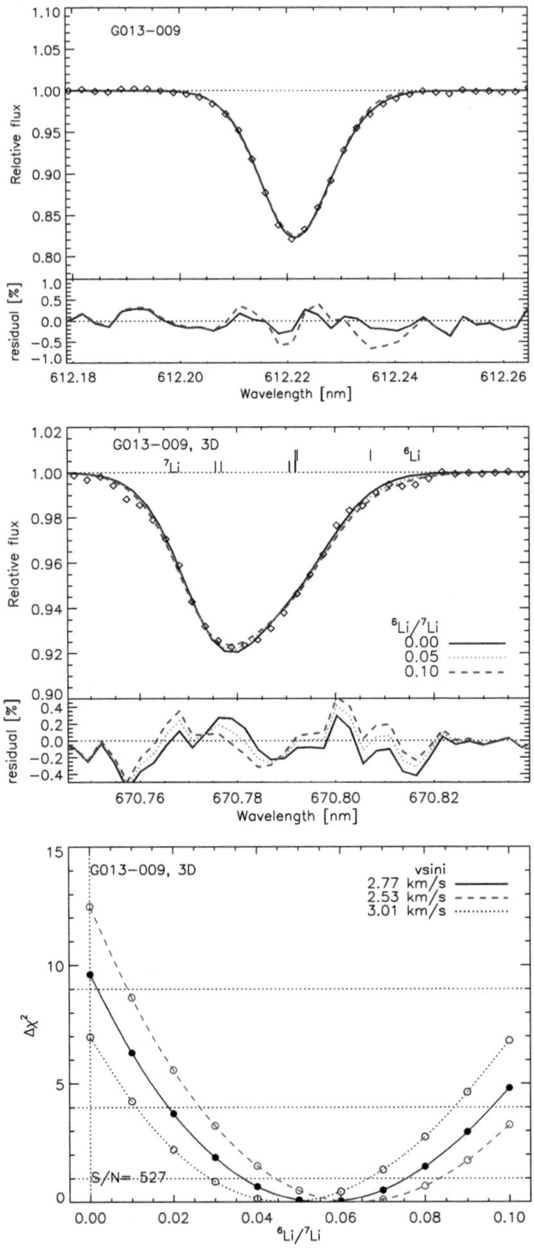

Fig. 1. *Top*: Observed Ca I 612.2 nm profile in G13-9 (*diamonds*) together with best-fit 1D (dashed line) and 3D (solid line) theoretical LTE profiles. *Middle*: Observed Li I 670.8 nm profile in G13-9 compared to 3D theoretical LTE profiles for three values of ^{6}Li/^{7}Li. *Bottom*: χ^2 variation of the 3D fit to the Li line of G13-9 with $v_{\rm rot}\sin i = 2.77 \pm 0.24\,{\rm km\,s^{-1}}$.

As an example of the profile method, we show in Fig. 1 the determination of ^{6}Li/^{7}Li in the halo turnoff star G13-9 ($T_{\rm eff}$ = 6300 K, log$g$ = 4.0, [Fe/H] = −2.3). The spectrum was obtained with VLT/UVES at a resolution of $R \sim 120\,000$ and a signal-to-noise of ~ 500. As seen from the top figure the 3D modelling provides a better fit to the Ca I comparison line than the 1D model. The middle figure shows the 3D fit to the Li line and at the bottom the corresponding χ^2 variation as a function of ^{6}Li/^{7}Li is shown. Note, that for each value of ^{6}Li/^{7}Li the other free parameters in the fit, i.e. the total Li abundance, the wavelength zero-point of the observed spectrum, and the continuum level, were allowed to vary to optimize the fit and thus minimize χ^2. Hence, the 1-, 2- and 3-sigma confidence limits of the determination correspond to $\Delta\chi^2 = 1$, 4 and 9, respectively. From Fig. 1, the Li isotopic ratio of G13-9 is determined to be ^{6}Li/^{7}Li = 0.056 ± 0.021. Using the 1D method with a Gaussian broadening we get ^{6}Li/^{7}Li = 0.048 ± 0.019. As discussed in [1] there is in fact no significant systematic difference of the ^{6}Li/^{7}Li ratio derived in 1D and 3D for a sample of 18 stars, showing that line asymmetries are not large enough to affect the determination significantly.

When estimating the error of ^{6}Li/^{7}Li from the χ^2-analysis we have assumed Poisson noise in the spectrum. Flat-fielding problems may, however, add to the error. CCD detectors used in echelle spectrographs suffer from fringing in the red part of the spectrum, and after flat-fielding a residual fringing pattern may remain as shown in Fig. 2. Across the Li line, the continuum varies with an amplitude of about 0.3 %. The pattern can be removed by dividing with the spectrum of a bright, early-type star, which has no lines in the Li region. If such rectification of the spectrum is not carried out, spurious detections of ^{6}Li may result.

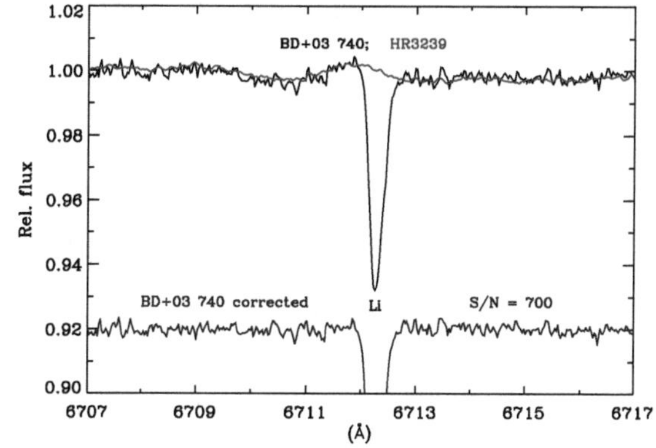

Fig. 2. Residual fringing problem in the UVES spectrum of the metal-poor turnoff star BD +03 740. After division with a 5-pixel smoothed spectrum of the B1.5V star HR 3239 observed on the same night as BD +03 740, a rectified spectrum (shifted to a continuum level of 0.92) is obtained. Note that the Li line in BD +03 740 is radial-velocity shifted by +200 km s^{-1}.

2 Lithium in Metal-poor Halo Stars

The first probable detection of ^{6}Li in a metal-poor halo star (HD 84937, ^{6}Li/^{7}Li $= 0.05 \pm 0.02$) was obtained by Smith et al. [16]. Later Hobbs & Thorburn [4] and Cayrel et al. [3] confirmed this result by independent observations and data analysis.

The most comprehensive study of ^{6}Li/^{7}Li in metal-poor stars was recently published by Asplund et al. [1]. For 9 of 24 stars, ^{6}Li appears to be detected at a significance level $\geq 2\sigma$. Fig. 3 shows the ^{6}Li abundances in these stars and HD 84937; for the remaining stars, 2σ upper limits are denoted with arrows. Furthermore, ^{7}Li abundances are shown with crosses.

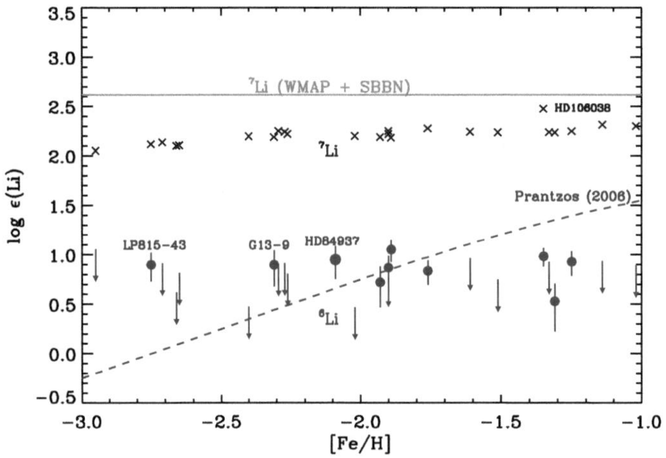

Fig. 3. ^{6}Li and ^{7}Li abundances in a sample of halo turnoff stars as derived by Asplund et al. [1]. The primordial ^{7}Li abundance predicted from the WMAP based baryon density in the Universe and standard big bang nucleosynthesis is indicated and the Galactic cosmic ray production of ^{6}Li according to Prantzos [12] is shown with a dashed line.

As seen from Fig. 3, the observed ^{7}Li "plateau" lies about 0.4-0.6 dex below the primordial Li abundance predicted from standard big bang nucleosynthesis using the baryon density in the Universe derived from WMAP data. According to standard stellar models ^{7}Li is not depleted in the turnoff stars observed, and non-standard models, e.g. models with rotational induced mixing [11], have difficulties in explaining a depletion of ~ 0.5 dex together with the small intrinsic dispersion (< 0.03 dex) in ^{7}Li among the plateau stars. Recently, it has been shown that models with turbulent diffusion may offer a solution [14], but the same models predict a ^{6}Li depletion of about 2 dex, which would imply an impossibly high ^{6}Li abundance in the gas that formed the halo stars. For the most metal-poor stars, the observed ^{6}Li abundance is already higher than predicted by models for the formation of Li by $\alpha + \alpha$ reactions and CNO spallation in Galactic cosmic rays [12].

The surprisingly high level of ^{6}Li abundances has led to suggestions of pre-Galactic formation of ^{6}Li in cosmic rays produced in connection with hierarchical structure formation [18] or by Pop. III stars [15]. Another interesting possibility is ^{6}Li formation by non-standard big-bang nucleosynthesis. The decay or annihilation of supersymmetric particles like the gravitino and neutralino may affect the nucleosynthesis. For the right combination of particle properties, a simultaneous production of ^{6}Li and a reduction of ^{7}Li appears achievable [7], thus explaining both the observed ^{6}Li and the low ^{7}Li.

As seen from Fig. 3, there is a hint of a decrease of ^{7}Li abundances for $[\mathrm{Fe/H}] < -2.5$. A similar trend and evidence for a cosmic scatter in the Li abundance of turnoff stars with metallicities below -3.0 is found by Nissen et al. [9] and Bonifacio et al. [2]. Piau et al. [10] suggest that this can be explained if part of the halo gas was processed through massive Pop. III stars, which destroy both Li isotopes completely. If the first low-mass stars are preferentially formed in the ejections of Pop. III supernovae they would then tend to have low and non-uniform ^{7}Li abundances. After some time, mixing with unprocessed primordial gas leads to a higher and more uniform ^{7}Li but still below the value predicted from std. big bang nucleosynthesis.

3 Search for ^{6}Li in Planet-hosting Stars

Models of low-mass stars with solar-like metallicities suggest that ^{6}Li is totally destroyed in their pre-main sequence phase. Hence, the apparent detection of ^{6}Li ($^6\mathrm{Li}/^7\mathrm{Li} = 0.126 \pm 0.014$) in HD 82943 (a metal-rich G0V star with two known Jupiter-mass planets) by Israelian et al. [5], was interpreted as evidence for a planet having been engulfed by the parent star. Reddy et al. [13] pointed out, however, that an unidentified weak blend at 6708.025 Å affects the derived ^{6}Li abundance significantly. The strength of the blend may be calibrated from the solar spectrum, and assuming that it is due to a low excitation Ti I line, Reddy et al. found that the profile of the Li line in HD 82943 is well fitted without any ^{6}Li. The upper limit for this star and another seven planet-hosting stars is $^6\mathrm{Li}/^7\mathrm{Li} \leq 0.03$.

In a new paper, Israelian et al. [6] have shown that the unidentified blend can not be a low excitation Ti I line, because it is then predicted to be stronger than observed in cooler stars. Assuming that the blend is a high excitation Si I line, and using a new UVES spectrum of HD 82943 with $S/N = 1000$, they derived $^6\mathrm{Li}/^7\mathrm{Li} = 0.05 \pm 0.02$. Given the uncertainty in the identification of the blend, this should not be considered as a final detection of ^{6}Li. It should also be noted that Mandell et al. [8] were unable to detect ^{6}Li in two lithium-poor stars with planets. Still, it would be interesting to search for ^{6}Li in a large sample of solar type stars. Presence of ^{6}Li due to engulfing of planets may be rare, because ^{6}Li could be destroyed at the bottom of the convective zone on a time scale that is relatively short compared to the stellar age. Hence, detection of ^{6}Li in just a few solar-type stars would be highly interesting.

4 The Need for New Instrumentation

In order to verify the detection of ^{6}Li and to study the possible decline and scatter of ^{7}Li at the lowest metallicities, there is high interest in determining Li isotopic abundances in stars with [Fe/H] < -3. With present days 8-10 m telescopes only a couple of such turnoff stars are bright enough ($V < 12$ mag) to be observed at a resolving power of $R \geq 10^5$ and $S/N \geq 500$. To reach a decent sample of very metal-poor stars, 30-50 m class telescopes will be needed. Furthermore, it would be important to have available high resolution spectrographs with higher wavelength stability and better flat-fielding properties than existing spectrographs like VLT/UVES, KECK/HIRES and SUBARU/HDS. As a first step it would be very valuable to get an efficient "super-HARPS" in the combined focus of the four VLT telescopes.

References

1. Asplund, M., Lambert, D.L., Nissen, P.E., et al. 2006, ApJ, 644, 229
2. Bonifacio, P., Molaro, P., Sivarani, T., et al. 2006, A&A (in press)
3. Cayrel, R., Spite, M., Spite, F., et al. 1999, A&A, 343, 923
4. Hobbs, L.M., & Thorburn, J.A. 1994, ApJ, 428, L25
5. Israelian, G., Santos, N.C., Mayor, M., & Rebolo, R. 2001, Nature, 411, 163
6. Israelian, G., Santos, N.C., Mayor, M., & Rebolo, R. 2003, A&A, 405, 753
7. Jedamzik, K. 2004, Phys. Rev. D, 70, 063524
8. Mandell, A.M., Ge, J. & Murray, N. 2004, AJ, 127, 1147
9. Nissen, P.E., Akerman, C., Asplund, M., et al. 2005, Proc. of IAU symp. 228, Eds. V. Hill, P. Francois, and F. Primas, p. 101
10. Piau, L., Beers, T.C., Balsara, D.S., et al. 2006, astro-ph/0603553
11. Pinsonneault, M. H., Steigman, G., Walker, T. P., et al. 2002, ApJ, 574, 398
12. Prantzos, N. 2006, A&A, 448, 665
13. Reddy, B.E., Lambert, D.L., Laws, C., et al. 2002, MNRAS, 335, 1005
14. Richard, O., Michaud, G., & Richer, J. 2005, ApJ, 619, 538
15. Rollinde, E., Vangioni, E., & Olive, K. 2006, ApJ, 651, 658
16. Smith, V.V., Lambert, D.L., & Nissen, P.E. 1993, ApJ, 408, 262
17. Smith, V.V., Lambert, D.L., & Nissen, P.E. 1998, ApJ, 506, 405
18. Suzuki, T.K., & Inoue, S. 2002, ApJ, 573, 168

Lithium Isotopic Abundances in Old Stars

Ana Elia García Pérez[1], Susumu Inoue[2], Wako Aoki[2], and Sean Ryan[1]

[1] University of Hertfordshire, College Lane, Hatfield AL10 9AB, UK
a.e.garcia-perez@herts.ac.uk,s.g.ryan@herts.ac.uk
[2] National Astronomical Observatory of Japan, Mitaka, Tokyo 181-8588, Japan
inoue@th.nao.ac.jp,aoki.wako@nao.ac.jp

Summary. Lithium is an element of great importance from a cosmological point of view. Observations of its light isotope ^{6}Li are very delicate and in the literature, detections for only a handful of metal-poor stars are found. High quality spectra ($S/N \sim 450$, $R \sim 100000$) of the Li I 670.8 nm resonance line of six metal-poor stars, taken at the Subaru telescope, are being analysed. The results based on the still early stage of the analysis suggest one possible detection and four possible upper limits (including the two most metal-poor stars in the sample) for ^{6}Li. The observed spectra of the remaining star suggest that there is very little ^{6}Li, if any at all.

1 Introduction

The atmospheres of metal-poor stars are good laboratories to investigate processes which occurred in the early Galaxy. These processes left their signatures printed in these atmospheres contents. The spectral analysis of metal-poor stars can sometimes be difficult, especially when dealing with detections of weak features. Nevertheless, these analysis represent an important way to retrieve some of the information contained within the stellar chemical composition. Most of the lithium observed in metal-poor stars is expected to be produced during Big Bang nucleosynthesis. However, lithium is a fragile element whose abundances could be depleted by mixing processes. These processes alter the initial composition and they do it at a different level depending on the element and on the isotope. Determinations of isotopic lithium abundances can set constraints on the possible amount of depletion and on the degree the observed abundance value of lithium in metal-poor stars (plateau value) might deviate from the primordial value. At the moment, the value derived from standard cosmology in combination with WMAP results disagrees with the plateau value.

The detection of ^{6}Li is still a challenge. High quality spectra and high precision analysis are necessary to detect the slight asymmetry, which is produced by the isotopic shift, in the line profile.

2 Data and Analysis

We have analysed the spectra of six metal-poor stars taken in May 2005 with the High Dispersion Spectrograph mounted at the 8.2-m Subaru telescope. Our intention, with this preliminary analysis, is to show why high precision spectroscopy is required for isotopic lithium abundance determinations, rather than to give final results. Observational data were reduced following the standard procedures, using IRAF. The reduced spectra are of very high quality; typical values reached for the signal-to-noise ratios and the resolving power are $S/N = 300-650$ and $R = 90000-100000$ around the Li I 670.8 line.

Note that isotopic abundance ratios determinations are not too sensitive to changes in the stellar parameters ($T_{\rm eff}$, $\log g$, [Fe/H], ξ). Hence for the purpose of this analysis, we have assumed values, with the exception of metallicity, from the literature ([2] and [3]). Metallicity values were derived spectroscopically from the equivalent width of a set of Fe I lines. The assumed stellar parameter values are given in Table 1.

Table 1. Stellar parameters and ^{6}Li/^{7}Li ratios for the best guess fits

Star	$T_{\rm eff}$	$\log g$	[Fe/H]	ξ	^{6}Li/^{7}Li
BD−04°3208	6338	4.00	−2.21	1.5	≤ 0.05
G 64-37	6318	4.16	−3.12	1.5	≤ 0.05
BD+02°3375	5855	4.16	−2.12	1.5	0.01
BD+20°3603	6092	4.04	−2.15	1.5	≤ 0.03
BD+26°3578	6239	3.87	−2.25	1.5	0.08
LP815-43	6514	4.23	−2.72	1.5	≤ 0.08

Once stellar parameters were specified, MARCS model atmospheres were computed and used in the synthesis of 1D-LTE spectra (BSYN) of the Li I 670.8 nm resonance line (and other lines of interest). The synthetic spectra were convolved with two functions which mimic the instrumental and the stellar line broadening respectively. The observed Th-Ar lamp lines were used to get the widths (FWHM=3.10–3.20 km/s) of the Gaussian describing the instrumental profile, while the observed Ca I line at 612.2 nm was used for the modelling of the macroscopic stellar broadening. A radial-tangential profile was assumed for the stellar broadening (widths= 2.50–4.20 km/s). The detection of the ^{6}Li isotope is based on the detection of the slight asymmetry, produced by the isotopic shift, in the red wing of the stronger ^{7}Li line. The observed and the synthetic line profiles were compared (see Fig. 1). ^{6}Li/^{7}Li ratios together with the total lithium abundances where changed until the best fit was reached.

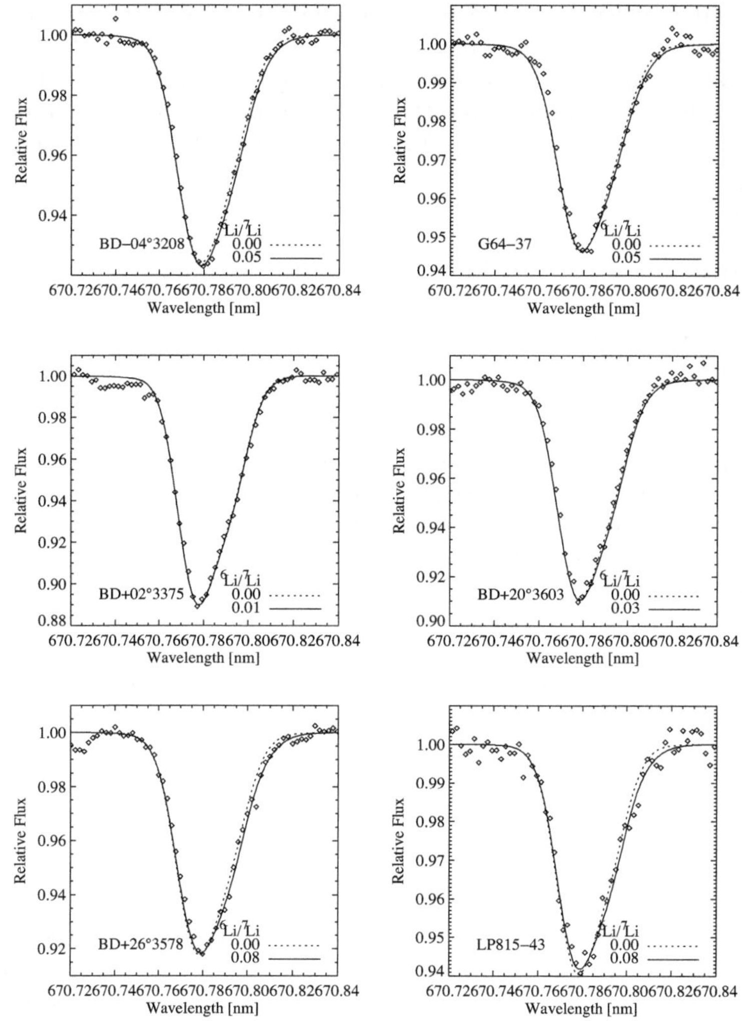

Fig. 1. Comparison of the observed spectra (diamonds) for the Li I 670.8 nm line with the synthetic spectra (solid and dashed lines).

3 Isotopic Abundance Ratios and Our Best Guess Fit

The analysis presented here is still at an early stage, hence we refrain of claiming any certain detection. Instead, we will refer to possible detections or possible upper limits according to our guess for the best fit. They will be confirmed by a proper statistical method like a χ^2-fit later (future publication). There are cases where the possible detection is more clear than in others. For some other stars, the spectral synthesis indicates that if the feature does exist, it is so weak that the high quality of the observed spectra is still not enough for the detection (eg. BD+20°3603, $S/N \sim 400$). Values for our best estimates so far are listed in Table 1, one possible detection and four possible upper limits. It is too soon for a proper discussion and for drawing final conclusions. However, we present the preliminary results (see Table 1) and discuss their limitations.

LP815-34 is an interesting target because of its low metallicity, the second lowest of the observed sample. [1] has reported a detection for this star. We are not yet in a good position to claim any detection, as our fits indicates that $0 \leq {}^6\mathrm{Li}/{}^7\mathrm{Li} \leq 0.08$ values are still compatible with the observations and the signal-to-noise ($S/N \sim 500$). The other metal-poor star, G 64-37 ([Fe/H]=−3.12), may represent a case where the broadening determination was not accurate enough. This would be not a surprise given that only one line was used for this purpose. The inclusion of more lines may improve the situation, but in such metal-poor star, it would be difficult to find suitable ones as the lines in the spectra of these stars are weak. We will investigate the effect of introducing two other free parameters, the wavelength zero-point and the continuum level, on our estimates.

In the case of the possible detections, we have BD+26°3578. This star was one of the first metal-poor stars for which a positive detection was published [4]. Recently, [1] has reported a value of ${}^6\mathrm{Li}/{}^7\mathrm{Li} = 0.01$ for this star, which is in disagreement with the value of 0.05 in [4] and even more with our estimate of 0.08. In order to make a more rigorous comparison, we should await the completion of our more detailed analysis. The other star in common with [1] is BD-04°3208 for which there is a good agreement. It will be very interesting to confirm whether BD+02°3375 has such a low ${}^6\mathrm{Li}$ content which could be a signature of depletion processes acting on the star. Irrespectively, this star is the one with the lowest T_{eff} value in the observed sample.

A more complete analysis will enable us to study in more detail the possible scenarios and mechanism of ${}^6\mathrm{Li}$ production, cosmic ray production by supernovae or by star formation shocks, stellar flares etc. If the results from our preliminary analysis are confirmed by our more detailed analysis, we would have a few interesting detections and some equally interesting non-detections. We have embarked on that task, and the results, along with an analysis of uncertainties, are going to appear in a later publication.

References

1. M. Asplund, D. L. Lambert, P. E Nissen, F. Primas and V. V. Smith, ApJ **644**, 229 (2006)
2. P. E. Nissen, F. Primas, M. Asplund, D. L. Lambert, A&A **390**, 235 (2002)
3. P. E. Nissen, Y. Q. Chen, M. Asplund, M. Pettini, A&A **415**, 993 (2004)
4. V. V. Smith, D. L. Lambert, P. E. Nissen, ApJ **506**, 405 (1998)

Nuno Santos, Claudio Melo and Alexandre C.M. Correia:
science is a serious business.

Accurate Quantitative Spectroscopy of OB Stars: the H, He and C Spectrum

M.F. Nieva[1,2] and N. Przybilla[1]

[1] Dr. Remeis Sternwarte Bamberg, Sternwartstr. 7, D-96049 Bamberg, Germany
[2] Observatório Nacional, Rua General José Cristino 77 CEP 20921-400, RJ, Brazil
nieva,przybilla @sternwarte.uni-erlangen.de

Summary. We present a state-of-the-art analysis technique able to reproduce the H, He and C spectra of OB-type stars simultaneously in the visual and the near-IR. Our spectrum synthesis relies on a hybrid non-LTE approach involving robust model atoms based on precise atomic data. Highly accurate atmospheric parameters free of systematic error are derived for a sample of randomly selected apparently slow-rotating stars in the solar vicinity. A mean carbon abundance of 8.33 ± 0.04 is indicated. Up to 40 features from C II–IV are considered. This includes the prominent C II $\lambda\lambda$4267 and 6578/82 Å lines, suitable for abundance determinations in fast-rotating and extragalactic objects at lower S/N. An excellent starting point for future highly-precise abundance analyses of other metals is established, allowing to put tight constraints on stellar and galactochemical evolution.

1 Introduction

Hydrogen and helium are of major interest in the astrophysical context, as they constitute practically all light-emitting plasma. Their lines are the strongest spectral features in OB stars and non-LTE effects play a dominant rôle in the line formation. Solving the entire measurable H and He spectrum is a pre-requisite for metal abundance analyses, in particular for one of the most abundant metals in the universe: carbon. Abundances derived from OB stars provide important constraints on stellar evolution and the chemical evolution of our own and other galaxies. In extragalactic applications (e.g. Magellanic Clouds) one desires to analyse the *strongest* features in the metal-line spectra (low S/N). In the case of ionized carbon these are the prominent lines of the two multiplets C II $\lambda\lambda$4267 Å and 6578/82 Å. These lines are highly sensitive to non-LTE effects and to the choice of stellar atmospheric parameters. So far, all studies from the literature failed to derive consistent abundances from these lines and also to establish the C II/III ionization equilibrium.

The present work aims to simultaneously investigate the entire measurable H, He and C spectrum and provide a solution to the classical non-LTE problem of carbon abundance determinations in OB stars. A reliable C II-IV model atom is developed and first applications on high-quality spectra are presented. Besides great care in the selection of atomic data, special emphasis is also given to an accurate determination of the atmospheric parameters, both in order to minimise systematic uncertainties.

2 Model Calculations

A hybrid approach is used for the non-LTE line-formation computations. These are based on line-blanketed, plane-parallel, homogeneous and hydrostatic LTE model atmospheres calculated with ATLAS9. Non-LTE synthetic spectra are computed with recent versions of DETAIL and SURFACE. These codes solve the coupled radiative transfer and statistical equilibrium equations and compute synthetic spectra using refined line-broadening data, respectively. Non-LTE line-formation calculations are performed for C II/III/IV, hydrogen and He I/II using state-of-the-art model atoms based on critically selected atomic data [4, 7, 8, 9].

3 Observations and Analysis

A first sample of six apparently slow-rotating Galactic B-type dwarfs and giants from OB associations and from the field in the solar vicinity is analysed. The observational database consists of very high S/N spectra obtained with FEROS on the 2.2m telescope at La Silla, Chile (ESO). The stellar parameters are derived from application of an extensive iterative method resulting in simultaneous fits to all measurable H, He [5] and C II-IV lines [6, 7]. The iteration is performed on effective temperature $T_{\rm eff}$ and surface gravity $\log g$ (goal: to achieve ionization equilibrium) as well as the micro-, macroturbulent and projected rotational velocities (from carbon line profiles), He and C abundances and different sets of atomic data. The final solution for each star is obtained when the statistical deviations of the averaged C abundance are minimised. By application of the procedure to all programme stars it was possible to calibrate the C II-IV model atom for the entire parameter range ($21\,500 \leq T_{\rm eff} \leq 32\,000$ K, $3.10 \leq \log g \leq 4.30$) and to obtain highly accurate C abundances for the sample at the same time.

4 Results for the Programme Stars

The synthetic spectra match all hydrogen and helium lines in the optical and (where available) also in the near-IR simultaneously for each of the six test stars covering a wide range of stellar parameters [5]. We also derive an unprecedently uniform C abundance for each star from numerous lines from different ionization stages. Non-LTE and LTE abundances vs. equivalent width W_λ are displayed in Fig. 1 for all individual lines, showing excellent consistency in non-LTE. Identifications of the C II $\lambda\lambda$4267, 6578/82, 6151 and 6462 Å lines are displayed. These lines are very sensitive to non-LTE effects. The final parameters as well as the carbon abundances are also given in Fig. 1. An immediate consequence of our careful analysis is a highly consistent value of carbon abundance free of systematic effects for all the stars of the sample.

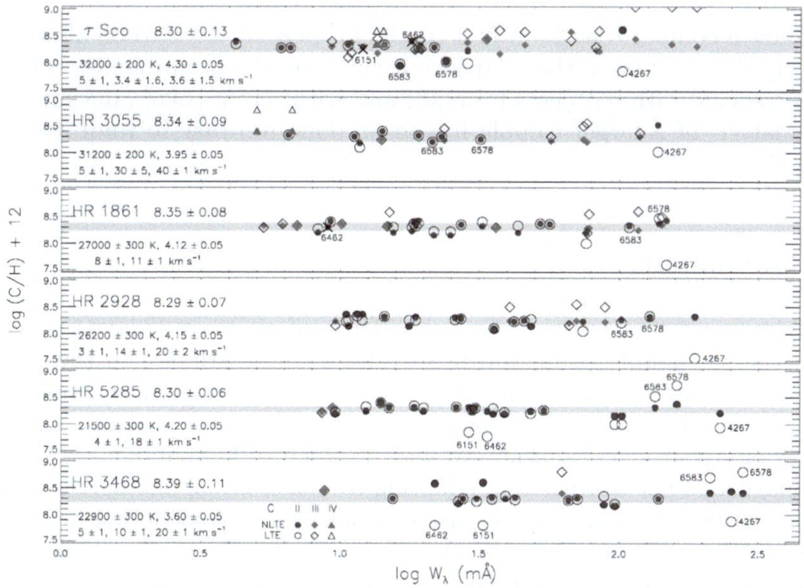

Fig. 1. Non-LTE and LTE abundances vs. equivalent width for C II-IV lines that could be measured in each spectrum. The ID of the stars and the derived mean carbon abundance is given in the upper left corner of each row, as well as derived basic stellar parameters ($T_{\rm eff}$, $\log g$, micro-, macroturbulent and projected rotational velocity). The grey rectangles correspond to 1σ-uncertainties of the stellar carbon abundance. Identification of lines with high sensitivity to non-LTE effects is displayed. Emission lines are marked by crosses: LTE calculations are not able to reproduce them even qualitatively.

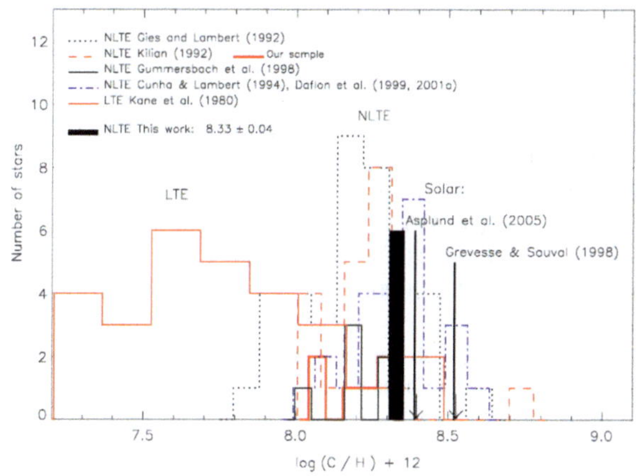

Fig. 2. Comparison of carbon abundances derived from our sample of early B-type stars in the solar vicinity with results from the literature and solar values.

We show a comparison of our first results for present-day carbon abundances of early B-type stars in the solar vicinity with results from the literature (our stars coincide with 6 objects from [3]) and with recent solar values in Fig. 2. An analysis of a larger sample has to be done in order to improve on the statistics.

5 Conclusions

A highly consistent mean carbon abundance of $\varepsilon(\mathrm{C}) = 8.33\pm0.04$ is derived from the six sample stars, which provides a tight estimate to the present-day C abundance from young stars in the solar vicinity. The atmospheric composition appears to be basically unaffected by chemical mixing in the course of stellar evolution, i.e. we find no trend of C abundances with evolutionary age. For comparison, adopting results from [3] one derives a mean $\varepsilon(\mathrm{C}) = 8.19\pm0.12$ from the same six stars, implying a considerable systematic shift and a significantly increased statistical scatter. More objects need to be analysed in order to confirm such homogeneous present-day (slightly) sub-solar carbon abundances – considering as references $\varepsilon(\mathrm{C})_\odot = 8.39\pm0.05$ [1] or $\varepsilon(\mathrm{C})_\odot = 8.52\pm0.06$ [2] – in nearby associations (HR 1861: Ori OB1; τ Sco, HR 5285: Sco-Cen) and in the field (the other stars). Despite the small sample size analysed so far, our highly accurate results indicate that the large scatter found for carbon abundances in early-type stars in previous work could be a consequence of systematic uncertainties introduced by the choice of inappropriate atomic data and stellar parameters.

References

1. M. Asplund, N. Grevesse, A. J. Sauval, *The Solar Chemical Composition*, 2005, in Cosmic Abundances as Records of Stellar Evolution and Nucleosynthesis, ed. T. G. Barnes III & F. N. Bash (San Francisco: ASP), 25
2. N. Grevesse & A. J. Sauval, 1998, Space Sci. Rev., 85, 161
3. J. Kilian, 1992, A&A, 262, 171
4. M. F. Nieva & N. Przybilla, 2006a, ApJ, 639, L39
5. M. F. Nieva & N. Przybilla, 2006b, A&A accepted (astro-ph/0608117)
6. M. F. Nieva & N. Przybilla, *Present-Day Carbon Abundances*, 2006c, in Nuclei in Cosmos IX, Proceedings of Science, PoS(NIC-IX)150 (astro-ph/060922)
7. M. F. Nieva & N. Przybilla, 2006d, in prep.
8. N. Przybilla, 2005, A&A, 443, 293
9. N. Przybilla, & K. Butler, 2004, ApJ, 609, 1181

High Resolution Spectroscopy of HgMn stars: A Time of Surprises

S. Hubrig[1], C.R. Cowley[2], F. González[3], F. Castelli[4]

[1] European Southern Observatory, Casilla 19001, Santiago, Chile shubrig@eso.org,
[2] University of Michigan, Ann Arbor, MI 48109-1042, USA cowley@umich.edu
[3] Complejo Astronómico El Leoncito, Casilla 467, 5400 San Juan, Argentina fgonzalez@casleo.gov.ar,
[4] Osservatorio Astronomico di Trieste, Trieste, Italy castelli@ts.astro.it

The origin of the abundance anomalies observed in late B-type stars with HgMn peculiarity is still poorly understood. Observationally, these stars are characterized by low rotational velocities and weak or non-detectable magnetic fields. The most distinctive features of their atmospheres are an extreme overabundance of Hg (up to 6 dex) and/or Mn (up to 3 dex) and a deficiency of He. Anomalous isotopic abundances have been reported in the past for the elements Hg, Pt and Tl. Observational evidence for large isotopic shifts in the infrared triplet of Ca II was presented in the last two years (Castelli & Hubrig 2004 [1], Cowley & Hubrig 2005 [2], Cowley et al., these proceedings). Shifts of up to +0.2 Å were found in a number of HgMn and magnetic Ap stars indicating the dominant isotope is the terrestrially rare ^{48}Ca.

As more than 2/3 of the HgMn stars are known to belong to spectroscopic binaries (Hubrig & Mathys 1995 [3]), the variation of spectral lines observed in any HgMn star is usually explained to be due to the orbital motion of the companion. Here we present the results of a high spectral resolution study of a few spectroscopic binaries with HgMn primary stars. We detect for the first time in the spectra of HgMn stars that for many elements the line profiles are variable over the rotation period (Hubrig et al. 2006 [4]). The strongest profile variations are found for the elements Pt, Hg, Sr, Y, Zr, Mn, Ga, He and Nd. The slight variability of He and Y is also confirmed by the study of high resolution spectra of another HgMn star, α And.

In Fig. 1 we show the behavior of the line profile of Hg II λ 3983.9 in the spectra of AR Aur at different rotation phases. Our preliminary modelling of abundance distributions of the elements Sr and Y over the stellar surface suggests that these elements are very likely concentrated in a fractured ring along the rotational equator (Hubrig et al. 2006 [4]). In Fig. 2 we present recent FEROS observations of variable line profiles of Y II λ 3982.5 in the spectra of the HgMn star HD 11753 and of Y II/Hg II lines in the spectra of the HgMn double-lined spectroscopic binary HD 27376.

The discovery of an inhomogeneous distribution of various elements in the atmospheres of HgMn stars challenges our understanding of the nature of HgMn stars. We believe that factors as the presence of a weak tangled

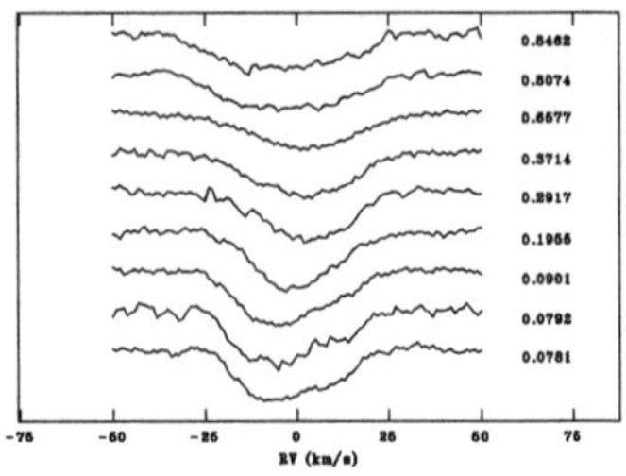

Fig. 1. Variations of the line profile of Hg II λ 3983.9 in the spectra of AR Aur phased with the rotation period P = 4.13 days.

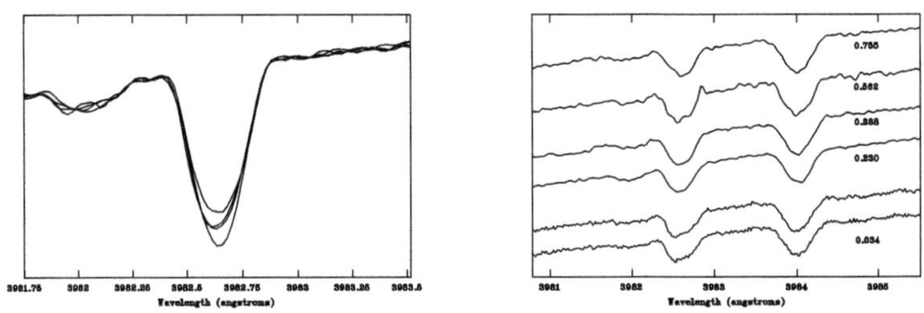

Fig. 2. Left: Profile variations of the Y II λ 3982.5 Å line in the spectra of HD 11753 at various rotation phases. Right: Profile variations of the Y II λ 3982.5 and Hg II λ 3983.9 lines in the spectra of HD 32964 over the rotation period.

magnetic field, tidal distortion, or the reflection effect can play a role in the development of anomalies in HgMn stars. Although diffusion due to gravitational settling and radiative levitation remain the most popular explanation for the HgMn star abundances, the selective accretion of interstellar material during the pre-main-sequence phase by HgMn binary systems seems to be a promising possibility for explaining some surface anomalies, given the presence of magnetic fields. Probably, all these mechanisms have to be taken into account in future studies of these stars.

References

1. Castelli F., Hubrig S. 2004, A&A 421, L1
2. Cowley C.R., Hubrig S. 2005, A&A 432, L21
3. Hubrig S., Mathys G. 1995, ComAp 18, 167
4. Hubrig S., González J.F., Savanov I., et al. 2006, MNRAS 371, 1953

High-resolution Spectroscopy of Faint Stars with Transiting Planets

N.C. Santos[1,2,3], C. Melo[4], F. Pont[3], T. Guillot[5], A. Ecuvillon[6], M. Mayor[3], S. Udry[3], D. Queloz[3], G. Israelian[6], F. Bouchy[7], C. Moutou[7]

[1] Centro de Astronomia e Astrofísica da Universidade de Lisboa, Portugal
nuno.santos@oal.ul.pt
[2] Centro de Geofísica de Évora, Portugal
[3] Observatoire de Genève, Switzerland
[4] European Southern Observatory, Chile
[5] Observatoire de la Côte d'Azur, France
[6] Instituto de Astrofísica de Canárias, Spain
[7] Laboratoire d'Astrophysique de Marseille, France

Summary. We present a series of studies based on high-resolution UVES/VLT spectroscopy of several faint stars for which a transit by a giant planet has been detected. The spectra were used to derive accurate stellar parameters and chemical abundances for the targets, as well as to estimate their ages and galactic cartesian coordinates. The results are discussed in the context of the chemical abundances of stars with known giant planets. The derived chemical abundances are also compared with those found for stars in the solar neighborhood.

1 Introduction

From 1995 to the present about 200 extra-solar giant planets have been discovered orbiting solar-type stars[8]. So far, most of the known extra-solar planets have been unveiled by the use of the Doppler radial-velocity technique. Alone, this only gives us information about the orbital parameters of the planets and their minimum masses, and no information is given about the planetary physical properties, like their real masses, radii, and mean densities. A complementary technique that can give us the possibility to study these quantities is the photometric transit technique. Until recently, however, in only a few cases it had been possible to measure the light dimming of the star as its planet crossed the stellar disk.

Fortunately, some major planet search programs using the photometric transit technique start now to deliver consistent results. Many candidates have been announced by surveys like OGLE (e.g. [2]), and in a few cases, the planetary nature was confirmed by follow-up radial-velocity measurements. The today's 14 known transiting extra-solar giant planets [1, 3, 4, 5, 7, 8, 6, 23, 10, 11, 12, 13] are finally giving us information about the properties

[8]For a continuously updated table see http://obswww.unige.ch/exoplanets

of planets orbiting other stars, and opening the possibility to confront the observed properties with those predicted by the models.

Many interesting issues have also been raised though. For example, one half of the transiting planets discovered by the OGLE campaign have orbital periods of the order of 1-2 earth days (these planets are dubbed "very-hot jupiters"). This result is in clear contrast with the known lower limit around 3-days previously observed for ("hot-jupiter") giant planets discovered by radial-velocity surveys. Whether this is the result of a detection bias or of some other physical process is now a matter of debate [22].

The discovery of new transiting planets, some of them orbiting distant (>1Kpc) stars (e.g. the OGLEs), motivated a project to study their host stars. The need for both high resolution and high S/N data to observe the faint (V>16) OGLE stars implied the use of a large aperture telescope with a high resolution spectrograph. The UVES spectrograph at the ESO VLT-UT2 Kueyen telescope is the best available instrument to conduct this program. A series of UVES spectra were thus obtained in two different programs (075.C-0185 and 076.C-0131) for the 5 known OGLE transiting planet-host stars (OGLE-TR-10, 56, 111, 115 and 132) as well as for TrES-1. The major results of the above mentioned studies are discussed below. These have also been published in a series of papers [18, 19, 21, 20, 9].

2 Results

2.1 Stellar Parameters and Refined Planetary Radii

Stellar parameters and chemical abundances were derived in LTE using the 2002 version of the code MOOG and a grid of Kurucz Atlas plane-parallel Kurucz model model atmospheres[9]. The whole procedure is described in [17], and is based on the analysis of 39 Fe I and 12 Fe II weak lines, imposing excitation and ionization equilibrium. In order to check the excitation temperatures derived from the Fe-line analysis, we compared the observed H_α line profiles with synthetic profiles for metal-rich dwarfs ([M/H]=0.3 and $\log g$=4.5) computed by R. Kurucz for different temperatures. The results of this comparison show that the H_α temperatures agree with the excitation temperatures.

The derived stellar parameters were used with the photometric transit curves to refine the derived stellar and planetary masses and radii [18, 9].

2.2 Chemical Abundances

A few authors (e.g. [15]) have discussed the possible fact that shorter period planets orbit stars that are, on average, even metal-richer than the average planet-hosts [14, 16, 17]. In this sense, and given their short period orbits,

[9] http://verdi.as.utexas.edu/moog.html and http://kurucz.harvard.edu/grids.html

the knowledge of the position of the new transiting candidates in the [Fe/H] distribution of stars with giant planets is extremely important.

The analysis of the UVES spectra revealed that the 6 studied stars are all metal-rich (0.06<[Fe/H]<0.37) [18, 9]. This is particularly relevant for the most distant targets (some of the OGLE stars are more than 2 Kpc from the sun), and suggest that the metallicity-giant planet correlation is valid also for stars out of the solar-neighborhood. On the other hand, and although a slight tendency exists, our data does not support that shorter period planets orbit even higher metallicity stars [18]. However, we caution that the small number of points available does not permit to give a definite answer to this issue.

The obtained spectra were also used to derive and discuss the chemical abundances for a large set of chemical elements, including C, O, Na, Mg, Al, Si, S, Ca, Sc, Ti, V, Cr, Mn, Co, Ni, Cu, and Zn [19]. The resulting abundances were compared with those found for stars in the solar neighborhood (e.g. [24]). We conclude that, besides being particularly metal-rich and with small possible exceptions, OGLE-TR-10, 56, 111, 113, 132, and TrES-1 are chemically indistinguishable from the field (thin disk) stars regarding their [X/Fe] abundances. This is particularly relevant for the most distant of the targets. We also did not find any correlation between the abundances and the condensation temperature of the elements, evidence that strong accretion of planetary-like material did not occur (e.g. [25]).

2.3 Stellar Ages

The reason why radial velocities surveys were not able to detect "very-hot jupiters" is under discussion. A possible explanation is that these close-in planets are short-lived, being evaporated in a short time-scale due to UV flux of their host stars. In this case, stars hosting transiting planets would be systematically younger than those in the radial velocity sample.

To test this hypothesis we used the available spectra to estimate ages for the target stars [21]. These were derived using several methods, including the lithium abundances and the measurement of a chromospheric activity index based on the flux within the core of the Ca II H & K lines. The results show that none of the stars seems to be younger than $\sim$1 Gyr. This conclusion has a strong impact on the derived evaporation rates of hot and very-hot jupiters [21]. Recent results favor indeed that the lack of very-hot jupiters in radial-velocity surveys is related to a statistical and sampling bias. Ground based photometric transit searches seem to be particularly sensitive to very short period planets [22].

2.4 Stellar Metallicity and Planet Internal Structure

Having precise stellar metallicities and observational radii for the planets, we searched for correlations between these two variables. To do this we derived the difference between the observed radii and the theoretical radii (a "radius anomaly"). This latter was computed using a simple model for the planetary structure (e.g. no core), and taking into account the stellar radiation. A correlation between this "radius anomaly" and the stellar metallicity was then searched.

The results, presented in Guillot et al. [20], provide strong evidence for a correlation between the stellar metallicity and the planetary internal structure. Indeed, metal-richer stars seem to be orbited by planets having a higher metal-content (bigger core?). The model-dependent "core" masses range from about 10 times the mass of the Earth for the planets orbiting the most metal-poor stars, to more than 100 $M_\oplus$ for the planets orbiting their metal-richer counterparts. This result may have important impact on the models of planet formation and evolution, and lends support to the core-accretion model for planetary formation.

N.C.S. would like to thank the support from FCT (Portugal), in the form of a fellowship (SFRH/BPD/8116/2002) and a grant (POCI/CTE-AST/56453/2004).

References

1. D. Charbonneau, T.M. Brown, D.W. Latham, M. Mayor: ApJ **529**, L45 (2000)
2. A. Udalski, B. Paczynski, K. Zebrun et al.: Acta Astr. **52**, 1 (2002)
3. R. Alonso, T.M. Brown, G. Torres et al.: ApJ **613**, L153 (2004)
4. M. Konacki, G. Torres, S. Jha, D. Sasselov: Nature **121**, 507 (2003)
5. F. Bouchy, F. Pont, N.C. Santos et al.: A&A **421**, L13 (2004)
6. F. Bouchy, S. Udry, M. Mayor et al: A&A **444**, L15 (2005)
7. F. Pont, F. Bouchy, D. Queloz et al.: A&A, **426**, L15 (2004)
8. F. Pont, F. Bouchy, C. Melo et al.: A&A, **438**, 1123 (2005)
9. F. Pont, M. Gillon, C. Moutou et al.: A&A, submitted (2006)
10. P. McCullogh, J. Stys, J. Valenti et al.: ApJ **648**, 1228 (2006)
11. F. O'Donovan, D. Charbonneau, G. Mandushev et al.: ApJ, submitted (2006)
12. G.A. Bakos, R.W. Noyes, G. Kovacs et al.: ApJ, in press (2006)
13. A. Collier Cameron, F. Bouchy, G. Hebrard et al.: MNRAS, submitted (2006)
14. G. Gonzalez: A&A **334**, 221 (1998)
15. A. Sozzetti: MNRAS **354**, 1194 (2005)
16. N.C. Santos, G. Israelian, M. Mayor: A&A **373**, 1019 (2001)
17. N.C. Santos, G. Israelian, M. Mayor: A&A **415**, 1153 (2004)
18. N.C. Santos, F. Pont, C. Melo et al.: A&A **450**, 825 (2006)
19. N.C. Santos, A. Ecuvillon, G. Israelian et al.: A&A **458**, 997 (2006)
20. T. Guillot, N.C. Santos, F. Pont et al.: A&A **453**, L21 (2006)
21. C. Melo, N.C. Santos, F. Pont et al.: A&A **460** 251 (2006)
22. S.B. Gaudi, S. Seager, G. Mallen-Ornelas: ApJ **623**, 472 (2005)
23. B. Sato, D.A. Fischer, G.W. Henry et al.: ApJ **633**, 465 (2005)
24. A. Bodaghee, N.C. Santos, G. Israelian, M. Mayor: A&A **404**, 715 (2003)
25. A. Ecuvillon, G. Israelian, N.C., Santos et al.: A&A **449**, 809 (2006)

First Resolved Narrow Line Profiles in Ultracool Dwarfs

Ansgar Reiners[1,2]

[1] Hamburger Sternwarte, Gojenbergsweg 112, 21029 Hamburg, Germany
areiners@hs.uni-hamburg.de
[2] Max-Planck-Institut for Solar System Research, Katlenburg-Lindau, Germany

Summary. High precision spectroscopy has produced a wealth of information on atmospheric and surface structure in hot and sun-like stars, and in cooler K-stars. In M-dwarfs, however, temperatures are so low that atomic lines have either vanished or are heavily broadened hiding information on surface velocity fields or spots. The spectra are dominated by molecular bands that are mostly useless for line profile analysis. This implies that a detailed investigation of surface velocity fields, the measurement of very low rotation velocities, and Doppler Imaging are very difficult although information on surface structure would be particularly interesting around temperatures where stars become fully convective.

I present the first high precision spectra of individual lines of the molecular FeH band near 1μm observed with CES/ESO ($R \sim 200\,000$). These lines resemble the first intrinsically resolved lines in M-dwarfs, which are not dominated by pressure broadening. These lines for the first time allow a detailed analysis of very low surface velocities (turbulence and rotation) and the investigation of line profile shape and variability. Furthermore, they allow a comparison to broadening profiles predicted by calculations of convective motions in M-star atmospheres.

1 Introduction

The measurement of very slow rotation requires high spectral resolution so that line broadening is not dominated by the instrumental profile. Although one can in principle correct for instrumental broadening if accurately known, extra broadening due to rotation becomes only visible if it is on the order of the instrumental broadening. In other words, slow rotation on the order of 1 km s^{-1} has very little influence on a line profile that is broadened by the limited resolution of 5 km s^{-1} width. On the other hand, it will be easily detectable if the instrumental profile is only of 1 km s^{-1} width. Furthermore, broadening effects other than rotation (convection, Zeeman splitting, etc.) may influence the line shape as well, a fact that is usually accounted for by taking an observed spectrum of a "slow" rotator as a template. As a result, the line profile of a "non-rotating" M-star has a FWHM on the order of 6-7 km s^{-1} at a resolution of $R \sim 50\,000$ and on the order of 2 km s^{-1} at a resolution of $R \sim 200\,000$. With the lower resolution, such a profile is sampled in only 1–2 resolution elements while at high resolution it is seen in about 5–10 resolution elements. Clearly, projected rotation velocities of $v \sin i \approx 2$ km s^{-1}

are difficult to detect in the former case but easily detectable in the latter, and a measurement of rapid rotation will be much more accurate at higher resolution.

A second problem arises in the spectra of M-dwarfs, their spectra being dominated by molecular absorption bands with hardly a single isolated line that reaches the continuum. The line wings reaching the continuum, however, is a crucial requirement for the measurement of slow rotation since it is the whole line that has to be analyzed; if the continuum level is not obvious it might only be part of the line that is analyzed (or used as a template) leading to too low a rotation velocity (too high if the problem lies only in the template line).

Finally, the observation of faint red M-dwarfs is a challenge particularly in optical wavelength regions where high SNR is difficult to reach. High SNR is crucial particularly if single lines of the target stars are to be compared to spun up versions of theoretical lines or lines in template stars.

2 Zeeman Splitting

Zeeman splitting for the first time was measured in ultracool dwarfs (spectral type later than M5) in molecular lines of FeH in [2]. This molecule provides a number of well isolated individual lines that show a variety of Zeeman sensitivities. The splitting of the FeH lines cannot yet be modelled theoretically, and the authors use the method of [1] to calibrate Zeeman splitting by using spectra of template stars with known magnetic flux.

Using this technique, [2] measure the mean magnetic field in a sample of 24 ultracool dwarfs with the finding that magnetic flux as strong as 4 kG occurs in such objects, and that activity as measured in Hα and X-ray emission scales with magnetic flux as it does in warmer stars. Thus, activity is of magnetic origin also in the predominantly neutral atmospheres of such cool objects.

3 Very Slow Rotation – The Rising Part of the Rotation Activity Connection in M-dwarfs

Data at a resolving power of $R \sim 200\,000$ was taken at the Coudé Echelle Spectrograph (CES) at the 3.6m telescope, ESO, La Silla. The most reliable way determining rotational line broadening is to utilize the spectrum of a star that shows virtually no rotation, i.e. $v \sin i$ is below the detection limit. The star with the narrowest profile in the sample is Gl 273, which is also the star with the lowest ratio L_X/L_{bol}. No rotation has been detected in this star before and there is no reason to assume that significant broadening due to rotation affects its line profile. However, the line profiles of the FeH

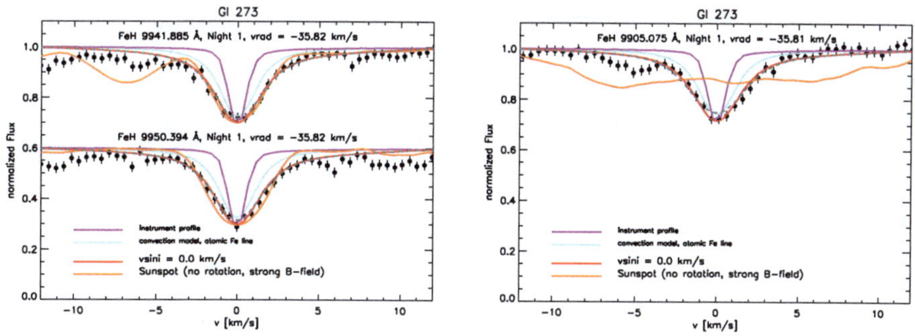

Fig. 1. FeH lines in Gl 273. Models are overplotted as indicatied in the figure. *Left:* Two magnetically insensitive lines. *Right:* A magnetically very sensitive lines. Note the strong Zeeman broadening in the Sunspot spectrum.

lines in Gl 273 are well resolved and they are significantly broader than the instrumental profile.

The two spectral lines $\lambda 9942.9$ and $\lambda 9950.4$ Å of Gl 273 are shown in Fig. 1 (left) in velocity units relative to the line center. Data are plotted as dots with error bars from the SNR. The instrumental profile measured from ThAr lines is overplotted, it is significantly smaller than the observed line showing that the line shape has been resolved for the first time. Unfortunately, radiative transfer calculations of the FeH lines are not yet reliably reproducing the observations so that it was not possible to compare the line shape to a fully consistent model of the two molecular lines. However, the main broadening mechanisms in such narrow lines are temperature broadening and convective velocities. The atomic weight of Fe is a good approximation of the weight of the FeH molecule, thus temperature broadening in Fe lines can be expected to be very similar to the effect in FeH lines. The strength of the FeH lines in early M stars is also comparable to some Fe lines, so that the line formation depths should also be similar in both species. Thus, line broadening due to convective motions are probably also of comparable strength.

In Fig. 1, the intrinsic profile of the Fe I line at $\lambda = 6518.37$ Å for $T_{\rm eff} = 2800$ K and $\log g = 5.0$ is overplotted. The profile was kindly provided by [3]. Atmospheric parameters correspond to an M6 atmosphere which is slightly cooler than our target objects, i.e. one would expect the lines to be slightly broader in the target spectra, although the difference is much smaller than what is measurable in the data. The convolution of the instrumental profile and the synthetic Fe I line is also overplotted (red line). It is immediately clear that our expectation of an observed Fe I line has a shape that very accurately fits the line shape observed in the two FeH lines in question. The similarity of the two FeH lines in Gl 273 to the shape expected for observed Fe lines means that within the uncertainties the FeH lines are consistent with zero rotation.

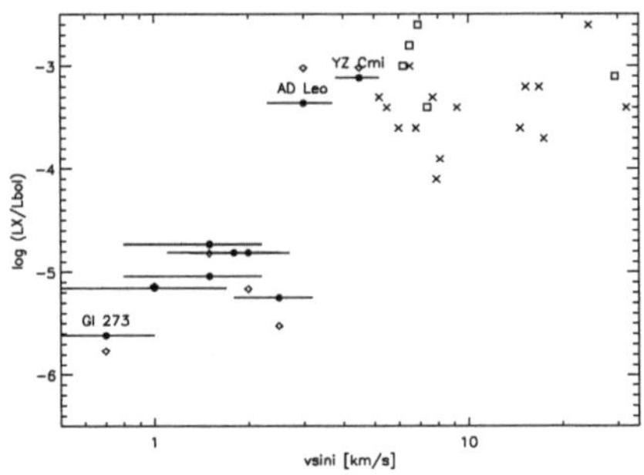

Fig. 2. Normalized X-ray emission L_X/L_{bol} plotted agains projected rotation velocity $v \sin i$ for the observed sample. The very slow rotators suggest a rising part of the rotation activity connection among M-dwarfs.

Among the sample, projected rotation velocities between 1 and 2.5 km s^{-1} are measured. The detected (projected) rotation velocities are not inconsistent with a rotation-activity connection in the sense that stars at $v \sin i \sim 2\,\mathrm{km\,s}^{-1}$ have higher $\log L_X/L_{bol}$ than the slowest rotators at only half that velocity. In Fig. 2, seven of the eight objects follow a rather tight trend that falls only slightly below the relation expected from an extrapolation of the rotation-activity connection found in warmer stars (note that Gl 205 and Gl 514 share the same point). The only star not consistent with that picture is Gl 526 with the highest rotation velocity ($v \sin i = 2.5\,\mathrm{km\,s}^{-1}$) at very low activity ($\log L_X/L_{bol} < -5$). Considering Gl 526 as an outlier, seven objects define a very close rotation-activity connection in M-dwarfs rotating slower than 3 km s^{-1}.

AR has received research funding from the European Commission's Sixth Framework Programme as an Outgoing International Fellow (MOIF-CT-2004-002544).

References

1. Reiners, A., & Basri, G., 2006, ApJ, 644, 497
2. Reiners, A., & Basri, G., 2007, ApJ, accepted, astro-ph/0610365
3. Wedemeyer S., & Ludwig, H.-G., 2006, priv.comm.

Physical Parameters of Evolved Stars in Clusters and in the Field from Line-depth Ratios

K. Biazzo[1,2], L. Pasquini[2], A. Frasca[1], L. da Silva[3], L. Girardi[4], A. P. Hatzes[5], J. Setiawan[6], S. Catalano[1], and E. Marilli[1]

[1] Osservatorio Astrofisico di Catania-INAF, Catania, Italy
`kbiazzo@oact.inaf.it`
[2] European Southern Observatory, Garching bei München, Germany
[3] Observatório Nacional-MCT, Rio de Janeiro, Brazil
[4] Osservatorio Astronomico di Padova-INAF, Padova, Italy
[5] Thüringer Landessternwarte, Tauterburg, Germany
[6] Max-Planck-Institute für Astronomie, Heidelberg, Germany

Summary. We present a high-resolution spectroscopic analysis of two samples of evolved stars selected in the field and in the intermediate-age open cluster IC 4651, for which detailed measurements of chemical composition were made in the last few years. Applying the Gray's method based on ratios of line depths, we determine the effective temperature and compare our results with previous ones obtained by means of the curves of growth of iron lines. The knowledge of the temperature enable us to estimate other fundamental stellar parameters, such as color excess, age, and mass.

1 Introduction

The study of stellar populations in our Galaxy and in its neighborhoods has received in the last years a big impulse, especially thanks to the use of large telescopes and to the detailed spectroscopic analysis performed on high-resolution spectra. In this context, open clusters, that are homogeneous samples of stars having the same age and chemical composition, are very suitable to investigate the stellar and Galactic formation and evolution. In spite of this, the data on stars belonging to open clusters are often insufficient to adequately constrain age, distance, metallicity, mass, color excess, and temperature. This is due to the fact that the main classical tool to study cluster properties is the color-magnitude diagram, which suffers of several uncertainties and intrinsic biases due to, for example, the uncertain knowledge of the chemical composition and the reddening of the stars. As a consequence, spectroscopic methods, being independent of the reddening, are very efficient to evaluate temperatures of stars in clusters.

Spectroscopic effective temperatures are usually determined imposing that the abundance of one chemical element with many lines in the spectrum (typically iron) does not depend on the excitation potential of the lines. Another method for determining effective temperature is based on line-depth ratios (LDRs). It has been widely demonstrated that the ratio of the depths of two lines having different sensitivity to temperature is an excellent measure of stellar temperatures with a sensitivity as small as a few Kelvin degrees in the most favorable cases ([14, 13, 12]).

In the present paper, we apply the LDR method to high-resolution UVES and FEROS spectra for deriving effective temperatures in nearby evolved field stars with very good Hipparcos distances and in giants of the intermediate-age open cluster IC 4651. For both the star samples, the temperature was already derived spectroscopically, together with the element abundances, with the curves of growth of absorption lines spread throughout the optical spectrum ([17, 18]). In addition, for the stars belonging to the open cluster IC 4651, we make the first robust determination of the average color excess, based on spectroscopic measurements.

2 Star Samples

We have analysed seventy-one evolved field stars and six giant stars belonging to the open cluster IC 4651. The sample of field stars was already analysed by [18] for the determination of radii, temperatures, masses and chemical composition. The stars in the intermediate-age cluster IC 4651 have been selected from the sample studied by [17] for abundance estimates.

The field stars data were acquired with the FEROS spectrograph ($R = 48\,000$) at the ESO 1.5m-telescope in La Silla (Chile), while the IC 4651 spectra were acquired with UVES ($R = 100\,000$) at the ESO VLT Kueyen 8.2m-telescope in Cerro Paranal (Chile). In both cases, the signal-to-noise ratio (S/N) was greater than 150 for all the spectra, which make them very suitable for the temperature determination described in Sec. 3.

3 Effective Temperature Determination

The wavelength range covered by FEROS and UVES spectrographs contain a series of weak metal lines which can be used for temperature determination with the LDR method. Lines from similar elements such as iron, vanadium, titanium, but with different excitation potentials (χ) have indeed different sensitivity to temperature. This is due to the fact that the line strength, depending on excitation and ionization processes, is a function of temperature and, to a lesser extent, of the electron pressure. For this reason, the better line couples are those with the largest χ-difference. In the range 6150 Å $\lesssim \lambda \lesssim$ 6300Å there are several lines of this type whose ratios of their depths have been exploited for temperature calibrations ([14, 13, 7, 5]), and for studies of the rotational modulation of the average effective temperature of magnetically active stars ([10, 4]) or for investigating the pulsational variations during the phases of a Cepheid star ([15, 3]). In particular, we choose 15 line pairs for which [5] already made suitable calibrations.

Field giant stars

The comparison between the temperatures obtained by us ($T_{\mathrm{eff}}^{\mathrm{LDR}}$) and those obtained by [18] ($T_{\mathrm{eff}}^{\mathrm{SPEC}}$) is plotted in Fig. 1. We find a very good agreement between $T_{\mathrm{eff}}^{\mathrm{SPEC}}$ and $T_{\mathrm{eff}}^{\mathrm{LDR}}$ in all the temperature range 4000–6000 K.

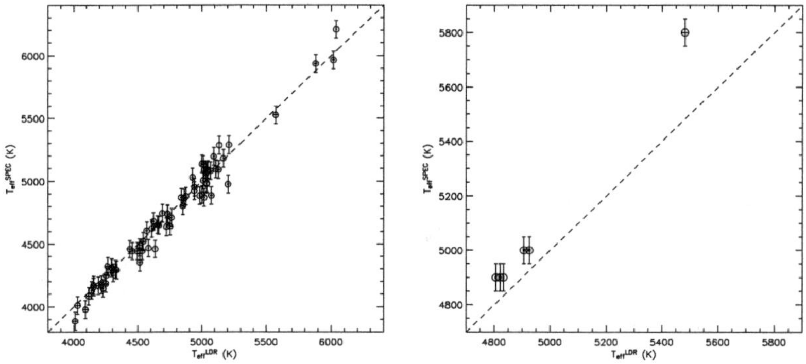

Fig. 1. Comparison between the temperatures obtained from LDR and curve-of-growth analyses for the field stars (*left*) and the giants belonging to IC 4651 (*right*).

Giant stars in IC 4651

$T_{\mathrm{eff}}^{\mathrm{LDR}}$ is systematically lower than $T_{\mathrm{eff}}^{\mathrm{SPEC}}$ by an amount typically between 70 and 90 K. The only exception is the star E95 for which the difference amounts to about 320 K. [17] already found for this star the largest difference between photometric and spectroscopic temperature among their giant star sample. However, the position on the HR diagram of this star corresponds to a subgiant and this could be the reason for the disagreement.

4 Color Excess of IC 4651

We can evaluate for each star the intrinsic color index $(B-V)_0$ by inverting for example the $(B-V)-T_{\mathrm{eff}}$ calibrations of [12] and [1, 2] with the aim to compute the color excess $E(B-V)$ of the cluster IC 4651. Thus, for the two temperature sets, $T_{\mathrm{eff}}^{\mathrm{LDR}}$ and $T_{\mathrm{eff}}^{\mathrm{SPEC}}$, we obtain $E(B-V) \approx 0.12$ and 0.16 for the Gray's calibration, and $E(B-V) = 0.13$ and 0.17 for the Alonso's calibration. It is worth noticing that there is not a large difference between color excesses obtained with the two calibrations. From a preliminary analysis, we find an improving of the agreement if we properly take into account the metallicity effects ([6]). Moreover, our color excess values are in good agreement with the results of $E(B-V)=0.13$ and $E(b-y)=0.091$ obtained by [9] and [17], respectively ($E(b-y)=0.72E(B-V)$, [8]). The present determination of $E(B-V)$ is a strong argument in favor of such a low reddening notwithstanding the distance of $\approx$ 900 pc estimated by [16] and the low galactic latitude of IC 4651 ($\simeq 9°$).

5 Conclusion

In this paper we have derived accurate atmospheric parameters for field evolved stars and giant stars in the open cluster IC 4651 by means of high-resolution spectra acquired with the ESO spectrographs FEROS and UVES.

For the field giant stars, we find a good agreement between temperatures computed by [18] with the curves-of-growth method and by ourselves with the LDR technique. For the giants in the intermediate-age open cluster IC 4651, we have determined the effective temperatures by means of the LDR method, that allowed us to compute the reddening. We find a rather low reddening towards the cluster, $E(B-V)\simeq 0.13$, that needs to be explained, given the high distance ($\simeq$ 900 pc) and the low galactic latitude of IC 4651.

We conclude that our technique is well suited to derive accurate effective temperatures and reddening of clusters with a nearly-solar metallicity. The determination of very precise temperatures is of great importance to derive stellar age and mass distributions ([11]), representing a powerful tool for stellar population studies in addition to those based on photometric data.

References

1. A. Alonso, S. Arribas & C. Martínez-Roger: A&A, **313**, 873 (1996)
2. A. Alonso, S. Arribas & C. Martínez-Roger: A&A, **140**, 261 (1999)
3. K. Biazzo, S. Catalano, A. Frasca & E. Marilli: MSAIS, **5**, 109 (2004)
4. K. Biazzo, A. Frasca, G.W. Henry et al.: ApJ, in press (2006)
5. K. Biazzo, A. Frasca, S. Catalano & E. Marilli: AN, submitted (2006)
6. K. Biazzo, L. Pasquini, L. Girardi et al.: A&A, in preparation (2006)
7. S. Catalano, K. Biazzo, A. Frasca, E. Marilli: A&A, **394**, 1009 (2002)
8. J.A. Cardelli, G.C. Clayton & J.S. Mathis: ApJ, **345**, 245 (1989)
9. O.J. Eggen: ApJ, **166**, 87 (1971)
10. A. Frasca, K. Biazzo, S. Catalano et al.: A&A, **432**, 647 (2005)
11. L. Girardi, L. da Silva, L. Pasquini et al.: in *Resolved Stellar Populations*, ed by D. Valls-Gaband, M. Chavez (ASP Conf. Ser.: San Francisco) in press
12. D.F. Gray: *The Observation and Analysis of Stellar Photospheres*, 3rd edn (Cambridge University Press 2005)
13. G.F. Gray & K. Brown: PASP, **113**, 723 (2001)
14. D.F. Gray & H.L. Johanson: 1991, ApJ, **103**, 439 (1991)
15. V.V. Kovtyukh & N.I. Gorlova: A&A, **358**, 587 (2000)

16. S. Meibom, J. Andersen, & B. Nordström: A&A **386**, 187 (2002)
17. L. Pasquini, S. Randich, M. Zoccali et al.: A&A **951**, 963 (2004)
18. L. da Silva, L. Girardi, L. Pasquini et al.: A&A, **458**, 609 (2006)

UVES and CRIRES Spectroscopy of AGB Stars: Technetium and the Third Dredge-up

Stefan Uttenthaler[1,2], Hans Ulrich Käufl[1], Josef Hron[2], Thomas Lebzelter[2], Maurizio Busso[3], and Mathias Schultheis[4]

[1] ESO Garching, Germany `suttenth;hukaufl@eso.org`
[2] Dept. of Astronomy, Univ. of Vienna, Austria `hron;lebzelter@astro.univie.ac.at`
[3] Dept. of Physics, Univ. of Perugia, Italy `busso@fisica.unipg.it`
[4] Observatoire de Besançon, France `mathias@obs-besancon.fr`

Summary. We present high-resolution UVES and CRIRES spectra of bright Oxygen-rich asymptotic giant branch (AGB) stars located in the outer galactic bulge. The UVES spectra were searched for absorption lines of the unstable element Technetium. This element is synthesised in deep layers of AGB stars, thus it is a reliable indicator of both recent s-process activity and third dredge-up. We aim to test theoretical predictions on the luminosity limit for the onset of third dredge-up. Additionally, our investigations have some implications for the age estimate of bulge AGB stars. The near-infrared spectra taken with the recently commissioned CRyogenic InfraRed Echelle Spectrograph CRIRES at ESO's VLT aim at studying the connection between the occurrence of Tc, Carbon isotopic ratios and the luminosity of the targets.

1 Introduction

The Asymptotic Giant Branch (AGB) phase forms the late stage of stellar evolution for initial masses between about 0.8 and 8 $M_{\odot}$. During the preceding stage on the Horizontal Branch, the star burns Hydrogen in a shell around the Helium burning core. Once the Helium is exhausted in the core, it burns in a shell around a Carbon-Oxygen core and the Early Asymptotic Giant Branch (E-AGB) phase begins. In the most luminous part of the AGB, the behaviour of a star is characterised by Thermal Pulses (TP). These are recurrent events of explosive He shell burning accompanied by changes in luminosity, temperature, pulsation period and internal structure (see, e.g., [1] for a review). Between TPs, heavy elements can be produced via the "slow neutron capture process" (s-process, see e.g. [9]) in the region between the hydrogen and the helium burning shells. The processed material is then brought to the stellar surface by the convective envelope that temporarily extends to these very deep layers. This event is called the third dredge-up (3DUP) and is the cause of the eventual metamorphosis of an Oxygen-rich M-star to a Carbon-rich C-star.

Along the s-process path in the table of nuclei lies the element Technetium (Tc) which has no stable isotopes. The isotope of Tc with the longest half-life time produced via the s-process is ^{99}Tc with $\tau_{1/2} = 2.1 \times 10^5$ years. This makes Tc a reliable indicator of the 3DUP, because due to the short life time any Tc we see in a star has been produced during its previous evolution on the TP-AGB and transported to the photosphere by convection.

In an earlier paper [5], we studied the Tc content of a sample of luminosity selected galactic field AGB stars (see also references in that work for earlier observations of Tc). The aim was to test theoretical predictions ([7, 1, 6], and references therein) on the minimum luminosity for 3DUP to occur. The results of that study suggest the theoretical minimum bolometric magnitude of $M_{\mathrm{bol}} = -3\overset{\mathrm{m}}{.}9$ to be in agreement with the observations.

Due to the uncertainties in distance of field AGB-stars (Hipparcos parallaxes for AGB stars have large systematic errors, as the AGB stellar diameters are of the order of several AUs) no definite conclusions could be drawn.

2 UVES Observations

To improve the situation we conducted observations of a sample of 27 luminosity selected AGB stars located in the outer galactic bulge (PG3 field) with the UVES spectrograph. Four stars in this sample were finally identified to contain Tc in their atmospheres.

The periods of the four Tc-rich stars are around or above 300 d. The two longest period stars both show Tc. This is in good agreement with the results for field stars [5], where an increasing fraction of Tc-rich stars for pulsation periods above 300 d is found.

The most important results of our investigation are summarised in Fig. 1. In this colour-magnitude diagram of our sample stars the theoretical minimum 3DUP luminosity, two isochrones from [2], and a section of an evolutionary track from O. Straniero (priv. comm.) are included. The evolutionary model track is similar to the $1.5M_\odot$ models presented in [7]. The filled symbols represent the four stars identified to contain Tc. For the construction of this diagram a correction of the depth-induced brightness scatter was performed by taking into account the difference in brightness to the logP - K relation of [3]. The only effect of this correction on the conclusions is that the Tc-rich stars become the brightest objects at a given $(J - K)_0$ colour. This can be understood if they are the most massive objects in the present sample: The Tc stars are the brightest objects at any time over the TP cycle at a given colour (= temperature).

As is evident, all of the Tc-rich stars fall above the theoretical 3DUP luminosity limit, drawn as dotted horizontal line. Thus, like in [5], theoretical predictions can be regarded to be in agreement with observations. However, since Tc is detected at all in bulge AGB stars, and from the agreement in luminosity between the evolutionary track and the observations (and the lack thereof for the "old" isochrones), an age of around 3 Gyrs can be estimated

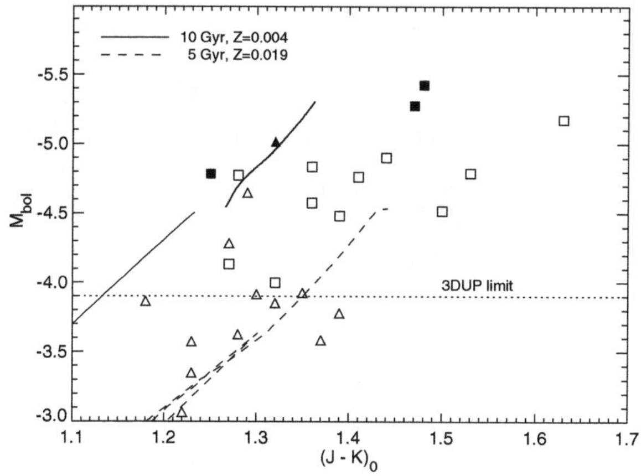

Fig. 1. Colour-magnitude diagram of the sample stars. Squares and triangles represent Mira and Semiregular variables, respectively. The filled symbols are stars with positive Tc detection. The dotted horizontal line shows the estimated minimum luminosity required for 3DUP (see [5]), the thin solid and dashed lines are isochrones from [2] (see legend for the adopted age/metallicity combinations), and the thick solid line is a section of an evolutionary model from O. Straniero between the last two thermal pulses. The evolution runs from the lower left (shortly after the TP) to the upper right (shortly before the next TP) along the track.

for these objects. A similar estimate was also presented by [4], in contrast to the bulge age estimate of 10 Gyrs derived by [10].

For more details on the data, the analysis and results we refer to [8].

3 CRIRES Observations

In the course of the commissioning of the CRyogenic InfraRed Echelle Spectrograph (CRIRES) in May and August 2006 high-resolution H- and K-band spectra of some of the sample stars were taken. For a more detailed presentation of this NIR spectrograph at ESO's VLT we refer to the contribution by H. U. Käufl et al. in these proceedings. The aim is to determine $^{12}C/^{13}C$ and C/O ratios and to correlate them with the occurrence of Tc and the luminosity of the targets. The desired values will be determined using state-of-the-art atmospheric models (MARCS code) and spectral synthesis techniques. The information acquired will clarify the usability of the mentioned ratios as a dredge-up indicator. Since in LMC and SMC giants no information on the occurrence of Tc can be gathered (due to the low flux of these objects in the blue spectral region) reliable substitute indicators have to be established. In the light of the age estimate for the targets discussed above, any abundance measurements are of high value to solve the obvious discrepancy.

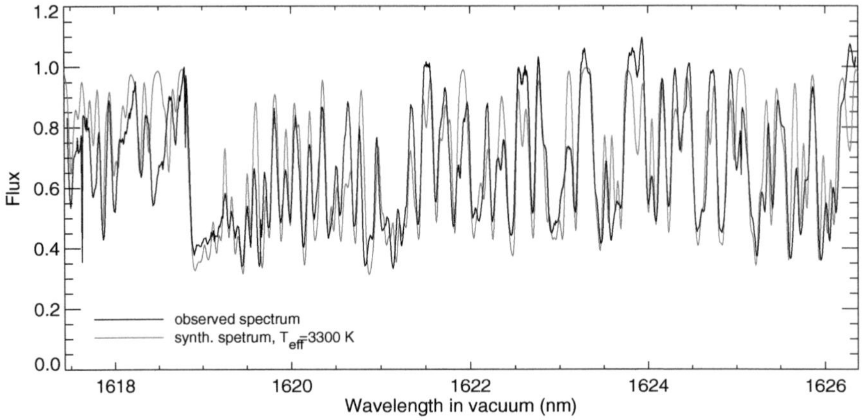

Fig. 2. H-band spectrum of a Tc-rich bulge star observed with CRIRES, and a synthetic spectrum based on a hydrostatic MARCS model atmosphere with $T_{\rm eff} =$ 3300 K. The band head at 1619 nm is due to $^{12}C^{16}O$, most of the other deep features are caused by the OH molecule.

The data analysis is still in progress, as a preliminary result we show in Fig. 2 a comparison between an observed CRIRES spectrum of a Tc-rich star and a synthetic spectrum based on a model atmosphere with $T_{\rm eff} = 3300$ K, solar metallicity, $log(g) = 0.0$, and solar C and O abundances.

Acknowledgements. We wish to thank B. Aringer, M. Gorfer, M. Messineo and O. Straniero for their diverse contributions to this work. Also, we would like to take this opportunity to thank the Paranal observatory staff and the CRIRES commissioning team for their engagement in taking the CRIRES spectra. TL acknowledges funding by the Austrian Science Fund FWF under project P18171.

References

1. M. Busso, R. Gallino & G. J. Wasserburg, ARA&A **37**, 239, (1999)
2. L. Girardi, A. Bressan, G. Bertelli et al., A&AS **141**, 371 (2000)
3. I. S. Glass & M. Schultheis, MNRAS **345**, 39–48 (2003)
4. M. A. T. Groenewegen & J. A. D. L. Blommaert, A&A **443**, 143 (2005)
5. T. Lebzelter & J. Hron, A&A **411**, 533–542 (2003)
6. M. Lugaro, F. Herwig, J. C. Lattanzio et al., ApJ **586**, 1305 (2003)
7. O. Straniero, A. Chieffi, M. Limongi et al., ApJ **478**, 332 (1997)
8. S. Uttenthaler, J. Hron, T. Lebzelter et al., astro-ph/0610500
9. G. Wallerstein, I. Iben, P. Parker et al., Rev. Mod. Phys. **69**, 995 (1997)
10. M. Zoccali, A. Renzini, S.Ortolani et al., A&A **399**, 931 (2003)

Characterisation of the Ursa Major Group

Matthias Ammler[1] and Eike W. Guenther[2]

[1] AIU Jena, Schillergäßchen 2-3, D-07745 Jena ammler@astro.uni-jena.de
[2] TLS Tautenburg, Sternwarte 5, D-07778 Tautenburg
guenther@tls-tautenburg.de

1 Introduction

The Ursa Major (UMa) group comprises the stars in the UMa cluster in the Big Dipper constellation and many co-moving stars spread over the whole sky. It is well-suited for many kinds of astrophysical studies since the members are close, bright, and have high proper motion. In addition the relatively young age of $\approx$ 300 Myrs [22] motivates evolutional studies and the search for low-mass companions.

These prospects are confronted with two complications. Firstly the real age of the UMa group is not yet well-established. Recent age estimates range from 200 Myrs [13] to 600 Myrs [12]. Secondly the kinematic membership status in the recent kinematic member lists of [16] and [12] agrees for only 7 out of 20 common stars.

The use of kinematic membership criteria becomes questionable but there are indications that spectroscopic features of activity allow to find mid- and late-type UMa group members [12, 9]. Knowledge on the stellar parameters, e.g. effective temperature and surface gravity, then is indispensable. [9] is the only recent work presenting homogeneously inferred accurate stellar parameters of mid- and late-type UMa group members. However this work is restricted to the few UMa group members within 25 pc so that we decided to extend this work beyond 25 pc and selected the 56 kinematic members from the lists of [16] (fulfilling both of Eggen's criteria) and [12] (kinematic membership assignment *Y*).

2 Observations, Data Reduction and Analysis

Échelle spectra covering the optical wavelength range with high resolving power ($\lambda/\Delta\lambda \gtrsim 60,000$) and high signal-to-noise ratio (mostly $\gtrsim 200$ near Hα) were obtained with *FOCES* on Calar Alto and the Coudé-Échelle spectrograph in Tautenburg. The data reduction follows the usual scheme. Special care was taken of the rectification of the relative continuum which is crucial for the derivation of effective temperatures from the wings of the Balmer lines [6, 15].

The analysis follows [8] and is based on LTE model atmospheres developed by T. Gehren [21]. The analysis is differential with respect to the Sun and therefore works for dwarf and subgiant stars with spectral types late-F to early-K, i.e. 19 out of the 56 kinematic UMa group members. We add analyses of 10 stars since 6 stars have already been analysed by the same methods [9, 14] and 3 stars are located on the southern sky and could not be observed. The stellar parameters are derived consistently from the observed spectra. Effective temperature is inferred by fitting synthetic Balmer line profiles to the observed Hα and Hβ profiles. Surface gravity is determined by means of the iron ionisation equilibrium or the wings of the Mg Ib triplet. Neither method works for four stars of the sample. In these cases, surface gravity is calculated from the *Hipparcos* parallax.

3 Results and Discussion

The resulting stellar parameters are tabulated in [2] and [3]. The typical precision achieved for effective temperature is ± 100 K, for surface gravity ± 0.15 dex and for iron abundance ± 0.07 dex. The parameters are exposed to several tests in order to judge the accuracy of the derived values. The most crucial test, in particular of the surface gravity, is the comparison of spectroscopic distance with *Hipparcos* distance [9]. The scatter of the data points from this work experienced in Fig. 1 results mainly from the application of Fuhrmann's methods to stars at the hot and the cool end of the range of applicability. Nevertheless the error bars are still consistent with the *Hipparcos* distances supporting the derived stellar parameters, especially the surface gravities. See [2] and [3] for notes on individual stars.

Being convinced about the accuracy of the inferred stellar parameters we turn to the properties of the UMa group as a whole. In here we concentrate on two main results, iron abundance and lithium depletion. Our results indicate a mean iron abundance of -0.03 ± 0.04 which is higher by 0.05 dex and 0.06 dex than previously determined means: -0.08 ± 0.09 [5] and -0.09 ± 0.02 [22], respectively. The UMa group iron abundance is quite different from the Hyades while it is at best marginally different from the Pleiades ($[\mathrm{Fe/H}]_{\mathrm{Hyades}} = 0.13 \pm 0.05$ and $[\mathrm{Fe/H}]_{\mathrm{Pleiades}} = -0.03 \pm 0.02$, [5]).

The amount of lithium absorption in the UMa group can be distinguished from both the Pleiades and the Hyades. Lithium absorption may be used as membership criterion at least for UMa group candidates cooler than the Sun and hotter than $T_{\mathrm{eff}} \approx 5000$ K. However this region is sparsely populated and definitely more data points have to be added in order to draw meaningful conclusions. At this point, the lithium equivalent widths confirm the previously determined age range of $200 - 600$ Myrs although a smaller intrinsic age spread is indicated.

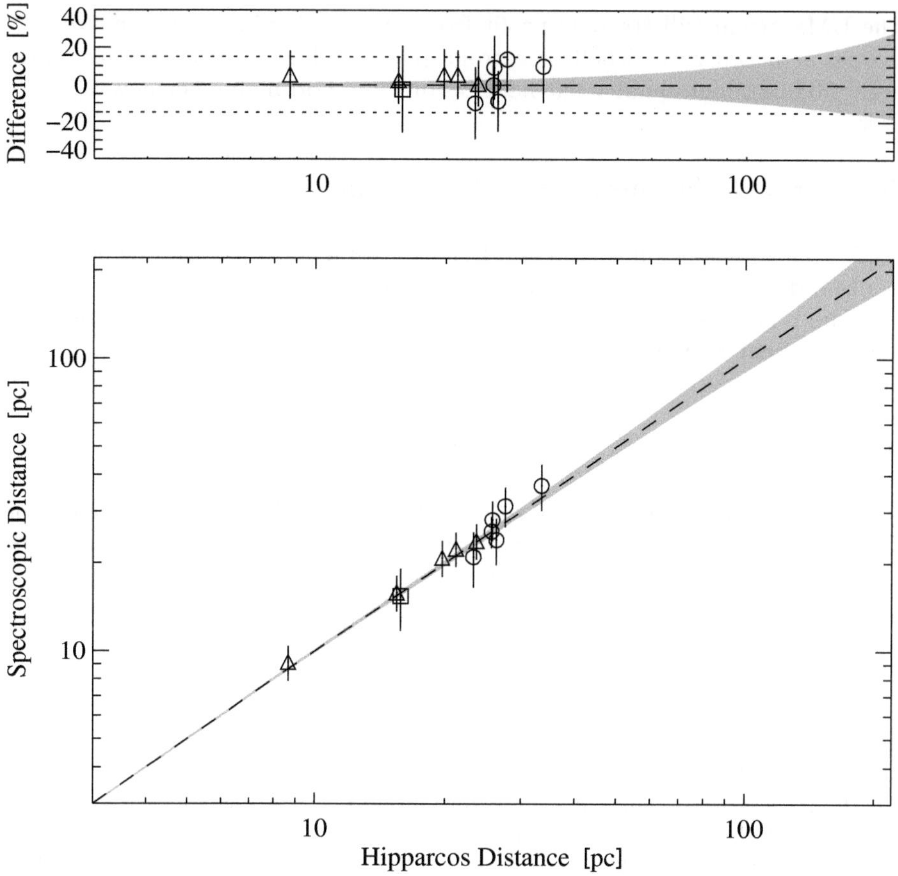

Fig. 1. Comparison of spectroscopic distances with *Hipparcos* distances – The long-dashed line represents perfect agreement with *Hipparcos*. The grey-shaded area displays the *Hipparcos* errors. The short-dashed lines in the upper panel represent the typical individual uncertainties of 10 − 15 % experienced by [9]. This work (circles) adds ten to the five analyses of kinematic members by [9] (triangles). The square represents the spectroscopic distance of HD 24916 A which was calculated from the parameters given by [14].

4 Requests

Surface gravity could not be derived for four stars of the sample possibly because of features of stellar youth which hamper the analysis. Better theoretical understanding of the line formation and the atmospheres in young stars is indispensable.

The 2m-class telescopes turned out to be hardly sufficient for the observations of mid- and late-type UMa group members beyond 25 pc. The study of the UMa group will strongly profit from present-day intermediate-size telescopes attached with Échelle spectrographs with high spectral resolution (at least 40,000), high-quality flat-field calibration, and high efficiency.

5 Acknowledgements and Remarks

M.A. thanks Klaus Fuhrmann and Brigitte König for providing the necessary tools and support for the quantitative spectral analysis.

The reader is referred to [2] and [3] for a full presentation of this work and pertinent references.

References

1. Ammler, M., Guenther, E.W., König, B. and Neuhäuser, R.: High-resolution spectroscopy of the UMa group. In: *Proceedings of 'The 13th Cambridge Workshop on Cool Stars, Stellar Systems and the Sun'*, ed by F. Favata, G.A.J. Hussain and B. Battrick (ESA), 2005, p.391
2. Ammler, M.: Characterisation of young nearby stars – The Ursa Major group. PhD Thesis, University of Jena, Germany (2006a)
3. Ammler, M. and Guenther, E.W.: A&A, to be submitted (2006b)
4. Barrado y Navascués, D., Deliyannis, C.P. and Stauffer, J.R.: ApJ **549**, 452 (2001)
5. Boesgaard, A.M. and Friel, E.D.: ApJ **351**, 467 (1990)
6. Fuhrmann, K.: Temperatur und Elementhäufigkeiten von F- und G-Sternen. PhD Thesis, LMU München (1993)
7. Fuhrmann, K., Pfeiffer, M., Frank, C., Reetz, J. and Gehren, T.: A&A **323**, 909 (1997)
8. Fuhrmann, K.: A&A **338**, 161 (1998)
9. Fuhrmann, K.: Astronomische Nachrichten **325**, 3 (2004)
10. Hatzes, A.P., Guenther, E.W., Endl, M., Cochran, W.D., Döllinger, M.P. and Bedalov, A.: A&A **437**, 743 (2005)
11. Kenworthy, M., Hofmann, K.-H., Close, L., Hinz, P., Mamajek, E., Schertl, D., Weigelt, G., Angel, R., Balega, Y.Y., Hinz, J. and Rieke, G.: ApJL **554**, L67 (2001)
12. King, J.R., Villarreal, A.R., Soderblom, D.R., Gulliver, A.F. and Adelman, S.J.: AJ **125**, 1980 (2003)
13. König, B., Fuhrmann, K., Neuhäuser, R., Charbonneau, D. and Jayawardhana, R.: A&A **394**, L43 (2002)
14. König, B., Guenther, E.W., Esposito, M. and Hatzes, A.: MNRAS **365**, 1050 (2006)
15. Korn, A.J.: Rectifying échelle Spectra – A Comparison Between UVES, FEROS and FOCES. In: *Scientific Drivers for ESO Future VLT/VLTI Instrumentation, Proceedings of the ESO Workshop held in Garching, Germany, 11-15 June, 2001*, ed by J. Bergeron, G. Monnet (Springer, Berlin Heidelberg New York 2002) pp 199–204

16. Montes, D., López-Santiago, J., Gálvez, M. C., Fernández-Figueroa, M.J., De Castro, E. and Cornide, M.: MNRAS **328**, 45 (2001)
17. Potter, D., Martín, E.L., Cushing, M.C., Baudoz, P., Brandner, W., Guyon, O. and Neuhäuser, R.: ApJL **567**, L133 (2002)
18. Perryman, M.A.C., Lindegren, L., Kovalevsky, J., Hoeg, E., Bastian, U., Bernacca, P.L., Crézé, M., Donati, F., Grenon, M., van Leeuwen, F., van der Marel, H., Mignard, F., Murray, C.A., Le Poole, R.S., Schrijver, H., Turon, C., Arenou, F., Froeschlé, M. and Petersen, C.S.: A&A **323**, L49 (1997)
19. Perryman, M.A.C., Brown, A.G.A., Lebreton, Y., Gomez, A., Turon, C., de Strobel, G.C., Mermilliod, J.C., Robichon, N., Kovalevsky, J. and Crifo, F.: A&A **331**, 81 (1998)
20. Pfeiffer, M.J., Frank, C., Baumueller, D., Fuhrmann, K. and Gehren, T.: A&AS **130**, 381 (1998)
21. Reile, C.: MA Thesis, LMU München (1987)
22. Soderblom, D.R. and Mayor, M.: AJ **105**, 226 (1993)
23. Terndrup, D.M., Stauffer, J.R., Pinsonneault, M.H., Sills, A., Yuan, Y., Jones, B.F., Fischer, D. and Krishnamurthi, A.: AJ **119**, 1303 (2000)

What Sara Ellison is saying at the conference dinner seems very convincing for Andreas Quirrenbach, but Michael Murphy is skeptical.

[C/O] Observations in Low-[Fe/H] Halo Stars

D. Fabbian[1], P. E. Nissen[2], M. Asplund[1], C. J. Akerman[3], M. Pettini[3]

[1] Research School of Astronomy & Astrophysics, The Australian National University, Mount Stromlo Observatory, Cotter Road, Weston ACT 2611, Australia damian@mso.anu.edu.au

[2] Institute of Physics and Astronomy, University of Aarhus, DK-8000 Aarhus C, Denmark

[3] Institute of Astronomy, University of Cambridge, Madingley Road, Cambridge, CB3 0HA, United Kingdom

Summary. We have observed 15 halo stars to determine the [C/O] behaviour at low [Fe/H]. Making use of our recent non-LTE calculations, which show that the high excitation C and O lines used in previous studies in the literature are affected by very significant departures from LTE, we aim to obtain accurate [C/O] ratios down to [Fe/H]~ -3.2, which will enable us to shed light on the possible presence of an upturn of [C/O] at low metallicities.

Oxygen is thought to originate from SNe generated at the death of short-lived massive stars and should thus increasingly influence the [C/O] trend at low metallicity, as seen in "standard" Galactic Chemical Evolution models without Pop. III stars. The origin of carbon is still debated, some authors suggesting that C comes mainly from low- and intermediate-mass stars, others arguing that massive stars are more important. Early observational results seemed to point at an almost flat trend at a level [C/O]~ -0.5 down to low [Fe/H] [5]. Recently, Akerman et al. [1] found an apparent [C/O] upturn to near-solar values at low metallicity, which was confirmed by Spite et al. [4]. However, the CH lines employed in this latter study are likely affected by large negative 3D corrections [2]. Akerman et al. (2004) argued that the increasing [C/O] trend with decreasing O abundance for [O/H] < -1, may signal C-rich ejecta from massive Pop. III SNe. Those authors warned that this tentative claim needed to be verified with more observations and theoretical investigations on non–LTE effects for C and O. Since at that stage no extensive study was available addressing the non–LTE effects for C in late-type stars down to low [Fe/H], they based their analysis on the assumption that such effects would balance out for the two elements.

We have embarked on a study of the non-LTE effects for both C [3] and O (Fabbian et al., in prep.). We find large negative non–LTE abundance corrections for the C I lines around 910 nm and the O I triplet at 777 nm, both of which were employed by Akerman et al. We are in the process of applying those results to their sample and to our new observational data, in this way targeting systematic errors and extending the sample of halo stars with accurately determined [C/O] values. To this purpose, high-resolution (R$\sim$ 50000) and high signal-to-noise (S/N$\sim 250-500$) observations were collected at the

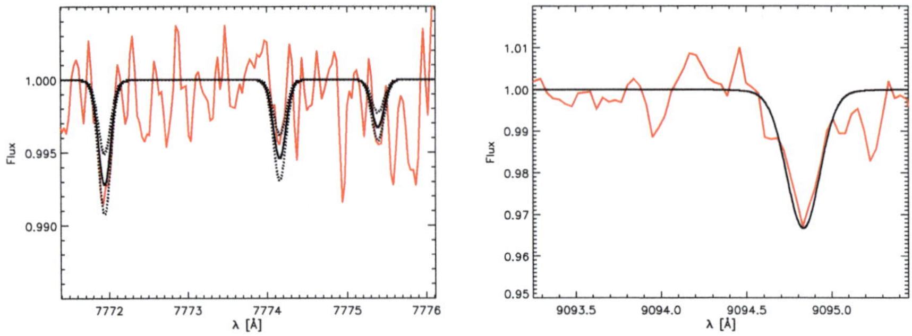

Fig. 1. Left panel: spectrum synthesis of the O I 777 nm triplet lines, for one of the lowest metallicity stars in our sample (CD-24_17504, at [Fe/H]~ -3.2). Observed and best-fit computed line profiles are shown (thin and thick solid lines respectively). The dotted lines mark the profiles obtained with O abundance decreased and increased respectively by 1 σ with respect to the best-fit computed LTE value of $\log \epsilon_O = 6.16$ (χ^2 fitting was used for these very weak lines). Right panel: observed and computed (with LTE abundance of $\log \epsilon_C = 5.71$, derived from equivalent width matching) line profiles for one of the high-excitation near-IR C I absorption features employed, as detected in CD-24_17504.

Very Large Telescope using UVES and at the Magellan telescope with the MIKE spectrograph. This will allow us to derive C and O in a sample of 15 metal-poor halo stars down to [Fe/H]~ -3.2, in addition to those stars observed by Akerman et al. The [C/O] thus determined is robust to errors in temperature but affected by departures from LTE. To tackle the systematic errors which might have been afflicting previous studies in the literature, we will make use of our detailed non–LTE calculations for C and O (showing for both elements more negative non–LTE effects at low metallicity than previously thought) on a star-by-star basis and thus determine the [C/O] ratio accurately down to very low metallicity. A detailed report on the results from this study will be published shortly (Fabbian et al., in prep.).

This work is partly funded by the Australian Research Council. DF also thanks the Astronomical Society of Australia and the conference organizers for their support.

References

1. C. J. Akerman, L. Carigi, P. E. Nissen, M. Pettini and M. Asplund: A&A **414**, 931 (2004)
2. M. Asplund: ARA&A **43**, 481 (2005)
3. D. Fabbian, M. Asplund, M. Carlsson, D. Kiselman: A&A **458**, 899 (2006)
4. M. Spite, R. Cayrel, B. Plez et al: A&A **430**, 655 (2005)
5. J. Tomkin, M. Lemke, D. L. Lambert, C. Sneden: AJ **104**, 1568 (1992)

Oxygen Abundances in Metal-poor Stars, from [OI], OI and IR OH Lines

B. Barbuy[1] and J. Meléndez[2]

[1] University of São Paulo, IAG, Rua do Matão 1226, São Paulo 05508-900, Brazil barbuy@astro.iag.usp.br
[2] Australian National University, Mount Stromlo Observatory, Weston ACT 2611, Australia jorge@mso.anu.edu.au

We derive Oxygen-to-iron ratios for 25 stars in the metallicity range $-3.0 <$ [Fe/H] < -1.2. A plateau oxygen-to-iron ratio [O/Fe] $\approx +0.4$ dex is found, which is independent of evolutionary state (dwarfs or giants), and for lower metallicities (down to [Fe/H] ≈ -3) the [O/Fe] ratio is about $+0.5$ dex.

1 Introduction

Oxygen abundances in a sample of 25 metal-poor stars ($-3.0 <$ [Fe/H] < -1.2) were determined from infrared OH lines in the H-band, based on new observations of seven metal-poor stars with the Phoenix high-resolution spectrograph at the 4m Kitt Peak telescope and our previous infrared spectra (Phoenix + 2.1 m Kitt Peak telescope and Nirspec + Keck II telescope). For most of the sample stars oxygen abundances were also determined from [OI] and OI lines, based on high resolution high S/N spectra obtained with the FEROS spectrograph at ESO or data available in the literature.

2 Observations

High-resolution infrared spectra were obtained from images taken at the 4m Kitt Peak telescope, using the Phoenix spectrograph (Hinkle et al. 2000, 2003), on 2001 March 12. Employing a 4-pixel slit a FWHM resolution of 50 000 was achieved. The IR spectra cover the wavelength range 1.55 - 1.56 μm. Optical spectra were obtained at the 1.52m telescope at ESO, La Silla, using the Fiber Fed Extended Range Optical Spectrograph (FEROS) (Kaufer et al. 2000). The total spectrum coverage is 356-920 nm with a resolving power of 48 000, covering a large number of FeI and FeII lines, as well as the forbidden and permitted oxygen lines.

3 Analysis

The new infrared spectroscopic observations were carried out for a sample of seven stars with metallicities in the range -3.0 < [Fe/H] < -1.2. The results from our previous work (Meléndez et al. 2001; Meléndez & Barbuy 2002; Barbuy et al. 2003) are revised, analyzing all sample stars in a consistent way employing Kurucz overshooting model atmospheres [3] for both solar and stellar spectra. In total, the sample consists of twenty five metal-poor (-3.0 < [Fe/H] < -1.2) dwarfs, subgiants and giants.

Using new IRFM temperature scale by Ramírez & Meléndez (2005a,b) temperatures T_{eff} were calculated, and they showed to be different by more than 40 K from the T_{eff} obtained with the previous calibrations; this was the case of the K dwarfs (and a few giants), because the new calibrations have a better coverage of metal-poor K dwarfs and should result in more reliable internal temperatures.

Trigonometric gravities were derived for stars having accurate Hipparcos parallaxes We also derived evolutionary gravities employing α-element enhanced isochrones by Kim et al. (2002). For cases where Hipparcos gravities have large errors the evolutionary gravities were preferred.

Metallicities were derived employing both FeI and FeII lines, with equivalent widths measured on high resolution optical spectra. Microturbulence velocities v_t were obtained by requiring no dependence of [Fe/H] against reduced equivalent width.

ATLAS9 and MARCS models are employed. Also, three different codes are checked: MOOG (Sneden 1973), PFANT (Cayrel et al. 1991; Barbuy et al. 2003) and Turbospectrum (Alvarez & Plez 1998).

3.1 FeII Oscillator Strengths

In this work, a revision of FeII oscillator strengths is presented, based on a combination of laboratory and theoretical determinations. The bulk of laboratory gf-values are probably correct in an absolute scale, but there are large uncertainties on a line-by-line basis. On the other hand, although the theoretical calculations are not always correct in an absolute scale, the theoretical relative line intensities within multiplets should be highly reliable. In Meléndez & Barbuy (2002) we exploited the advantages of both laboratory and theoretical methods, adopting relative line intensities within a given multiplet from theoretical calculations, whereas the absolute transition probabilities for each multiplet were determined from laboratory measurements. For the present work we revised the gf-values of FeII lines. One example of the improvement of gf-values is shown in Fig. 1, where we derive solar abundances, using Biémont et al. (1991)'s and revised values.

4 Conclusions

The agreement between the oxygen abundance derived from the forbidden [OI] and infrared OH lines is very impressive, with a mean zero difference

[3] http://kurucz.harvard.edu

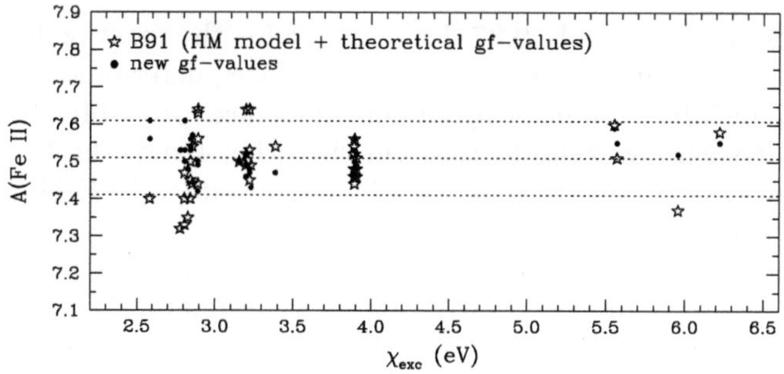

Fig. 1. Fe abundances for the Sun, using gf-values from Biémont et al. 1991, and revised values from the present work.

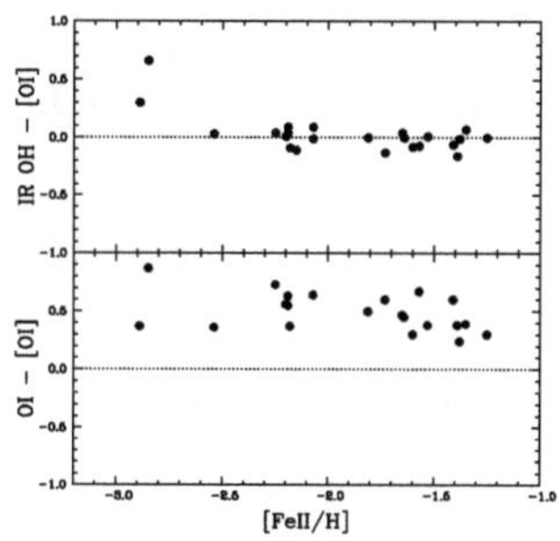

Fig. 2. Δ[O/Fe] between the oxygen-to-iron obtained from the IR OH lines and the [OI]630nm line (upper panel), and between the permitted OI triplet and the [OI]630nm line (lower panel).

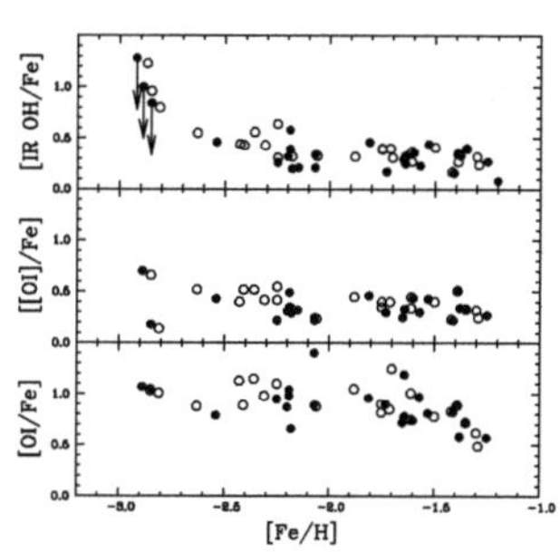

Fig. 3. [O/Fe] vs. [Fe/H] based on IR OH (upper panel), forbidden [OI]630nm (middle panel) and permitted OI triplet lines (lower panel).

(star-by-star scatter of only 0.07 dex), except for the most metal-poor stars with [Fe/H] ≈ -3, where the oxygen abundances determined from IR OH lines are artificially enhanced by about +0.5 dex, probably due to strong 3D effects at such low metallicities. On the other hand, the O abundances derived from the OI triplet are artificially enhanced (in LTE 1D analysis) at all probed metallicities by about +0.5 dex.

Solar and stellar abundances were determined in a consistent way employing both the Kurucz and OSMARCS model atmospheres. We obtain that the sample stars with metallicities $-2.5 <$ [Fe/H] < -1.2 show a plateau oxygen-to-iron ratio [O/Fe] $\approx +0.4$ dex, which is independent of evolutionary state (dwarfs or giants), and for lower metallicities (down to [Fe/H] ≈ -3) the [O/Fe] ratio is about +0.5 dex, for a solar oxygen abundance of ϵ(O) = 8.77. These results confirm previous findings by e.g. Barbuy (1988), García-Perez et al. (2006) and Meléndez et al. (2006). For metallicities lower than [Fe/H]<-3.0, Cayrel et al. (2004) indicate a somewhat higher oxygen-to-iron ratio. [4]

References

1. Alvarez, R., Plez, B. 1998, A&A, 330, 1109
2. Barbuy, B. 1988, A&A, 191, 121
3. Barbuy, B., Perrin, M.-N., Katz, D., Coelho, P., Cayrel, R., Spite, M., Van't Veer-Menneret, C. 2003, A&A, 404, 661
4. Cayrel, R., Perrin, M.-N., Barbuy, B., Buser, R. 1991, A&A, 247, 108
5. Cayrel, R., Depagne, E., Spite, M., Hill, V., Spite, F. et al. 2004, A&A, 416, 1117
6. Biémont, E., Baudoux, M., Kurucz, R.L., Ansbacher, W., Pinnington, A.E., 1991, A&A, 240, 539
7. García-Pérez, A.E., Asplund, M., Primas, F., Nissen, P.E., Gustafssson, B. 2006, A&A, 451, 621
8. Hinkle, K.H., Joyce, R.R., Sharp, N., Valenti, J.A. 2000, SPIE, 4008, 720
9. Hinkle, K.H., Joyce, R.R., Hedden, A., Wallace, L., Engleman, R. Jr. 2001, PASP, 113, 548
10. Kaufer, A., Stahl, O., Tubbesing, S. et al. 2000, Proc. SPIE, 4008, 459
11. Kim, Y.-C., Demarque, P., Yi, S.K., Alexander, D.R., 2002, ApJS, 143, 499
12. Meléndez, J., Barbuy, B. & Spite, F. 2001, ApJ, 556, 858
13. Meléndez, J. & Barbuy, B. 2002, ApJ, 575, 474
14. Meléndez, J., Shchukina, N.G., Vasiljeva, I.E., Ramírez, I. 2006, astro-ph/0601256
15. Ramírez, I., Meléndez, J. 2005a, ApJ, 626, 446
16. Ramírez, I., Meléndez, J. 2005a, ApJ, 626, 465
17. Sneden, C. 1973, ApJ, 184, 839

[4]**Acknowledgements: B.B. acknowledges CNPq and Fapesp.**

Sulphur Abundances in Metal-poor Stars

P.E. Nissen[1], C. Akerman[2], M. Asplund[3], D. Fabbian[3], and M. Pettini[2]

[1] Department of Physics and Astronomy, University of Aarhus, DK-8000 Aarhus C, Denmark pen@phys.au.dk
[2] Institute of Astronomy, University of Cambridge, Madingley Road, Cambridge, CB3 0HA, UK
[3] Research School of Astronomy and Astrophysics, Mount Stromlo Observatory, Cotter Road, Weston ACT 2611, Australia

Summary. We report on sulphur abundances in halo stars as derived from near-IR UVES spectra. The importance of removing telluric lines and residual CCD fringing patterns by using early B-type stars as calibrators is emphasized. Comparison of data from the weak $\lambda 8694.6$ and the stronger $\lambda 9212.9, 9237.5$ pair of S I lines provides important constraints on non-LTE effects. We do not confirm the high sulphur abundances reported by others for some metal-poor stars; our results instead indicate that sulphur behaves like other typical α-capture elements with a plateau at [S/Fe] $\sim +0.3$ dex in the Galactic halo.

1 Introduction

Despite several recent works on the abundance of sulphur in Galactic stars, the trend of [S/Fe] vs. [Fe/H] for halo stars is still being debated. Some studies [8] [9] show that [S/Fe] is approximately constant at a level of +0.3 dex in the metallicity range $-3 <$ [Fe/H] < -1 as also predicted from chemical evolution models of the Milky Way, with Type II SNe being the dominant source of element production up to [Fe/H] ~ -1. Other investigations [5] [10] suggest, however, an increasing trend of [S/Fe] towards lower metallicities with [S/Fe] reaching +0.8 dex at [Fe/H] ~ -2. Such high values of [S/Fe] may be explained if SNe with very large explosion energies, so-called hypernovae, make a substantial contribution to the nucleosynthesis of elements in the early Galaxy [7].

In a recent paper, Caffau et al. [2] provide even more puzzling data on [S/Fe] in halo stars. For the metallicity range $-2.2 <$ [Fe/H] < -1.0 both high [S/Fe] $\sim +0.8$, and low [S/Fe] $\sim +0.3$ are found, suggesting a bimodal distribution of [S/Fe]. If real, it would complicate the interpretation of abundance data for damped Lyman-alpha systems (DLAs), for which sulphur is of particular importance as a volatile element that is not depleted onto dust.

2 Observations of Sulphur Lines

The difficulty of determining S abundances in metal-poor halo stars stems from the fact that the strongest S I lines occur in a spectral region which is severely affected by telluric absorption lines. These lines may be removed by dividing with the spectrum of an early-type B star observed together with the program star. The technique is illustrated in Fig. 1 of [8]. In order to have the method working properly it is important to observe with a stable instrument profile throughout the night. This can be a problem if the seeing is variable and the star is imaged directly onto the slit - especially if the seeing disk is less than the width of the slit. Use of an image slicer or scrambling by a fiber help to minimize the problem.

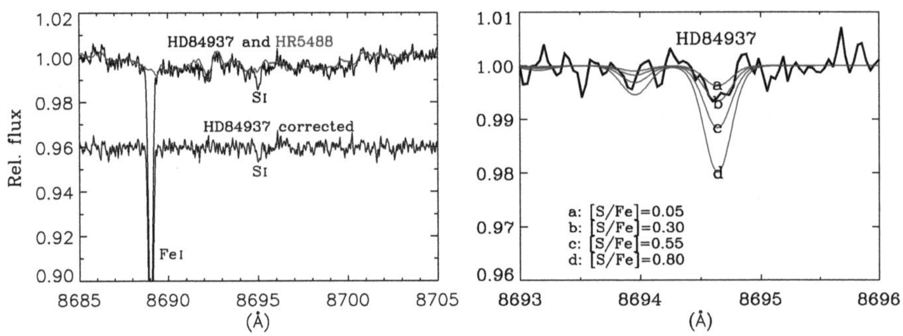

Fig. 1. *Left*: Residual fringing problem in our UVES spectrum of the metal-poor turnoff star HD 84937. After division with a 5-pixel smoothed spectrum of the B2III star HR 5488 a rectified spectrum (shifted to a continuum level of 0.96) is obtained. *Right*: Synthesis of the λ8694.6 S I line in the spectrum of HD 84937.

Another problem is CCD fringing in the near-IR part of the spectrum. As discussed by Korn & Ryde [6] this makes it difficult to determine precise S abundances of halo stars from the weak λ8694.6 S I line. B-type stars may, however, also be used to correct for residual fringing remaining after flat-fielding as illustrated in Fig. 1 here.

3 Sulphur Abundances from UVES Spectra

In [8], VLT/UVES spectra were used to derive sulphur abundances for 34 stars. Recently, we have extended the program with 12 turnoff stars having $-3.3 <$ [Fe/H] < -1.9. Effective temperatures were derived from the Hβ line profile, and iron abundances from Fe II lines.

Three bright halo stars claimed by others to have high [S/Fe] are on our program. For HD 84937, Takada-Hidai et al. [10] derive a ratio [S/Fe] $\simeq 0.6$ from the λ8694.6 line as observed with KECK/HIRES, whereas we get [S/Fe] $\simeq 0.3$ from the same line (see Fig. 1). Assuming LTE we obtain [S/Fe] $= 0.36$ and 0.27, respectively, from the λ9212.9, 9237.5 S I lines in our spectrum of HD 84937. For HD 140283, Takeda et al. [11] derive [S/Fe] ~ 1.0 from the λ8694.6 line in a UVES POP database [1] spectrum. We barely detect this line in our $S/N \sim 500$ spectrum of HD 140283; the 2-σ upper limit is [S/Fe] < 0.6, and from the $\lambda 9212.9, 9237.5$ lines of HD 140283 we derive [S/Fe] = 0.25 and 0.24, respectively. Finally, Caffau et al. [2] obtain [S/Fe] = 0.84 for HD 181743 from the $\lambda 8694.6$ line in their UVES spectrum. We cannot detect this line in our $S/N \sim 300$ spectrum, but estimate an upper limit [S/Fe] < 0.5. For both $\lambda 9212.9$ and $\lambda 9237.5$ we derive [S/Fe] = 0.3 as shown in Fig. 2. Note, that the large differences of [S/Fe] cannot be explained in terms of different values of $T_{\rm eff}$, $\log g$, [Fe/H] or the adopted solar abundances. Possibly, the S abundance derived by others from the very weak $\lambda 8694.6$ line is overestimated due to fringing residuals in their spectra.

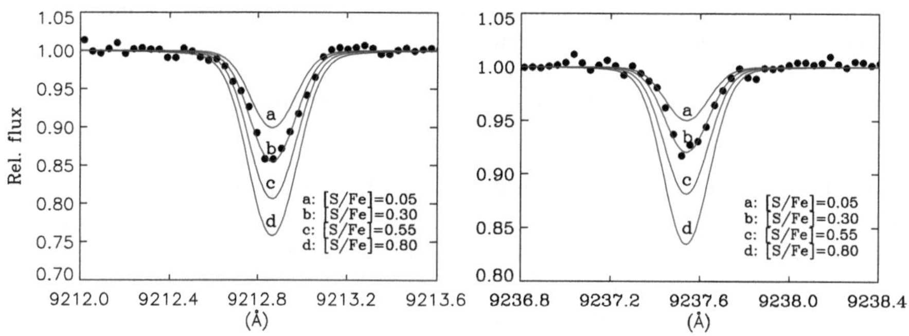

Fig. 2. Synthesis of the $\lambda 9212.9$ and $\lambda 9237.5$ S I lines in HD 181743.

According to Takeda et al. [11], the $\lambda 9212.9, 9237.5$ S I lines are significantly affected by deviations from LTE. The weak $\lambda 8694.6$ line is much less affected. As seen from Fig. 3, there is a similar good agreement between the two sets of S abundances if LTE or non-LTE with $S_H = 1$ is adopted. Here, S_H is a scaling factor applied to the classical formulae of Drawin [4] for hydrogen collisions. The case of $S_H = 0.1$ seems, however, to be excluded.

Fig. 3 (right) shows our LTE [S/Fe] values vs. [Fe/H]. As seen, the halo stars have a near-constant [S/Fe] $\sim +0.3$ with perhaps a tendency of an upturn below [Fe/H] $= -3$. If the non-LTE corrections of Takeda et al. with $S_H = 1$ are adopted, the halo trend becomes even more flat around an average value of [S/Fe] $\sim +0.2$ dex.

4 Conclusions

The present work indicates that sulphur in halo stars shows the same kind of overabundance with respect to iron as other α-capture elements. The high sulphur abundances ([S/Fe] > 0.6) reported in several recent works are not confirmed. Looking ahead, it would be interesting to follow the trend of [S/Fe] to lower metallicities than [Fe/H] $\simeq -3$, but extremely large telescopes will be

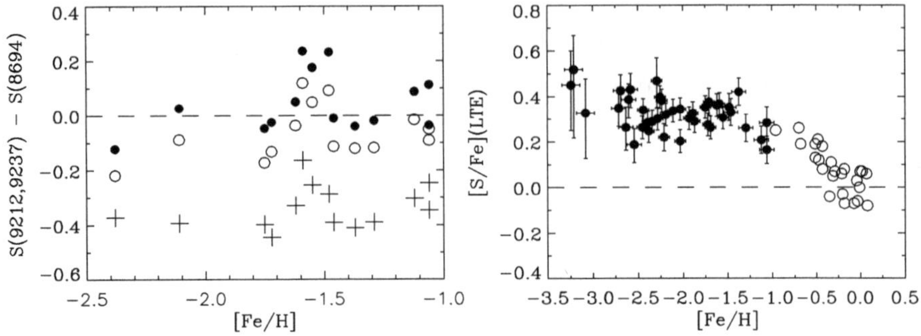

Fig. 3. *Left*: Comparison of S abundances derived from the S I λ9212.9, 9237.5 Å pair and the λ8694.6 line. Filled circles refer to LTE. Open circles include non-LTE corrections according to [11] with the hydrogen collisional parameter $S_H = 1$, and crosses refer to $S_H = 0.1$. *Right*: [S/Fe] vs. [Fe/H] for our sample of halo stars (filled circles) supplemented with disk stars (open circles) from [3].

needed to observe the weak S lines with sufficient precision. Further testing of non-LTE effects by observing S I lines at 1046 nm and the forbidden [S I] line at 1082 nm would be valuable. Finally, it is important to expand the set of measurements of sulphur and zinc (another volatile element not depleted onto dust) in DLAs in order to improve our knowledge of the star formation history of these chemically unevolved galaxies at high redshifts [12].

On the instrumental side, effort should be devoted to developing better flat-fielding methods so as to avoid the problem of residual fringing in near-IR spectra. It would also be interesting to investigate if telluric lines can be modelled with sufficient accuracy to allow their removal without wasting time on observing bright early-type stars as calibrators.

References

1. Bagnulo, S., Jehin, E., Ledoux, C., et al. 2003, ESO Messenger, 114, 10
2. Caffau, E., Bonifacio, P., Faraggiana, R., et al. 2005, A&A, 441, 533
3. Chen Y.Q., Nissen P.E., Zhao G., Asplund M. 2002, A&A, 390, 225
4. Drawin, H.W. 1968, Z. Physik, 211, 404
5. Israelian G., Rebolo R. 2001, ApJ, 557, L43
6. Korn, A.J., Ryde, N. 2005, A&A, 443, 1029
7. Nakamura T., Umeda H., Iwamoto K.I. et al. 2001, ApJ, 555, 880
8. Nissen, P.E., Chen, Y.Q., Asplund, M., Pettini, M. 2004, A&A, 415, 993
9. Ryde, N., Lambert, D.L. 2004, A&A, 415, 559
10. Takada-Hidai, M., Takeda, Y., Sato, S., et al. 2002, ApJ, 573, 614
11. Takeda, Y., Hashimoto, O., Taguchi, H., et al. 2005, PASJ, 57, 751
12. Wolfe, A.M., Gawiser, E., & Prochaska, J.X. 2005, ARA&A, 43, 861

Isotopic Abundances of Eu, Ba, and Sm in Metal-Poor Stars

Ian U. Roederer[1], Chris Sneden[1], James E. Lawler[2], Jennifer S. Sobeck[1], Catherine A. Pilachowski[3], and John J. Cowan[4]

[1] University of Texas, Dept. of Astronomy, Austin, Texas, 78712, USA
`iur@astro.as.utexas.edu`
[2] University of Wisconsin, Dept. of Physics, Madison, Wisconsin, 53706, USA
[3] Indiana University, Dept. of Astronomy, Bloomington, Indiana, 47405, USA
[4] University of Oklahoma, Dept. of Physics and Astronomy, Norman, Oklahoma, USA 73019

Summary. We have examined the isotopic mix of the heavy neutron-capture elements Eu, Ba, and Sm in three metal-poor stars with different enrichment histories. Isotopic abundances are more fundamental than elemental abundances as probes of the contributions from the rapid and slow nucleosynthesis reactions. We show preliminary results from our study, the first to examine isotopic abundances of three elements simultaneously in the same star.

1 Introduction

Nearly all heavy elements in the Universe are produced in stars by the rapid (r)- and slow (s)-processes during the late stages of stellar evolution. Understanding the nature of these processes requires detailed knowledge of their products, which can be observed today in metal-poor stars. Most neutron (n)-capture species have multiple stable isotopes, and therefore the elemental abundances are usually a composite of several isotopes; a more fundamental analysis involves measuring isotopic abundances of these species. Europium (Eu), Barium (Ba), and Samarium (Sm) are ideal species for an isotopic analysis because relatively large energy shifts exist between stable isotopes, and these species have specific measures of the r- and s-process contributions to the isotopic mix, as shown in Table 1. Previous measurements of isotopic abundances of heavy n-capture elements using line profile fits have been made for Eu ([15], [1], [2], and [8]) and Ba ([11], [9]). In this study we examine the isotopic mix of Eu, Ba, and Sm in three metal-poor stars with different enrichment histories.

2 Observations and Analysis

We know that our three target stars have very different enrichment histories from previous studies of their elemental and molecular abundances. We hypothesize that the different isotopic abundances should reflect these different

Table 1. Isotopic Mixes for Pure r- and s-process Material

Species	Pure r	Pure s	Notes	Refs
Ba	$f_{odd} = 0.46$	$f_{odd} = 0.11$	$f_{odd} = (N(^{135}\text{Ba}) + N(^{137}\text{Ba}))/N(\text{Ba})$	[3, 9]
Sm	$f_{odd} = 0.36$	$f_{odd} = 0.09$	$f_{odd} = (N(^{147}\text{Sm}) + N(^{149}\text{Sm}))/N(\text{Sm})$	[13, 10]
Eu	$f_{151} = 0.47$	$f_{151} = 0.54$	$f_{151} = N(^{151}\text{Eu})/N(\text{Eu})$	[15, 2]

enrichment histories. HD 122563 is a very metal-poor ([Fe/H]$= -2.7$) giant that has a clear underabundance of n-capture elements [6], [7]. HD 175305 is a metal-poor ([Fe/H]$= -1.5$) giant that is enriched in r-process material [4], [5]. HD 196944 is a very metal-poor ([Fe/H]$= -2.3$) "lead" star that is enriched in s-process material [17], [16]. We have acquired new observations of these three stars with the 2.7m Harlan J. Smith Telescope at McDonald Observatory using the 2dcoudé spectrograph. To resolve the isotope shifts and hyperfine splitting, very high resolution and high S/N are required; our data has $R \sim 130,000$ and S/N$\sim$160–1000. Standard reductions were performed using *IRAF* routines.

To measure the isotopic abundances, the observed absorption lines are fit by synthetic spectra. The synthetic spectra, computed with the LTE code MOOG [14], account for the isotope shifts and hyperfine splittings that broaden these absorption lines. Atmospheric parameters from previous studies of these stars were checked with our spectra and either adopted or modified slightly; however, the isotopic abundances are very insensitive to the atmospheric parameters [1].

3 Results

3.1 Europium

In HD 175305, we have observations of three Eu II lines (4129Å, 4205Å, and 4435Å), and one is shown in Figure 1. The mean isotopic abundance of these three lines is $f_{151} = 0.48 \pm 0.05$, in excellent agreement with the expected pure-r-process mix of $f^r_{151} = 0.47$. In HD 196944 and HD 122563 we have observations of only the 4205Å line, and we find $f_{151} = 0.54 \pm 0.08$ and $f_{151} = 0.50 \pm 0.08$ in these stars, respectively. These results suggest a mixed r- and s-process origin of the Eu in HD 122563 and an s-process origin for the Eu in HD 196944, although our current uncertainties cannot exclude an r-process origin at the 1σ uncertainty level. We anticipate that the uncertainties quoted here will shrink somewhat as we identify and minimize sources of systematic error.

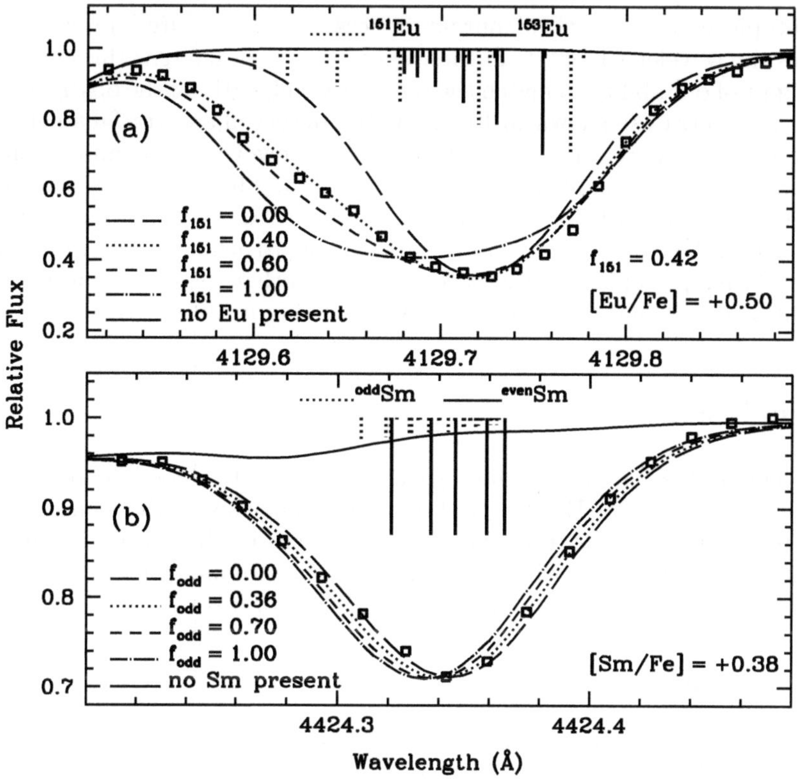

Fig. 1. Isotopic abundance syntheses of one Eu II line (panel **a**) and one Sm II line (panel **b**) in the r-process-enriched star HD 175305. The mean Eu isotopic abundance from three lines is $f_{151} = 0.48 \pm 0.05$, in very good agreement with an r-process mix. Given the small differences in the Sm isotopic mixes, neither a pure r-process mix (= 0.36; shown) nor a pure s-process mix (= 0.09; not shown) can be excluded on the basis of this observational data

3.2 Barium

The Ba II 4554Å line is very strong in all three of our stars, with equivalent widths ranging from 97 mÅ to 190 mÅ. Our synthetic spectra provide a poor fit to the center of this saturated line, and we attribute this to the saturation. Some previous studies (e.g. [12]) have noted that the Ba 4554Å line profile can be sensitive to NLTE effects and have effectively modeled this strong line using NLTE line profiles. We will attempt this approach in the near future.

3.3 Samarium

We have examined the Sm II 4424Å line in our three stars. Although we can fit the line profile well with our synthesis in HD 175305, the result of changing the isotopic mix from one of pure-r-process to one of pure-s-process only changes the wavelength offsets of these syntheses by $\sim$ 0.001Å; this currently is the limit of our ability to match the relative wavelengths of the observed and synthetic spectra. As shown in Figure 1, the observed spectra falls between the pure-r- and pure-s-process isotope mixes; however, we cannot exclude either mix on the basis of this observational data. This Sm line is weaker in HD 196944 and HD 122563, and even our high-S/N data cannot measure the isotopic abundances. We are currently attempting to achieve greater precision in the relative placement of the observed and synthetic spectra, and the use of additional Sm lines for isotopic analysis is being investigated.

4 Conclusions

Given a favorable star, our results suggest that it might be possible to measure isotopic abundances of Eu, Ba, and Sm simultaneously in the same star. If so, it is advantageous to use isotopic abundances of these three n-capture elements in conjunction with elemental abundances of many heavy n-capture elements to constrain the nucleosynthetic history of a star.

Acknowledgments

This research has been funded by U. S. National Science Foundation grants AST 03-07495 and AST 06-07708 to C. S.

References

1. W. Aoki, S. Honda, T. C. Beers, & C. Sneden: ApJ, **586**, 511 (2003)
2. W. Aoki, S. G. Ryan, N. Iwamoto, T. C. Beers, J. E. Norris, H. Ando, T. Kajino, G. J. Mathews, & M. Y. Fujimoto: ApJ, **592**, L67 (2003)
3. C. Arlandini, F. Käppeler, K. Wisshak, R. Gallino, M. Lugaro, M. Busso, & O. Straniero: ApJ, **525**, 886 (1999)
4. D. L. Burris, C. A. Pilachowski, T. E. Armandroff, C. Sneden, J. J. Cowan, & H. Roe: ApJ, **544**, 302 (2000)
5. J. J. Cowan, C. Sneden, T. C. Beers, J. E. Lawler, J. Simmerer, J. W. Truran, F. Primas, J. Collier, & S. Burles: ApJ, **627**, 238 (2005)
6. S. Honda, W. Aoki, T. Kajino, H. Ando, T. C. Beers, H. Izumiura, K. Sadakane, & M. Takada-Hidai: ApJ, **607**, 474 (2004)
7. S. Honda, W. Aoki, Y. Ishimaru, S. Wanajo, & S. G. Ryan: ApJ, **643**, 1180 (2006)
8. I. I. Ivans, J. Simmerer, C. Sneden, J. E. Lawler, J. J. Cowan, R. Gallino, & S. Bisterzo: ApJ, **645**, 613 (2006)
9. D. L. Lambert & C. Allende Prieto: MNRAS, **335**, 325 (2002)
10. J. E. Lawler, E. A. Den Hartog, C. Sneden, & J. J. Cowan: ApJS, **162**, 227 (2006)

11. P. Magain: A&A, **297**, 686 (1995)
12. L. Mashonkina & G. Zhao: A&A, **456**, 313 (2006)
13. J. Simmerer, C. Sneden, J. J. Cowan, J. Collier, V. M. Woolf, & J. E. Lawler: ApJ, **617**, 1091 (2004)
14. C. Sneden: Carbon and Nitrogen Abundances in Metal-Poor Stars. PhD Thesis, Univ. of Texas, Austin (1973)
15. C. Sneden, J. J. Cowan, J. E. Lawler, S. Burles, T. C. Beers, & G. M. Fuller: ApJ, **566**, L25 (2002)
16. S. Van Eck, S. Goriely, A. Jorissen, & B. Plez: Nature, **412**, 793 (2001)
17. L. Začs, P. E. Nissen, & W. J. Schuster: A&A, **337**, 216 (1998)

Part II

QSO Absorption Lines

Gas-phase Deuterium Abundances, Near and Far

Edward B. Jenkins

Princeton University Observatory, Princeton, NJ, USA ebj@astro.princeton.edu

Summary. Deuterium abundances in gaseous media reveal information on the baryon density of the universe Ω_b and the processing of chemical elements by stars in galaxies. By observing D in its atomic form in regions that are not heavily populated by molecules, we avoid complications that arise from chemical fractionation when deuterated molecules are produced. Atomic D can be observed by a variety of means: (1) absorption features in the Lyman series, (2) fluorescent Balmer emission from photon-dominated regions associated with H II regions, and (3) emission from the 92-cm hyperfine transition. For the first of these, there is a rich history of investigations of gaseous matter in our Galaxy, and in the recent past D/H has been measured in very distant systems that appear in absorption in the spectra of quasars. Associated with these observations are a number of important technical considerations, which are explored here. The value of D/H in the local part of our Galaxy is about 23 ppm, which is slightly lower than 27 ppm derived from either the distant, low-metallicity gas systems or a prediction based on primordial nucleosynthesis yield for a value of Ω_b determined from the temperature fluctuations in the cosmic microwave background radiation. Thus, the destruction of D by stellar processing in our Galaxy reduces D/H by less than about 20%, but many individual measurements far below 23 ppm indicate that significant amounts of D may be hidden on the surfaces of dust grains. A bewilderingly large dispersion of D/H seen in he distant systems may be resolved or at least better understood by the use of more powerful telescopes and spectrographs in the future.

1 The Importance of Deuterium Abundances

We start with the fundamental motivations for determining the abundances of deuterium in gaseous media within the universe. The two objectives outlined in this section address some basic questions that we were prepared to encounter beforehand, but a third, unexpected consideration arose after the observations were completed. This new factor will be discussed later in §3.2.

1.1 The Density of Baryons Ω_b in the Universe

In large part, the almost universal acceptance of the hot big bang model of cosmology, the so-called "standard model", is strongly reinforced by the near agreement of two observational outcomes that are driven by radically different physical principles manifested at very different epochs: (1) the existence of a universal black-body cosmic microwave background radiation (CMBR) which shows angular temperature fluctuations manifested at an age of about 4×10^5 yr, the interpretation of which is consistent with a reasonable set of ΛCDM cosmological parameters [23]; and (2) the approximate agreement of various light element abundances with the predicted nucleosynthesis yields, which span 8 orders of magnitude, which were set by the initial expansion and cooling within the first few minutes after the universe was created [1]. For the latter, the physical conditions are communicated to us according to the widely accepted principles on how atomic nuclei were formed and partly destroyed by nuclear reactions during an early phase of expansion and cooling. In this context, the ratio of deuterium to hydrogen from big bang nucleosynthesis, $(\mathrm{D/H})_{\mathrm{BBN}}$, is the most sensitive indicator of the baryon-to-photon ratio, η, which in turn can be translated into the baryon density of the present-day universe.

While a measurement of η based on determinations of D/H by itself is of fundamental importance, a comparison with the value of η deduced from fluctuations in the CMBR measured by the *Wilkinson Microwave Anisotropy Probe* (WMAP) allows us to test our concepts of physics extrapolated to extreme conditions and check that both the character of the universe's expansion and the manifestation of elementary particle reactions are consistent with our applications of present-day physical principles. For instance, any change with cosmic time in what we regard as the fundamental physical "constants" (G, α, μ etc.) might produce a discord between our interpretation of processes that occurred a few minutes after the big bang and the imprint on the background radiation at the much later time when the universe became transparent.

1.2 Galactic Chemical Evolution and Mixing in the ISM

Our quest to measure $(\mathrm{D/H})_{\mathrm{BBN}}$ with an accuracy that is useful for cosmology presents a special challenge because, over time, the ratio is likely to have been reduced from its original value by the conversion of D to ^{3}He when some fraction of the material cycled through stars (astration). Of course, looking at the problem from a different viewpoint, we can regard this not as a "challenge," but rather as an "opportunity."

The study of chemical evolution in our Galaxy encompasses the history of creation and destruction of broad, but very different classes of elements. Elements heavier than those created during BBN arise in large part either from fusion processes within stellar interiors or the explosive synthesis during supernova explosions. It is a complex story involving physical details that are uncertain and an intricate web of reactions which, for different types of objects, have yields that are known only approximately [5].

By contrast, for deuterium the problem should be far more straightforward. Deuterium is believed to be simply created during BBN and destroyed as gaseous material is cycled through stars. For this reason, a study of deuterium abundances should give the most straightforward indications from chemical evolution models that must incorporate various possible functional forms for the stellar birth rate with time, initial stellar mass functions, mass ejection rates from stars of different mass and in different stages of evolution, stellar activity as a function of galactocentric distance, and so forth. Moreover, deuterium abundances allow us to test models that propose significant infall of gas from the Local Group, i.e., material that has probably experienced very little processing through stars.

2 Methods of Observing Atomic Deuterium

There are three fundamental ways to observe the atomic form of D within diffuse gases in space. Deuterium is most easily detected by its absorption in the Lyman series in the spectrum of a UV-emitting continuum source (an early-type star or a quasar), but the measurements for material at zero redshift require the use of spectrographs and telescopes in space. The Lyα transition can be observed from the ground for systems at redshifts $z_{\rm abs} > 1.5$, and the entire Lyman series can be seen for $z_{\rm abs} > 2.4$. Fluorescent emission in the Balmer series has been observed for photon dominated regions next to H II regions [6, 7], but unfortunately the determination of D/H is difficult because the observed intensity is very strongly dependent on the temperature of the dominant illuminating star(s) [13]. In principle, the D counterpart of Hα recombination radiation should be observable, but in practice it is too weak and broad to detect [13]. Finally, one can observe the emission by the extremely weak hyperfine transition at the radio wavelength of 92 cm, a transition that is analogous to the one at 21 cm for H. This is a very difficult observation; despite a long series of attempts to detect this line, a successful outcome arose only very recently, yielding D/H = 15 and 23 ppm in two directions near the Galactic anticenter [19]. Unfortunately, these results may be undermined by uncertainties in the amount of H and D tied up in molecules, together with the possible saturation of the 21-cm line emission within many small clumps of gas that are not resolved by the telescope beam.

3 Lyman Series Absorptions

3.1 Fundamental Technical Considerations

Hydrogen and deuterium display the simplest spectra that we can imagine, ones that we all learned about as undergraduates. The deuterium transitions are separated from their hydrogen counterparts by $-82\,\mathrm{km\ s^{-1}}$, principally because the reduced mass of the electron is slightly larger. The Lyman series represents a progression with accurately known f-values that starts with Lα

and advances to successively weaker lines toward the higher series members. An interesting feature of this system is that the strengths of the absorption optical depths averaged over frequency remains constant and equal to the continuum opacity just beyond the limit [24], i.e., the line strengths are proportional to their spacing.

Absorption systems with a velocity spread (b parameter) equal to about $15\,\mathrm{km\,s^{-1}}$ are ideal for study, since larger velocity dispersions make the D lines blend with their nearby, much stronger H lines, and narrower velocity spreads make the most easily observed lines too saturated for obtaining reliable values of column density. In the context of investigating distant absorption systems, one can either study strong lines for systems that have weak to moderately strong absorption in the Lyman limit, the so-called Lyman Limit Systems (LLS), or much weaker Lyman series lines in the Damped Lyman Alpha (DLA) systems with $\log N(\mathrm{H\ I}) > 20.3$. One might initially think that the two choices are equivalent to each other, but they are not. One important obstacle in separating the absorption of D from that of H is a broad damping wing created by the very much stronger H line. For equivalent values of $Nf\lambda$, the higher series members which would be viewed for a DLA have relatively narrow damping wings because their absorption strengths scale in proportion to $Nf\lambda\gamma$, and the decay rates γ of the upper states are smaller. A chief drawback of DLAs is that they are fewer in number, compared to LLSs. Fortunately, a recent survey of quasar absorption lines using results from the *Sloan Digital Sky Survey* has revealed a large number of DLAs which ultimately can be accessed with telescopes with very large apertures [17].

3.2 A Brief History of D/H Measurements

There have been two major fronts in the determinations of D/H, one being investigations of sightlines in our Galaxy and the other being the observation of very distant systems seen quasar spectra. The former achieved an earlier start, because UV telescopes and spectrographs in orbit were built before the required 10-meter class telescopes on the ground were available.

Our Galaxy

The first detection of atomic deuterium in space was reported by Rogerson & York [20], who used the *Copernicus* satellite. Their measurement combined with other ones using *Copernicus* and the *Hubble Space Telescope* all gave values for D/H centered at about 15 ppm, but with some scatter from one case to the next. At the time, it was not clear if this scatter was real or simply a product of underestimated errors [11]. Definitive results came from the *Interstellar Medium Absorption Profile Spectrograph* (IMAPS), which could observe the far-UV transitions at a resolution of $3\,\mathrm{km\,s^{-1}}$ – a value far better than the resolution of *Copernicus*. Observations with IMAPS made it clear that D/H varied by at least a factor of 3 within the ISM of our Galaxy [8, 22].

A campaign to observe many different target stars with the *Far Ultraviolet Spectroscopic Explorer* (FUSE) gave further strength to the proposition that the abundances of D relative to H varied from one location to the next [12, 14]. Initially, it was thought that the variations might be explained by incomplete mixing of the ISM following the some localized processing through stars or perhaps after episodes of dilution from infalling, pristine gas from outside the Galaxy. However, the lack of any anticorrelation between the abundances of deuterium and those of heavier elements disfavored these ideas. In fact, the opposite seemed to hold: reductions in the deuterium abundances seemed to correlate with the depletions onto dust grains of various refractory elements, such as Ti [18], Fe, and Si [10].

Shortly before the abundance correlations were evident, Draine [3, 4] suggested that D atoms could be preferentially depleted from the ISM onto the surfaces of carbonaceous dust grains because the C–D bonds are stronger than the C–H ones by an energy difference of 0.083 eV ($\equiv$ 970 K). He estimated that this sequestering of D could be significant enough to create appreciable reductions in the gas-phase abundances. Observations with a spectrograph on the *Infrared Space Observatory* that showed a weak emission feature at 4.65 μm that could be produced by deuterated polycyclic aromatic hydrocarbons (PAHs) seem to support this idea [16]. After accounting for this effect, Linsky et al. [10] proposed that the intrinsic Galactic D abundance (after astration) is about equal to 23 ppm, based on the largest observed values of D/H. When compared to an estimate for $(D/H)_{BBN}$ of about 26-27.5 ppm based on the value of Ω_b determined from the WMAP measurements, one concludes that astration reduced the amount of D in our Galaxy by less than about 20%, which is nearly consistent with some models for Galactic chemical evolution [21].

Distant Galactic Systems

In principle, the quest to measure $(D/H)_{BBN}$ should be more transparent for distant gas systems that have a metallicity much lower than that of our Galaxy, since the astration should be substantially less, and the abundance of dust should be much lower. For the results obtained so far (see [9] and earlier determinations listed therein, plus two more recent measurements [2, 15]) a logarithmic average for D/H yields 27 ppm, which is somewhat greater than the results for our Galaxy (see above) but very close to $(D/H)_{BBN}$ derived from the WMAP results. Unfortunately, the dispersion of the individual outcomes appears to be much larger than that expected from the quoted errors, an effect that does not seem to be reflected by any concrete trend with metallicity, as indicated by [O/H] or [Si/H] [9].

4 The Value of New Advances in Spectroscopy

The investigations of D/H for distant systems discussed in §3.2 above have been carried out with extraordinary care to reduce the effects of systematic errors arising from uncertainties in continuum levels, interference from other lines, and complications in velocity substructure that might not have been modeled accurately. Nevertheless, improvements can be expected from new spectrographs that provide better wavelength resolution to more accurately portray velocity structures and reduce the blending from unwanted lines. With telescopes that have larger apertures, we should have the capability to probe DLAs in front of fainter quasars and thus, by the sheer numbers of different observations, obtain a better understanding of possible spurious effects from hydrogen interlopers (at $-82\,\mathrm{km\ s}^{-1}$ with respect to a main component) and uncertain continua. Finally, it would be beneficial to obtain better observations of emission from deuterated PAHs in the infrared, to confirm the initial findings discussed in §3.2.

References

1. Boesgaard, A.M., Steigman, G.: A.R.A.A. **23** , 319 (1985)
2. Crighton, N.H.M. et al.: M.N.R.A.S. **355** , 1042 (2004)
3. Draine, B.T.: Interstellar Dust. In: *Origin and Evolution of the Elements*, vol, ed by Mc William, A., Rauch, M. (Cambridge Univ. Press, Cambridge 2004) pp. 317-335
4. —: Can dust explain variations in the DH ratio? In: *Astrophysics in the Far Ultraviolet*, vol, ed by Sonneborn, G., Moos, H.W., Andersson, B.-G. (Astr. Soc. Pacific, San Francisco 2006) pp. 58- 69
5. Gibson, B.K., Loewenstein, M., Mushotzky, R.F.: M.N.R.A.S. **290** , 623 (1997)
6. Hébrard, G. et al.: Astr. Ap. **354** , L79 (2000)
7. — : Astr. Ap. **364** , L31 (2000)
8. Jenkins, E.B. et al.: Ap. J. **520** , 182 (1999)
9. Kirkman, D. et al.: Ap. J. (Suppl.) **149** , 1 (2003)
10. Linsky, J.L. et al.: Ap. J. **647** , 1106 (2006)
11. McCullough, P.R.: Ap. J. **390** , 213 (1992)
12. Moos, H.W. et al.: Ap. J. (Suppl.) **140** , 3 (2002)
13. O'Dell, C.R., Ferland, G.J., Henney, W.J.: Ap. J. **556** , 203 (2001)
14. Oliveira, C.M., Moos, H.W., Chayer, P., Kruk, J.W.: Ap. J. **642** , 283 (2006)
15. O'Meara, J.M. et al.: Ap. J. (Letters) **649** , L61 (2006)
16. Peeters, E. et al.: Ap. J. **604** , 252 (2003)
17. Prochaska, J.X., Herbert-Fort, S., Wolfe, A.M.: Ap. J. **635** , 123 (2005)
18. Prochaska, J.X., Tripp, T.M., Howk, J.C.: Ap. J. (Letters) **620** , L39 (2005)
19. Rogers, A.E.E. et al.: Ap. J. (Letters) **630** , L41 (2005)
20. Rogerson, J.B., York, D.G.: Ap. J. (Letters) **186** , L95 (1973)
21. Romano, D. et al.: M.N.R.A.S. **369** , 295 (2006)
22. Sonneborn, G. et al.: Ap. J. **545** , 277 (2000)
23. Spergel, D.N. et al.: astro-ph/0603449 (2006)
24. Sugiura, M.Y.: J. Physique **8** , 113 (1927)

Comprehensive Abundance Measurements in Damped Lyα Systems

M. Dessauges-Zavadsky[1], J. X. Prochaska[2], S. D'Odorico[3], F. Calura[4], and F. Matteucci[4,5]

[1] Geneva Observatory, Ch. des Maillettes 51, 1290 Sauverny, Switzerland
[2] UCO/Lick Observatory, University of California, Santa Cruz, CA 95064, USA
[3] ESO, Karl-Schwarzschildstrasse 2, 85748 Garching bei München, Germany
[4] Astronomy Department, University of Trieste, Via G. B. Tiepolo 11, 34131 Trieste, Italy
[5] INAF, Trieste Observatory, Via G. B. Tiepolo 11, 34131 Trieste, Italy

Summary. The high quality, high-resolution UVES/VLT and HIRES/Keck spectra of 10 quasars allowed us to analyze in detail 11 intervening damped Lyα systems at $z_{\rm abs} = 1.7 - 2.5$. We detected 54 metal-line transitions and obtained the column density measurements of 30 ions from 22 elements, B, C, N, O, Mg, Al, Si, P, S, Cl, Ar, Ti, Cr, Mn, Fe, Co, Ni, Cu, Zn, Ge, As, and Kr. Thanks to these comprehensive elemental abundances, we could for the first time study in detail the abundance patterns of a wide range of elements, the chemical variations of the ISM clouds, the star formation history, and the age of these high-redshift galaxies.

1 Introduction

Studies of elemental abundances have undergone a remarkable acceleration in the flow of data over the last few years. The main motivation common to all this observational effort is to understand the physical processes at play in the formation and evolution of galaxies. To achieve this goal, the chemical information can be used as one of the possible means at our disposal to link the properties of high-z galaxies with those of present-day galaxies.

The progress of abundance measurements at high redshift has significantly increased with the advent of the 10-m class telescopes and the high-resolution spectroscopy, and with the studies of damped Lyα systems (DLAs). Up to 26 elements are now detected in DLAs [1], and DLAs hence are real physical laboratories with which we are able to study the nucleosynthesis in the early Universe and the physical properties of these high-z galaxies.

While an evolutionary link between the emission-selected galaxies at $z = 2 - 3$ and the bulge component of the galaxy population today is strongly suggested, an analogous connection for DLAs is still a matter of debate, because of the difficult detection of DLA galaxy counterparts. Low redshift deep imaging is revealing a mixing bag for the DLA counterparts between spiral, irregular, and low surface brightness galaxies. We propose here to constrain the nature of high-z DLA galaxies with their chemical abundances.

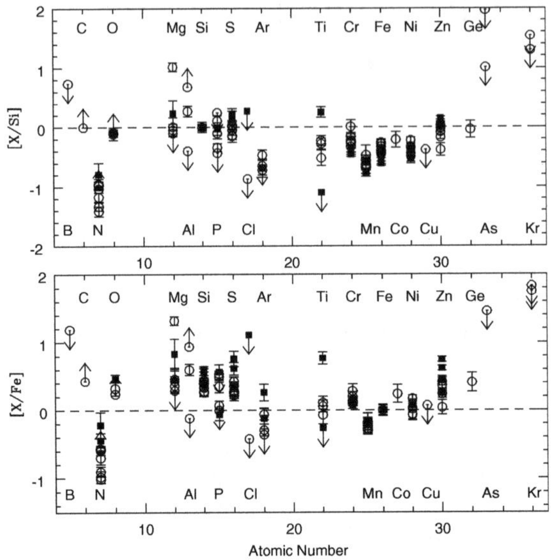

Fig. 1. Global gas-phase relative abundances [X/Si] (*upper panel*) and [X/Fe] (*lower panel*) for our sample of 11 DLAs. Our sample of data is composed of 4 DLAs analyzed in [2] (filled squares) and of 7 DLAs studied in [3] (circles).

2 Chemical Variations

2.1 Global Gas-phase Relative Abundances

The global relative abundances of DLAs allow to test their nucleosynthesis enrichment, their dust depletion level, and their ionization effects. Fig. 1 shows the relative abundances as a function of the atomic number of the 11 DLAs in our sample. We present both the [X/Si] and [X/Fe] abundance ratios for the entire set of 22 detected elements X. The general trend which comes out is the *uniformity*. Indeed, the RMS dispersions reach only 2–3 times higher values than the statistical errors for the majority of elements.

This uniformity is remarkable given that the DLAs in our sample have H I column densities ranging from 2×10^{20} to 4×10^{21} cm^{-2} and metallicities from 1/55 to 1/5 solar, and that their redshifts are spread over 2 Gyr. Moreover, large variations of 0.3–0.5 dex are observed when comparing the relative abundances in the Milky Way, the Small and Large Magellanic Clouds, and the dwarf spheroidals (e.g. [4]). This suggests that the effects of nucleosynthesis enrichment, dust depletion, and ionization are low in DLAs.

2.2 Relative Abundances of ISM Clouds

Few studies have investigated the relative abundances of ISM clouds within DLAs [5, 6], where variations are highly expected. The cloud-to-cloud analysis

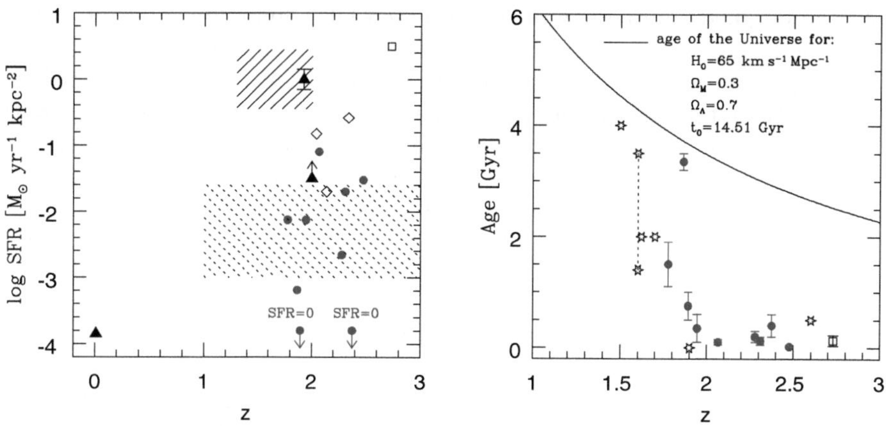

Fig. 2. *Right panel.* SFR per unit area versus redshift. Our DLA results are shown by circles. The dotted area represents the SFRs obtained for DLAs by [9], and the dashed area the SFRs obtained for emission-selected galaxies from GDDS [11]. The triangles correspond to 3 DLAs for which we have an estimation of the SFR directly from emission lines, the square to the Lyman-break galaxy MS 1512-cB58 [10], and the diamonds to the γ-ray burst host galaxies which seem to have intermediate SFRs. *Left panel.* Age versus redshift for the DLA galaxies studied (circles) and for galaxies analyzed via integrated spectra (stars; [12]).

of 7 DLAs in our sample reveals that 5 of them show statistically significant abundance ratio variations higher than 0.2 dex at more than $3\,\sigma$. The sources of these variations are either the differential dust depletion or the ionization, but there is no evidence for variations due to nucleosynthesis enrichment.

These results suggest that the ISM clouds within DLAs have different physical properties, but have the same origin and enrichment history. As a consequence, this could suggest the presence of efficient mixing in protogalaxies, and the moderate scatter observed in the dust depletion levels may indicate that the gas in DLAs is neither very dense nor cold but rather diffuse and warm. This has also important implications on the CDM hierarchical theory which describes a galaxy as multiple merging clouds bound to individual dark matter halos. Indeed, the challenge here is to explain how it is that the ISM protogalactic clouds may have a unique enrichment history, while they do not share a common gas reservoir and merge over a large timescale.

3 Star Formation Histories

Given these properties, what are the star formation histories (SFHs) of the DLA galaxies? We consider here two chemical evolution models identified as the "spiral model" and the "dwarf irregular model" according to the type of galaxies they do match best. These models are the same as those described in

[8]. The detailed comparison of the observed DLA abundance patterns with these models allows to determine their SFH and their age [7].

The general result that comes out is that the DLAs seem to be associated either with outer regions of spiral galaxies or with dwarf irregulars both characterized by *low star formation efficiencies.* In Fig. 2 (left panel), we show the derived star formation rates (SFRs) per unit area of our sample of DLAs. They all have moderate SFRs, between $-3.2 < \log \mathrm{SFR} < -1.4$ $M_\odot$ yr^{-1} kpc^{-2}, that are in good agreement with the interval of SFR values obtained by [9] (see the dotted region) and are similar to the SFR measured in the Milky Way ISM. However, they are 10–1000 times lower than the SFRs of the emission-selected galaxies at the same redshifts (e.g. [10, 11]). In Fig. 2 (right panel), we show the determined ages of the DLAs. Observed at $z_{\rm abs} = 1.7 - 2.5$, some of them seem to be very young galaxies with ages between 50–250 Myr likely experiencing their first star formation episodes, but others seem to be already old with ages longer than 1 Gyr indicating that galaxies have formed at $z > 10$.

4 Future Prospects for DLAs with ELTs

With high-resolution spectrographs on ELTs, there is no doubt that new elements, currently undetected, will become accessible as well as fundamental molecules like CO and HD in addition to H_2. However, will our knowledge on the nature of DLA galaxies improve? The answer is very likely *no.* We really need direct detections of DLA galaxy counterparts! Only the conjunction of absorption and emission properties of DLA galaxies will bring a major step in (i) the understanding of their link both with present-day and other high-z galaxies and (ii) a proper probe of the chemical evolution of galaxies.

References

1. J.X. Prochaska, J.C. Howk, A.M. Wolfe: Nature **423**, 57 (2003)
2. M. Dessauges-Zavadsky, F. Calura, J.X. Prochaska et al: A&A **416**, 79 (2004)
3. M. Dessauges-Zavadsky, J.X. Prochaska, S. D'Odorico et al: A&A **445**, 93 (2006)
4. M. Shetrone, K.A. Venn, E. Tolstoy et al: AJ **125**, 684 (2003)
5. J.X. Prochaska: ApJ **582**, 49 (2003)
6. E. Rodríguez, P. Petitjean, B. Aracil et al: A&A **446**, 791 (2006)
7. F. Matteucci: *The Chemical Evolution of the Galaxy*, Dordrecht: Kluwer Academic Publisher, Astrophysics and Space Science Library, vol 253 (2001)
8. F. Calura, F. Matteucci, G. Vladilo: MNRAS **340**, 59 (2003)
9. A.M. Wolfe, J.X. Prochaska, E. Gawiser: ApJ **593**, 215 (2003)
10. M. Pettini, C.C. Steidel, K.L. Adelberger et al: ApJ **528**, 96 (2000)
11. S. Savaglio, K. Glazebrook, R.G. Abraham et al: ApJ **602**, 51 (2004)
12. G.A. Bruzual: *Revista Mexicana de Fisica*, astro-ph/0202178 (2002)

Molecular Hydrogen at High Redshift and the Variation with Time of the Electron-to-proton Mass Ratio, $\mu = m_e/m_p$

P. Petitjean[1], C. Ledoux[2], R. Srianand[3], P. Noterdaeme[2], and A. Ivanchik[4]

[1] Institut d'Astrophysique de Paris, 98bis Bd Arago, 75014 - Paris, France
petitjean@iap.fr
[2] European Southern Observatory, Alonso de Córdova 3107, Casilla 19001, Vitacura Santiago, Chile
[3] IUCAA, Post Bag 4, Ganesh Khind, Pune 411 007, India
[4] Ioffe Physical Technical Institute, Polytekhnicheskaya 26, 194021 Saint-Petersburg, Russia

Summary. We search for the signature of molecular hydrogen in damped Lyman-α systems at high redshift. In the course of this survey, we constructed a representative sample of 18 DLA/sub-DLA systems with log N(H I) > 19.5 at high redshift ($z_{\rm abs} > 1.8$) and with metallicities relative to solar [X/H] > -1.3 (with [X/H] = log N(X)/N(H)$-$log(X/H)$_\odot$ and X either Zn, S or Si). We show that the presence of molecular hydrogen at high redshift is strongly correlated with metallicity. The comparison of H_2 transition wavelengths observed at high redshift to those measured in the laboratory can be used to constrain the variation with time of the proton-to-electron mass ratio, μ. Using the two best cases, we obtain the most stringent limit on the variation of μ over the last 12 Gyrs ever obtained, $\Delta\mu/\mu = 1.65\pm0.74\times10^{-5}$.

1 Introduction

High-redshift Damped Lyman-α systems (DLAs) detected in absorption in QSO spectra are characterized by their large neutral hydrogen column densities, N(H I) $\geq 2\times10^{20}$ cm^{-2}, similar to what is measured through local spiral disks. The corresponding absorbing clouds are believed to be the reservoir of neutral hydrogen in the Universe (Prochaska et al. 2005). Though observational studies of DLAs have been pursued over more than two decades, important questions are still unanswered, such as (i) the amount of in-situ star-formation activity in DLAs, (ii) the connection between observed abundance ratios and the dust content, and (iii) how severe is the bias due to dust obscuration in current DLA samples. One way to tackle these questions is to derive the physical conditions of the gas from observation of its molecular content. Early searches for molecular hydrogen in DLAs, though not systematic, have led to either low values of or upper limits to the molecular fraction of the gas. A major step forward in understanding the nature of DLAs through their molecular hydrogen content has recently been made possible by the unique high-resolution and blue-sensitivity capabilities of UVES at the VLT. In the course of the first large and systematic survey for H_2 at high redshift, we searched for H_2 in DLAs down to a detection limit of typically $N(H_2) = 2 \times 10^{14}$ cm^{-2} (Ledoux et al. 2003).

2 High Metallicities and H_2 Molecules

In Fig. 1 we plot the molecular fraction, $\log f = \log 2N(H_2)/(2N(H_2) + N(\text{H I}))$, versus metallicity, $[X/H] = \log N(X)/N(H) - \log(X/H)_\odot$, for our representative sample of DLAs with $[X/H] > -1.3$ (18 measurements, see Petitjean et al. 2006) and measurements by Ledoux et al. (2003) for $[X/H] < -1.3$ (23 measurements). The $\log f$ distribution is bimodal with an apparent gap in the range $-5 < \log f < -3.5$. Note that this jump in $\log f$ was already noticed by Ledoux et al. (2003) and is similar to what is seen in our Galaxy (Savage et al. 1977; see also Srianand et al. 2005). It is apparent that the

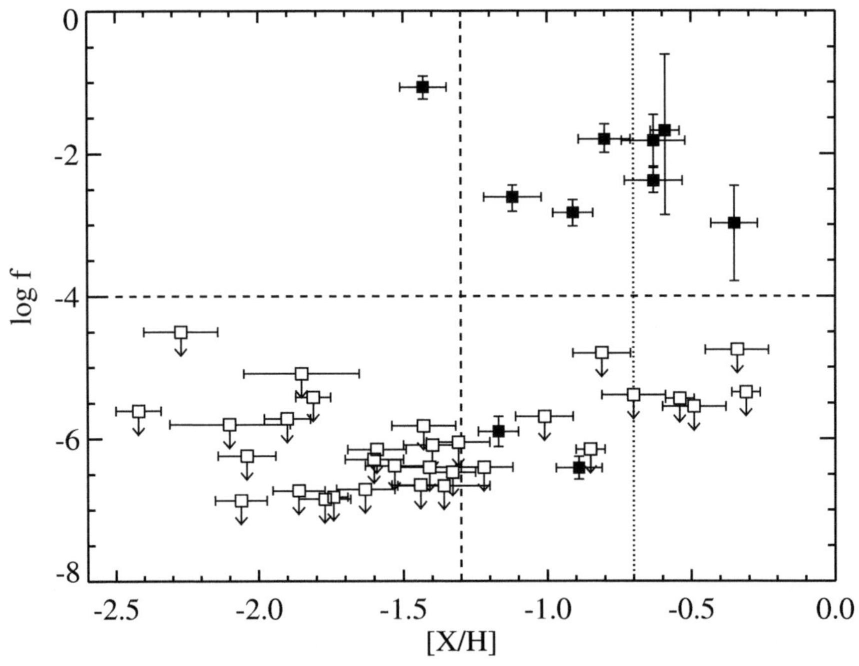

Fig. 1. Logarithm of the molecular fraction, $f = 2N(H_2)/(2N(H_2) + N(\text{H I}))$, versus metallicities, $[X/H] = \log N(X)/N(H) - \log(X/H)_\odot$ with X either Zn, S or Si, in DLAs. Filled squares indicate systems in which H_2 is detected. The dotted line shows the median of the high-metallicity sample ($[X/H] = -0.7$).

fraction of systems with molecular fraction log $f > -4$ increases with increasing metallicity. It is only ~5% for [X/H] < -1.3 when it is ~39% for [X/H] > -1.3. This fraction is even higher, 50%, for [X/H] > -0.7 which is the median metallicity for systems with [X/H] > -1.3. In addition, all systems with [X/H] < -1.5 have log $f < -4.5$.

We conclude that metallicity is an important criterion for the presence of molecular hydrogen in DLAs. This may not be surprising as the correlation between metallicity, and the depletion of metals onto dust grains observed for the whole DLA population (Ledoux et al. 2003) implies that higher metallicity means higher dust content and therefore higher H_2 formation rate. More generally, Ledoux et al. (2006a) have shown that a correlation exists between metallicity and velocity width in DLAs. If the latter kinematic parameter is interpreted as reflecting the mass of the dark-matter halo associated with the absorbing object, then DLAs with higher metallicity are associated with objects of higher mass where star formation could be enhanced. All this makes it arguable that, in DLAs, star-formation activity is probably correlated with the molecular fraction (Hirashita & Ferrara 2005). It is therefore of prime importance to survey a large number of DLA systems to define their molecular content better.

3 The Variation of μ

Once H_2 absorption features are detected at high-z, one can constrain the cosmological variation of the electron-to-proton mass ratio. Indeed, wavelengths of electron-vibro-rotational lines depend on the reduced mass of the molecule, with the dependence being different for different transitions. The measured wavelength λ_i of a line formed in the absorption system at the redshift z_abs can be written as

$$\lambda_\mathrm{i} = \lambda_i^0(1 + z_\mathrm{abs})(1 + K_\mathrm{i}\Delta\mu/\mu) \qquad (1)$$

where λ_i^0 is the laboratory (vacuum) wavelength of the transition, and $K_\mathrm{i} = \mathrm{d}\ln\lambda_i^0/\mathrm{d}\ln\mu$ is the sensitivity coefficient calculated for the Lyman and Werner bands of molecular hydrogen (e.g. Varshalovich & Potekhin 1995).

The systems in which the measurement is done must be carefully chosen because of the numerous systematics induced by heavily blended lines. We have chosen the two absorbers where the absorption lines can be modeled by a single Gaussian function, at $z_\mathrm{abs} = 2.59473$ and 3.02490 in the spectra of the quasars, respectively, Q 0405−443 and Q 0347−383,. We have re-observed the quasars to obtain the best data possible today. We measure the position of 76 H_2 absorption lines and search for any correlation between the relative positions of H_2 absorption, $\zeta_\mathrm{i} = (z_\mathrm{i} - \bar{z})/(1+\bar{z})$, and the sensitivity coefficients K_i. Results are shown in Fig. 2, using recent laboratory wavelengths by Philip et al. (2004). A correlation is present at the 2.5 σ limit that has been confirmed by Reinhold et al. (2006).

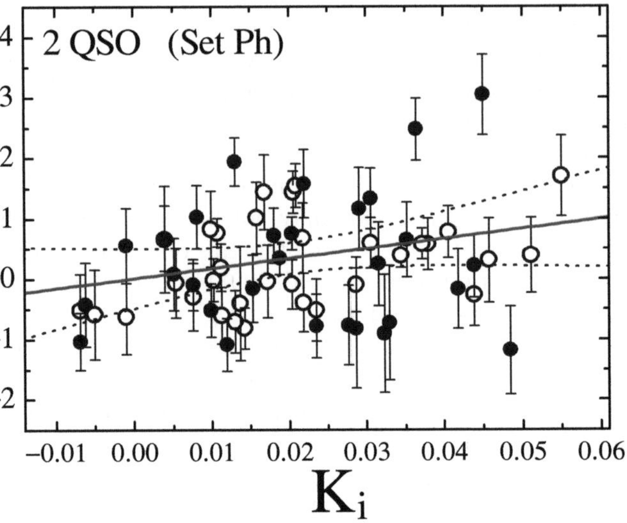

Fig. 2. The relative position of the lines, $\zeta_i = (z_i - \bar{z})/(1+\bar{z})$, is plotted versus the sensitivity coefficient (see Eq. 1). The regression analysis gives $\Delta\mu/\mu = 1.65 \pm 0.74 \times 10^{-5}$. Open and filled symbols are for two different quasars.

References

1. Hirashita, H., & Ferrara, A. 2005, MNRAS, 356, 1529
2. Ivanchik, A. et al., 2005, A&A, 440, 45
3. Ledoux, C., et al., 2006, A&A, 457, 71
4. Ledoux, C., Petitjean, P. & Srianand, R. 2003, MNRAS, 346, 209
5. Petitjean, P., Ledoux, C., Noterdaeme, P., & Srianand, R. 2006, A&A, 456, L9
6. Philip, J. et al. 2004, Can. J. Chem., 82, 713
7. Prochaska, J. X., et al., 2005, ApJ, 635, 123
8. Reinhold, E. et al., 2006, Phys. Rev. Lett., 96, 151101
9. Savage, B. D., Bohlin, R. C., Drake, J. F., & Budich, W. 1977, ApJ, 216, 291
10. Srianand, R., et al., 2005, MNRAS, 362, 549
11. Varshalovich, D., & Potekhin, A. 1995, Space Sci. Rev., 74, 259

Spectroscopy of QSO Pairs

Sara L. Ellison

University of Victoria, Dept. Physics & Astronomy, 3800 Finnerty Rd, V8V 1AP, British Columbia, Canada. sarae@uvic.ca

Summary. I review some of the applications of spatially resolved spectroscopy of multiple QSO lines of sight, including gravitationally lensed quasars, binaries and projected pairs. I focus on size constraints that have been determined for classes of absorbers that range from the intergalactic medium up to damped Lyman alpha systems. I also briefly discuss some insights that can be gained into the QSO environment and its physical properties by using projected pairs with disparate redshifts.

1 Introduction and Motivation

During this meeting we have heard about a number of the triumphs of quasar absorption line spectroscopy as a probe of high redshift galaxies and the intergalactic medium (IGM). For example, there are now over 100 damped Lyman alpha (DLA) systems for which chemical abundances have been measured, and with errors that typically rival that of stellar measurements in our own Galaxy. In a handful of cases, the suite of species detected includes rarely (at high z) detected lines from elements such as boron, cobalt and krypton, allowing us to model nucleosynthetic patterns, dust depletion and star formation histories (see the contribution by Dessauges-Zavadsky in this volume). In addition to atomic species, there is a growing compilation of molecular hydrogen detections which provide further insight into the physical conditions in the interstellar medium (ISM, see contribution by Petitjean). Cosmological applications of optical quasar absorption line spectroscopy include an independent determination of the density of baryons in the universe, Ω_B, via measurements of the primordial deuterium abundance and several possible avenues for searching for variations in fundamental constants (contributions by Jenkins, Murphy, Chand, Levshakov and Thompson).

Despite this glowing report card of successes, the technique of quasar absorption line spectroscopy has its limitations. Few absorbers have their counterpart galaxies identified, so although we have extensive chemical profiles for them, we are often lacking basic information concerning their luminosities, sizes and morphologies. A single line of sight through a galaxy yields an integrated measure of gas phase abundances, so we have no information on chemical homogeneity or gradients within the ISM. Finally, the often complicated velocity profiles can be difficult to interpret and can be explained by a range of models that includes rotation, inflow and outflow. Many of these limitations can be addressed when, rather than studying absorbers in a single line of sight, we utilize pairs of QSOs. Multiple sightlines may be in the form of gravitational lenses (typical transverse separations of up to a few arcseconds), binaries at the same redshift (typical transverse separations of a few arcseconds) and projected pairs filling the larger separations. Until recently, the number of multiple sightlines was relatively small, with only a subset bright enough for follow-up precision spectroscopy. Large samples of QSOs such as the 2dF and SDSS have changed this situation with several hundreds of pairs and multiples now known (e.g. [11]). In this proceedings contribution I review some of the results that are emerging from recent work, focussing on determinations of absorber size, chemical homogeneity and the proximity effect of ionizing sources.

2 Absorber Sizes and Coherence Scales

2.1 The Lyman Alpha Forest

The Lyman α forest represents absorption due to the IGM, both in the over-dense filaments and the under-dense 'voids' that constitute the cosmic web. Very early measurements of Lyα cloud sizes, which showed them to be coherent over Mpc scales, were instrumental in establishing their nature as intergalactic structures. Although clustering has been measured along single lines of sight using the 2-point correlation function with correlations found up to ~ 300 km/s (~ 1 Mpc), such measurements may be affected by redshift distortions and peculiar velocities. Measuring transverse sizes circumvents this uncertainty. Early work with lensed QSOs showed that the Lyα forest is highly coherent on scales of a few tens of kpc, implying actual sizes considerably larger than this number. More recently, a number of studies (e.g. [6], [7]) have shown that the IGM is coherent over scales of 1–2 Mpc at $z \sim 2$, although precise sizes for the absorber depend on the assumed geometry of the 'cloud'. This indicates that despite earlier concerns, redshift distortions and line of sight velocity effects do not seriously affect the clustering scale obtained from the 2 point correlation function. In the future, larger databases of high resolution, high S/N spectra of the Lyα forest can also be used to perform the Alcock-Paczynski test, yielding an independent measurement of Ω_Λ.

2.2 CIV Absorbers

The high column density Lyα clouds, N(HI) > 15, often show associated CIV absorption. An outstanding question is whether this CIV is associated with galaxy haloes or the truly diffuse IGM. On the one hand, [2] find a strong correlation between Lyman break galaxies and CIV absorption which implies haloes $\sim$ 80 kpc in size. On the other hand, [14] have shown that whilst there is certainly a strong signature from galactic haloes amongst the CIV absorbers, there is also a component which is not associated with nearby galaxies. This can be tested by measuring the transverse coherence scale in QSO pairs. Currently, CIV has been traced on transverse scales up to $\sim$ 100 kpc, mostly using lensed QSOs. Figure 1 shows that although the CIV equivalent widths vary by up to $\sim$ 50% on these scales, there is still some large scale coherence, consistent with the findings of [2], see also Figure 2. But is there coherence on larger scales as we may expect if some CIV is associated with the IGM? Whilst wide (arcminute) separation pairs exhibit a clustering of CIV on Mpc scales (e.g. [6]), the velocity shifts between lines of sight can be significant, indicating that the clustering is due to galaxies, rather than tracing coherent gas structures. In order to answer our question using QSO pairs, we first need to start filling in the parameter space in Figure 1. We have very few constraints on the coherence scale of CIV absorbers over the order of magnitude between 100 kpc and 1 Mpc. However, the SDSS and 2dF have provided us with a large sample of pairs with separations of tens of arcseconds with which to address this. In addition, high resolution, high S/N spectra are required in order to trace weak CIV absorbers. Whilst it is becoming well-established that the bulk of strong CIV absorbers are associated with galaxy haloes, it is the low column density cousins that may be associated with the true IGM.

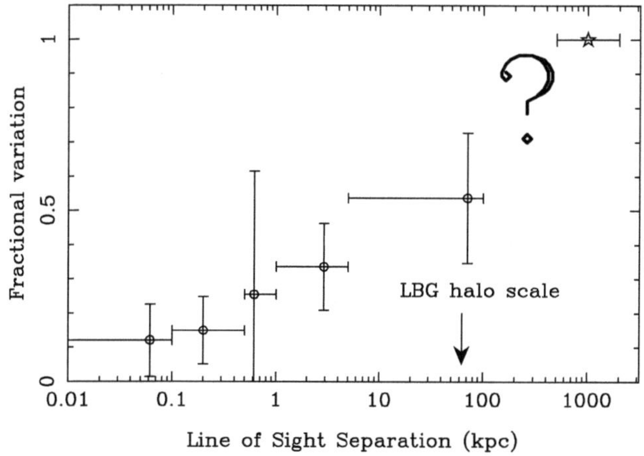

Fig. 1. Fractional difference in CIV equivalent width versus proper transverse separation for QSO pairs, adapted from [9]. Open points are mostly based on measurements from lensed QSOs. The large open star indicates that *high equivalent width* CIV absorbers in the wide separation (500 kpc – 2 Mpc) projected pairs of [6] are not generally seen in multiple lines of sight. Current observations of multiple lines of sight leave us with a gap between scales of 100 to 500 kpc. Also marked is the size scale of LBG CIV halos found by [2].

2.3 MgII Absorbers

For the last decade, we have worked with the picture that *high equivalent width* MgII absorbers are associated with large, luminous galaxies with typical sizes $\sim$ 40 kpc. This picture was established by the pioneering work of Steidel, Bergeron and collaborators. Figure 2 summarizes this canonical picture, which was established at $z \sim 1$ by matching absorbers with galaxies.

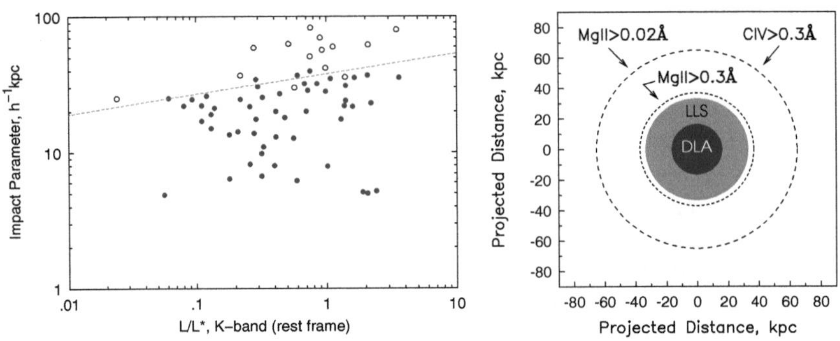

Fig. 2. Left panel: filled points show galaxies which cause strong MgII absorption in a nearby QSO. Open circles are galaxies with no corresponding MgII. The dashed line indicates that the size of the MgII halo is a simple function of galaxy luminosity and that the gas must have a filling factor of approximately unity. Right panel: Canonical picture of absorber sizes. Figure taken from [5].

A modification of this picture is apparently required by the work of [9] who presented spatially resolved STIS spectra of the $z = 3.91$ QSO APM08279+5255. The very small separations of the lensed images means that MgII along the line of sight is probed over proper transverse scales of 0.5 – 3 kpc. Significant fractional differences ($\sim$ 50%) in MgII EWs were found even on these small scales, whereas CIV is highly coherent on the same scales (e.g. Figure 1). Indeed, as had been previously found by [15] and [16] individual low ionization absorption components can rarely be traced over even pc scales. A pairwise analysis of *low equivalent width* MgII systems by [9] indicated a coherence scale of $\sim$ 2 kpc. This can be reconciled with the larger sizes derived by [18] if we consider a halo that is comprised of many pc-scale cloudlets with a filling factor of approximately unity in the inner few tens of kpc of the galaxy, but with lower values at larger distances. The classical strong MgII absorbers are therefore those that intersect many cloudlets, i.e. through the inner few tens of kpc where the filling factor is very high. A weak MgII absorber is observed when the line of sight intersects the galaxy at higher impact parameters and consequently only traverses a few of these cloudlets. This picture is supported by an observed anti-correlation between equivalent width and galaxy-QSO separation ([4]).

2.4 Damped Lyman Alpha Systems

The first estimates of the sizes of DLAs came from observations of absorption towards extended radio sources (e.g. [10]) leading to sizes of 10 – 15 kpc. Relatively little has been added to this view from QSO pairs due to the combined rarity of DLAs and close QSO pairs. In fact, only 4 lensed QSOs with DLAs are currently known with transverse proper separations ranging from $\sim$ 150 pc to 10 kpc. For the 3 pairs with separations $<$ 5 kpc, damped absorption is seen in both lines of sight, providing a lower limit on the size. It is a clear priority to identify and study many more QSO pairs with separations of a few to a few tens of kpc. Despite the paucity of data, the DLA seen in both lines of sight towards HE0512-3329 ([13]) is an interesting case study. This is the only case where spatially resolved chemical abundances have been measured in a DLA. Unfortunately, only Mn and Fe have measured column densities and it is difficult to disentangle whether the 0.5 dex differences in the lines of sight are of nucleosynthetic origin or due to differential dust depletion.

3 The Properties of QSOs and Their Environments

The well-known proximity effect seen in QSO spectra is the deficit of Lyα forest lines near to the redshift of the quasar due to enhanced ionizing radiation. Conversely, there is evidence that optically thick absorbers, such as Lyman limit systems (LLS) and DLAs are more common when $z_{\rm abs} \sim z_{\rm em}$ (e.g. [8], [17]). It has been suggested that this is due to clustering of galaxies around QSOs, a phenomenon that is well established from imaging surveys (e.g. [1]). Projected QSO pairs with disparate redshifts offer an independent way to test this idea, as shown in Figure 3. It has been shown ([11]) that in 50% of projected pairs with separations $<$ 150 kpc a LLS or DLA is seen at the redshift of the closer QSO, whereas only 2% would be expected based on the established number density of intervening absorbers. Although this qualitatively supports the idea of galaxy clustering around QSOs, the excess of galaxies observed near to QSOs in the transverse direction (i.e. with pairs) is a factor of 4–20 higher than found in single lines of sight ([12]). In some rare cases, the proximity of a QSO to a galaxy provides insight into the physical properties of the QSO itself. [3] detected fluorescent Lyα emission associated with a QSO pair similar to that shown in Figure 3 and discuss how such objects can constrain the radiative structure of the QSO and their lifetimes. This is currently the only example of fluorescence from a pair, despite the

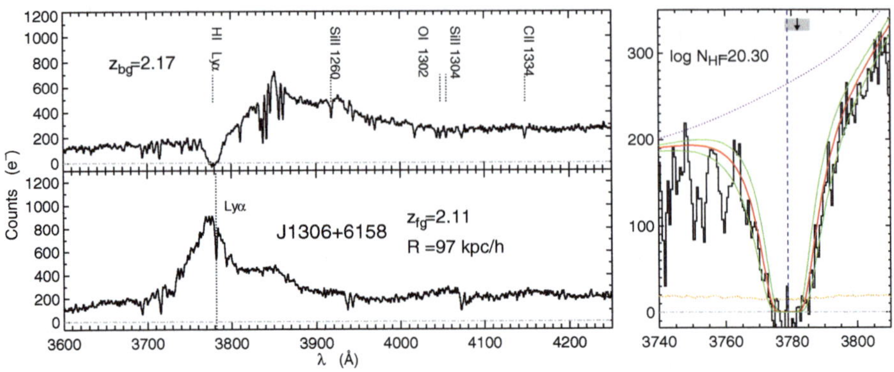

Fig. 3. [12] and [11] have studied projected pairs with disparate redshifts. In this Figure (taken from their paper), we see a DLA at the redshift of the lower z QSO imprinted onto the spectrum of the more distant QSO.

growing sample studied from the SDSS. This is surprising given the high luminosities of some of the QSOs observed, and may indicate that the ionizing radiation is highly anisotropic.

References

1. K. Adelberger, C. Steidel: ApJ **627**, L1 (2005)
2. K. Adelberger et al: ApJ **629**, 636 (2005)
3. K. Adelberger et al: ApJ **637**, 74 (2006)
4. C. W. Churchill et al.: ApJ **534**, 577 (2000)
5. C. W. Churchill et al.: MgII absorption through intermediate redshift galaxies. In: *Probing Galaxies through Quasar Absorption Lines* ed by P. Williams, C.-G. Shu, B. Menard (CUP 2005) pp 24–41
6. F. Coppolani et al.: MNRAS **370**, 1804 (2006)
7. V. D'Odorico et al.: A&A in press (2006)
8. S. L. Ellison et al: A&A **383**, 91 (2002)
9. S. L. Ellison et al: A&A **414**, 79 (2004)
10. C. Foltz et al.: ApJ **335**, 35 (1988)
11. J. F. Hennawi et al.: AJ **131**, 1 (2006)
12. J. F. Hennawi & J. X. Prochaska: ApJ **655**, 735 (2007)
13. S. Lopez et al.: ApJ **626**, 767 (2005)
14. M. Pieri, J. Schaye, A. Aguirre: ApJ **638**, 45 (2006)
15. M. Rauch et al.: ApJ **515**, 500 (1999)
16. M. Rauch et al.: ApJ **576**, 45 (2002)
17. D. Russell, S. L. Ellison, C. R. Benn: MNRAS **367**, 412 (2006)
18. C. Steidel: The Nature and Evolution of Absorption-Selected Galaxies. In *QSO Absorption Lines*, ed by G. Meyaln (Springer-Verlag 1995) pp 139.

Small-scale Structure of High-redshift OVI Absorption Systems

S. Lopez[1], S. Ellison[2], S. D'Odorico[3], and T.-S. Kim[4]

[1] Departamento de Astronomía, Universidad de Chile, Casilla 36-D, Santiago, Chile
[2] University of Victoria, Dept. Physics & Astronomy, Elliott Building, 3800 Finnerty Rd, Victoria, V8P 1A1, British Columbia, Canada
[3] European Southern Observatory, Karl-Schwarzschild-Str. 2, 85748 Garching-bei-München, Germany
[4] Astrophysikalisches Institut Potsdam, An der Sternwarte 16, D-14482, Potsdam, Germany

Summary. We present results of the first survey of high-redshift ($< z >\sim 2.3$) OVI absorption systems along parallel lines-of-sight toward two lensed QSOs.

1 Outline and Results

The Lyα forest frequently exhibits associated metal absorption lines, but the source of metal enrichment is not fully understood yet. At low redshift, $z <\sim 0.3$, a population of OVI absorbers discovered by HST and FUSE have revealed predominantly collisionally ionized gas at $T >\sim 10^5$ K (e.g., [4]), the so-called warm-hot intergalactic medium (WHIM). At higher redshifts some of this gas might in principle be detected in form of OVI. Whether a WHIM at high-redshift exists, how those metals have been released to the IGM, and what keeps the gas ionized, all are questions that are intimately bound to our understanding of galaxy formation and evolution.

Here we present a survey of OVI absorption systems in spatially-resolved VLT UVES spectra of two lensed QSOs [2]. Ten intervening OVI systems were detected in the range $z = 2.087$ to 2.633. Other species like CIV were often detected at the same redshift. We analyzed differences and similarities across the lines of sight in the line parameters of various species to extract physical parameters of the absorbing gas. We complemented our coherence scale limits derived from lensing with models of one and two phase ionized media. The combination of constraints from these two directions allowed us to derive approximate sizes (under certain assumptions of the multi-phase structure) and relevant abundances of C, N and O. Our conclusions are:

1. Every OVI system detected in one line-of-sight (LOS) shows OVI also in the adjacent LOS. On the kpc scale, OVI appears featureless (with a null fractional change in column density and no velocity offsets), while CIV in the same systems shows structure (Fig. 1).

2. A two-phase photoionization model in which CIV arises both in a low and a high-ionization regime and OVI only in the latter (ionization parameter $U \approx 10^{-0.6}$), with characteristic sizes of a few tens and a few hundreds of kpc, respectively, successfully explains the various LOS differences.
3. Photoionization requires [C/O]≈ -0.7, an abundance ratio that is consistent with gas processed in galaxies recently before the absorption epoch. This is consistent with correlations between galaxies and metal systems at $z = 2-3$ [1][3] showing that OVI is always found at distances up to a few hundred kpc from an observed galaxy.

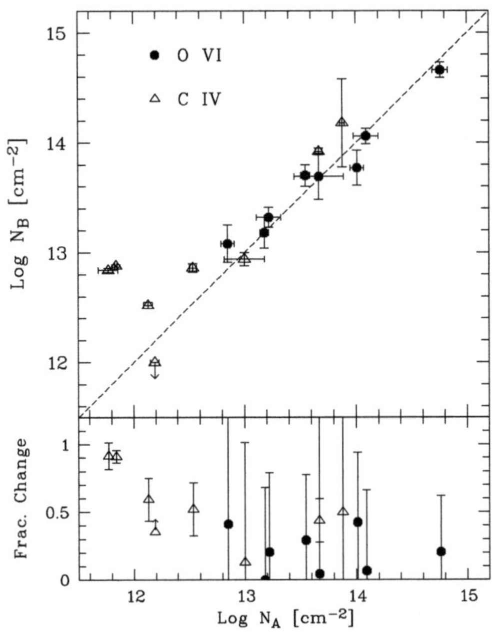

Fig. 1. Comparison of column densities between the A and B lines of sight (upper panel), and fractional change between A and B, $|N_\mathrm{A} - N_\mathrm{B}|/\max(N_\mathrm{A}, N_\mathrm{B})$ for OVI and CIV in OVI systems. Only intervening systems are shown in the plot.

References

1. Adelberger, K., Shapley, A. E., Steidel, C. C., Pettini, M., Erb, D., Reddy, N. A., 2005, ApJ, 630, 50
2. Lopez, S., Ellison, S., D'Odorico, S., & Kim, T.-S., 2006, A&A submitted.
3. Simcoe, R. A., Sargent, W. L. W., Rauch, M., Becker, G., 2006, ApJ, 637, 648
4. Tripp, T. M., Savage, B. D., Jenkins, E., 2000, ApJ, 534, L1

Hot Halos around High-Redshift Galaxies

Andrew J. Fox[1], Patrick Petitjean[1], Cédric Ledoux[2], and R. Srianand[3]

[1] Institut d'Astrophysique de Paris, 98bis Blvd Arago, 75014, Paris, France fox@iap.fr, ppetitje@iap.fr
[2] European Southern Observatory, Alonso de Córdova 3107, Casilla 19001, Vitacura, Santiago 19, Chile cledoux@eso.org
[3] IUCAA, Post Bag 4, Ganesh Khind, Pune 411 007, India anand@iucaa.ernet.in

1 Introduction and Methods

Damped Lyman-α (DLA) systems, defined as those QSO absorbers with H I column densities $> 2 \times 10^{20}$ cm^{-2}, are known to represent the largest reservoirs of neutral gas in the redshift range 0–5 [5]. In addition, DLAs also exhibit a multi-phase interstellar medium with molecular [2] and ionized phases present. Si IV and C IV absorption in the ionized phase of DLAs has been studied before [4]. We have begun the first survey for O VI in DLAs.

We analyze a VLT/UVES dataset that consists of 35 DLAs with coverage of O VI at S/N>5. Finding O VI among these data is difficult, due to the high density of blends from the Lyα forest. However, by using the doublet ratio between O VI λ1037 and λ1031, and looking for corresponding absorption in C IV, Si IV, and N V, we report O VI detections in 12 DLAs, three of which are at <5 000 km s^{-1} from the quasar and so may be associated systems. Three of the intervening DLA systems are shown in Figure 1.

2 Results and Conclusions

- The C IV and O VI profiles are typically twice as broad as the neutral gas profiles, with ionized components seen at up to 500 km s^{-1} from the neutral gas.
- The component line widths imply both warm and hot ionized phases exist, traced by narrow C IV and broad O VI components, respectively.
- The column densities of hot hydrogen in the O VI-bearing gas are of similar order to the column densities of H I in the neutral gas.
- If the fraction of oxygen atoms in the five-times-ionised state is $\ll$0.2, the O VI-bearing hot haloes can contain a significant fraction of the "Missing Metals" at high redshift [1]. This would occur if the temperature in the O VI-containing gas were near 10^6 K, and most oxygen atoms were ionised up to O VII or O VIII [3].
- Accretion and galactic outflows are the likely processes that can generate O VI halos around DLAs.

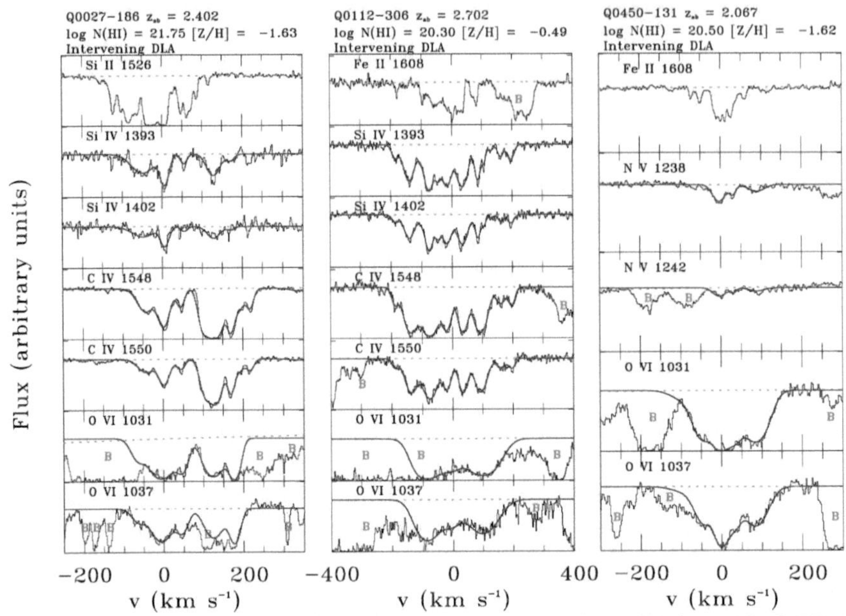

Fig. 1. VLT/UVES data of high ion absorption in three DLAs with O VI. The smooth line in each panel is our Voigt profile fit to the data, which can be seen underneath. Blends are marked with 'B'.

A full version of this survey is to be published in A&A. We thank the organizers of this meeting for a highly enjoyable conference.

References

1. Ferrara, A., Scannapieco, E., & Bergeron, J. ApJ, **634**, L37 (2005)
2. Ledoux, C., Petitjean, P., & Srianand, R. MNRAS, **346**, 209 (2003)
3. Sutherland, R. S., & Dopita, M. A. ApJS, **88**, 253 (1993)
4. Wolfe. A. M., & Prochaska, J. X. ApJ, **545**, 591 (2000)
5. Wolfe. A. M., Gawiser, E., & Prochaska, J. X. ARA&A, **43**, 861 (2005)

Part III

Fundamental Constants

Astrophysical Probes of Fundamental Physics

C.J.A.P. Martins

CFP, Universidade do Porto, Rua do Campo Alegre 687, 4169-007 Porto, Portugal
DAMTP, University of Cambridge, Wilberforce Rd., Cambridge CB3 0WA, U.K.
C.J.A.P.Martins@damtp.cam.ac.uk

Summary. I review the theoretical motivation for varying fundamental couplings and discuss how these measurements can be used to constrain a number of fundamental physics scenarios that would otherwise be inacessible to experiment. As a case study I will focus on the relation between varying couplings and dark energy, and explain how varying coupling measurements can be used to probe the nature of dark energy, with important advantages over the standard methods. Assuming that the current observational evidence for varying α and μ is correct, a several-sigma detection of dynamical dark energy is feasible within a few years, using currently operational ground-based facilities. With forthcoming instruments like CODEX, a high-accuracy reconstruction of the equation of state may be possible all the way up to redshift $z \sim 4$.

1 Theoretical Expectations

The deepest question of modern physics is whether or not there are fundamental scalar fields in nature. They are a key ingredient in the standard model of particle physics (cf. the Higgs particle, which is supposed to give mass to all other particles and make the theory gauge-invariant), but after four decades of particle physics searches there is still no evidence that nature has any use for them. Yet in recent years we have come to realize that the early universe is an ideal place to search for scalar fields, if they exist at all, and there have been some possible hints for them in various contexts. The field of astrophysical searches for varying couplings has been particularly active in recent years, as can be seen by the extensive series of new results and ongoing or forthcoming projects presented at this conference. Observations suggest that the recent universe is dominated by an energy component whose gravitational behaviour is quite similar to that of a cosmological constant. This could of course be the right answer, but the observationally required value is so much smaller than what would be expected from particle physics that a dynamical scalar field is arguably a more likely explanation. Theoretical motivation for such a field is not hard to find. In string theory, for example, dimensionful parameters are expressed in terms of the string mass scale and a scalar field vacuum expectation value.

Now, the slow-roll of this field (which is mandatory so as to yield negative pressure) and the fact that it is presently dominating the universe imply (if the minimum of the potential vanishes) that the field vacuum expectation value today must be of order m_{Pl}, and that its excitations are very light, with $m \sim H_0 \sim 10^{-33}$ eV. But a further consequence of this is seldom emphasized [1]: couplings of this field lead to observable long-range forces and to time-dependence of the constants of nature (with corresponding violations of the Einstein Equivalence Principle). A spacetime varying scalar field coupling to matter mediates a new interaction. If the recent evidence for varying couplings [2, 3] is explained by a dynamical scalar field, this automatically implies the existence of a new force. A series of space missions (ACES, μSCOPE, STEP) will improve on current bounds on the Einstein Equivalence Principle by as many as 6 orders of magnitude. These must find violations if the current data is correct [4]. Joint analyses of varying coupling and Equivalence Principle measurements will shortly provide key tests to a number of fundamental paradigms, such as string theory (and may well be our only opportunity to find evidence for it).

Moreover, in theories where a dynamical scalar field is responsible for varying α, the other gauge and Yukawa couplings are also expected to vary. Specifically, in GUTs there is a relation between the variation of α and that of the QCD scale, Λ_{QCD}, implying that the nucleon mass will vary when measured in units of the Planck mass. Similarly, one would expect variations in the Higgs vacuum expectation value (VEV), v, leading to changes in all particle mass scales including the electron mass. We therefore expect variations of the proton-to-electron mass ratio, $\mu = m_p/m_e$. Measurements of μ have been suggested a long time ago [5]. Typically one has [6]

$$\frac{\dot{\mu}}{\mu} \sim \frac{\dot{\Lambda}_{QCD}}{\Lambda_{QCD}} - \frac{\dot{v}}{v} \sim R\frac{\dot{\alpha}}{\alpha}; \tag{1}$$

the latter equality should be seen as the first term in a Taylor series, but given the expected level of variations the approximation should be good enough for most purposes. The value of R is model-dependent (indeed, even its sign is not determined *a priori*), but large values and negative values are naively expected for GUT models in which modifications come from high-energy scales. The large proportionality factors arise simply because the strong coupling constant and the Higgs VEV run (exponentially) faster than α. Note that with current data [2, 3, 7] one infers $R \sim -4$. Be that as it may, the wide range of α-μ relations implies that simultaneous measurements of both are a powerful discriminating tool between competing models: we can in principle test GUT scenarios without ever needing to detect any GUT model particles, say at accelerators.

2 From α and μ to $w(z)$

A crucial goal of modern observational cosmology is characterizing the properties of dark energy, and in particular to look for dynamical behaviour. A

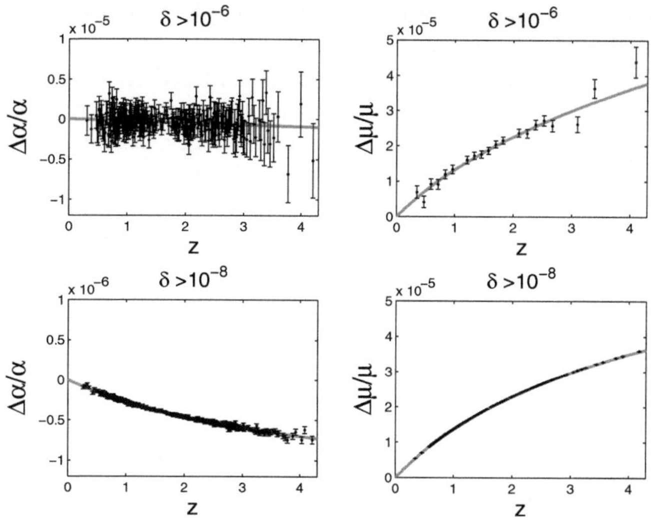

Fig. 1. Simulated datasets expected for α and μ in the near future (top panels) and with CODEX [12, 13] (lower panels), assuming a particular dark energy model.

simple property is its equation of state, $w = p/\rho$, and considerable effort is being put into trying to measure it. Current methods of choice are type Ia supernovae and (more recently) weak lensing. However, the question arises as to whether these are indeed the best tools for the task at hand. It has been known for some time [8] that supernova measurements are limited as a probe of the dark energy equation of state, especially if it is varying with redshift. Analysis of current and future constraints on the dark energy equation of state, from the various standard approaches and parametrized in the usual way [9], shows that a convincing detection of time variation of w is quite unlikely even with hypothetical future space-based experiments such as DUNE or JDEM (in any of its many versions). This is expected since any dynamical field providing the dark energy must be slow-rolling at the present time, and for slow variations there will always be a constant w model that produces nearly identical results over the redshift range where dark energy is dynamically important. This point has also recently been made in [10], and even the Dark Energy Task Force report [11] (an otherwise very naive document) revelas the shortcomings of the standard approaches.

Luckily, better (and cheaper) alternatives are available. A potentially effective tool for probing dynamical dark energy has been suggested previously in [14, 15], though not yet studied in detail: probing varying couplings is a key test to these models, and in particular the varying couplings can be used to infer the evolution of the scalar field, and thus to determine its equation of state. This is analogous to reconstructing the 1D potential for the classical motion of a particle once its trajectory has been specified. Note that this reconstruction method requires only the calculation of first derivatives of noisy data, while the standard methods rely on claculating second derivatives of noisy data. Previous efforts only considered the variation of α, but variations of μ may be easier to detect than those of α (if R is indeed large), and thus provide tighter constraints on dark energy, although the number of such measurements is currently much smaller than the α dataset. (The main reason for this is the difficulty in finding molecular Hydrogen clouds.) One of the goals of our work [16] is to encourage further measurements of μ, and significant such efforts are already in progress.

Having impreved measurements of both α and μ is extremely useful for various reasons. With both observables, the reconstruction will be a lot easier, not to mention less model-dependent. One has the advantage of a much larger lever arm in terms of redshift, since such measurements can be made up to redshifts of $z \sim 4$. Naively one might think that this is not a big advantage, since dark energy is only dynamically relevant at relatively low redshift, and even the DETF report [11] explicitly claims that there's no advantage in probing high redshifts. However, this completely misses the point that the additional redshift coverage probes the otherwise unaccessible z range where scalar field dynamics is expected to be fastest, deep in the matter era. This not only make the detection of any possible dynamics easier, but also reduces (and possibly elliminates) the model-dependence that is unavoidable in the standard methods (where parametrisations like $w = w_0 + w_a(1 - a)$ are dangerously naive). Last but not least, this method provides direct evidence distinguishing dynamical dark energy from a cosmological constant, which given the current data may be very challenging for the standard cosmological tests. Figs. 1 and 2 show an example of our recent work [16] displaying the benefits of a reconstruction using data on both couplings.

Finally, let us point out the fact that, because the reconstruction method requires calculating (first) derivatives of data, it is important to have a good redshift coverage. Therefore it is also important to have a method of measurement that can be applied to a large range of redshifts without changing systematics. Such a method does exist for measuring α: it is the Sunyaev-Zel'dovich effect [17], whose redshift-independence makes it ideal. Current data already provides interesting bounds, and an improvement of several orders of magnitude is expected in the coming years, when throusands of clusters will be at hand for this task, for example from the Planck Surveyor.

3 Conclusions

The prospects for further, more accurate measurements of fundamental constants are definitely bright, as has been highlighted at this conference. The methods described above and other completely new ones that may be devised thus offer the real prospect of an accurate mapping of the cosmological evolution of the fine-structure constant, $\alpha(z)$, and the proton to electron mass

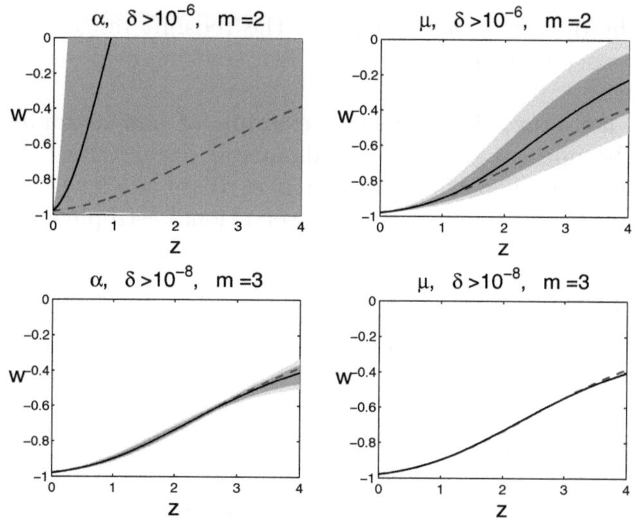

Fig. 2. The reconstruction of the equation of state and its error band is shown for the datasets of Fig. 1. The dashed line represents the dark energy equation of state corresponding to the potential used to generate the simulated data and the solid line corresponds to the reconstruction's best fit. The dark and light region are the 1σ and 2σ confidence levels.

ratio, $\mu(z)$. This may well prove to be the most exciting area of research in the coming years. The worse that can happen to cosmology is the scenario where a number of cosmological parameters are fixed by WMAP, then nothing new happens until Planck comes along and merely adds one digit to the precision of each already-known parameter. After that cosmology may well be dead: there will be little incentive to pushing research further to figure out what the next digit is. However, if in the meantime violations of the Equivalence Principle and/or varying fundamental constants are unambiguously confirmed, then one will (finally) have evidence for the existence of new physics—most likely in the form of scalar fields—in nature (which one may legitimately hope that Planck is able to probe) and an entirely new era begins.

The possibility of using varying couplings to reconstruct the equation of state of dark energy is particularly promising, especially if one obtains further measurements of μ. Let us emphasize that this method can applied now, using existing ground-based facilities. Assuming that the current observational evidence for varying couplings (as discussed during this conference) is correct, a several-sigma detection of dynamical dark energy could be obtained with only a few hundred hours of observation on a VLT-class telescope—an extremely modest investment given the potential gains. Let us stress once again the crucial advantage of a much larger lever arm in terms of redshift, since such measurements can easily be made up to redshifts of $z \sim 4$: this is perhaps the only way one can probe the redshift range where the field evolution is expected to be fastest (if it is a tracking field)—that is, deep in the matter era.

Last but not least, this is also an example of how astrophysical observations can be optimal probes of fundamental physics. Such astrophysical probes will become increasingly common in years to come, and hope this provides early encouragement for the observational astrophysics community. The early universe is the best possible laboratory for fundamental physics, and in this era of precision astrophysics and cosmology, astrophysics has the observational tools that can provide a unique impulse to the fundamental physics of this century. The opportunity is there, and one should take it.

Acknowledgments

I am grateful to Pedro Avelino, Nelson Nunes ans Keith Olive for an enjoyable collaboration that led to [16]. I also acknowledge stimulating conversations on the topic of this article with Nissim Kanekar and Rodger Thompson. This work was funded in part by FCT (Portugal), in the framework of the POCI2010 program, supported by FEDER. Specific funding came from grant POCI/CTE-AST/60808/2004.

References

1. S. M. Carroll: Phys. Rev. Lett. **81**, 3067 (1998)
2. M. T. Murphy, J. K. Webb, and V. V. Flambaum: M.N.R.A.S. **345**, 609 (2003)
3. E. Reinhold et al.: Phys. Rev. Lett. **96**, 151101 (2006)
4. T. Damour: Astrophys. Sp. Sci. **283**, 445 (2003)
5. R. I. Thompson: Astrophys. Lett. **16**, 3 (1975)
6. X. Calmet and H. Fritzsch: Phys. Lett. B **540**, 173 (2002)
7. N. Kanekar et al.: Phys. Rev. Lett. **95**, 261301 (2005)
8. I. Maor, R. Brustein, and P. J. Steinhardt: Phys. Rev. Lett. **86**, 6 (2001)
9. A. Upadhye, M. Ishak, and P. J. Steinhardt: Phys. Rev. D **72**, 063501(2005)
10. A. Liddle et al.: astro-ph/0610126 (2006)
11. A. Albecht et al.: astro-ph/0609591 (2006)
12. L. Pasquini et al.: Report OWL-CSR-ESO-00000-0160 (2005)
13. P. Molaro, M. T. Murphy, and S. Levshakov: astro-ph/0601264 (2006)
14. D. Parkinson, B. A. Bassett, and J. D. Barrow: Phys. Lett. B **578**, 235 (2004)
15. N. J. Nunes and J. E. Lidsey: Phys. Rev. D **69**, 123511 (2004)
16. P. P. Avelino, C. J. A. P. Martins, N. J. Nunes, and K. Olive: Phys. Rev. D **74**, 083508 (2006)
17. E. S. Battistelli et al.: in preparation (2006)

Revisiting VLT/UVES Constraints on a Varying Fine-structure Constant

M. T. Murphy[1], J. K. Webb[2], and V. V. Flambaum[2]

[1] Institute of Astronomy, University of Cambridge, Madingley Road, Cambridge CB3 0HA, UK; `mim@ast.cam.ac.uk`

[2] School of Physics, University of New South Wales, Sydney, NSW 2052, Australia; `jkw@phys.unsw.edu.au`, `flambaum@phys.unsw.edu.au`

Summary. Current analyses of VLT/UVES quasar spectra disagree with the Keck/HIRES evidence for a varying fine-structure constant, α. To investigate this we introduce a simple method for calculating the minimum possible uncertainty on $\Delta\alpha/\alpha$ for a given quasar absorber. For many absorbers in Chand et al. (2004) and for the single-absorber constraint of Levshakov et al. (2006) the quoted uncertainties are smaller than the minimum allowed by the UVES data. Failure of this basic consistency test prevents reliable comparison of the UVES and HIRES results.

1 Introduction

The 'many-multiplet' (MM) method is the most precise technique for constraining cosmological changes in the fine-structure constant, α, from QSO absorption spectra [1]. We have previously described self-consistent MM evidence from 143 Keck/HIRES absorbers for a smaller α over the redshift range $0.2 \le z_{\rm abs} \le 4.2$ at the fractional level $\Delta\alpha/\alpha = (-0.57 \pm 0.11) \times 10^{-5}$ [2]. Clearly, independent analyses of spectra from different spectrographs are desirable to refute/confirm this. First attempts with VLT/UVES spectra, e.g. [3, 4], generally found null results with quoted uncertainties $< 0.1 \times 10^{-5}$. To investigate these claims, we introduce a simple method for calculating the minimum possible uncertainty on $\Delta\alpha/\alpha$ from a given absorption system.

2 A Simple Measure of the Limiting Precision on $\Delta\alpha/\alpha$

The velocity shift, v_j, of transition j due to a small relative variation in α, $\Delta\alpha/\alpha \ll 1$, is determined by the q-coefficient for that transition,

$$\omega_{z,j} \equiv \omega_{0,j} + q_j \left[(\alpha_z/\alpha_0)^2 - 1 \right] \quad \Rightarrow \quad \frac{v_j}{c} \approx -2\frac{\Delta\alpha}{\alpha}\frac{q_j}{\omega_{0,j}}, \tag{1}$$

where $\omega_{0,j}$ & $\omega_{z,j}$ are the rest-frequencies in the lab and at redshift z, α_0 is the lab value of α and α_z is the shifted value measured from an absorber at z. The MM method is the comparison of measured velocity shifts from several transitions (with different q-coefficients) to compute the best-fit $\Delta\alpha/\alpha$. The linear equation (1) implies that the error in $\Delta\alpha/\alpha$ is determined only by the q-coefficients (assumed to have negligible errors) and the total velocity uncertainty, integrated over the absorption profile, from each transition, $\sigma_{\mathrm{v},j}$:

$$\delta(\Delta\alpha/\alpha)_{\mathrm{lim}} = \sqrt{S/D}\,, \tag{2}$$

for $S \equiv \sum_j \left(\frac{\sigma_{\mathrm{v},j}}{c}\right)^{-2}$ and $D \equiv S\sum_j \left(\frac{2q_j}{\omega_{0,j}}\right)^2 \left(\frac{\sigma_{\mathrm{v},j}}{c}\right)^{-2} - \left[\sum_j \frac{2q_j}{\omega_{0,j}} \left(\frac{\sigma_{\mathrm{v},j}}{c}\right)^{-2}\right]^2$. This expression is just the solution to a least-squares fit of $y = a + bx$ to data (x_i, y_i), with errors only on the y_i, where the intercept a is also allowed to vary; this mimics the real situation in fitting absorption lines where the absorption redshift and $\Delta\alpha/\alpha$ must be determined simultaneously.

The quantity $\sigma_{\mathrm{v},j}$ is commonly used in radial-velocity searches for extra-solar planets, e.g. [5], but is not normally useful in QSO absorption-line studies. Most metal-line QSO absorption profiles display a complicated velocity structure and one usually focuses on the properties of individual velocity components, each of which is typically modelled by a Voigt profile. However, it is important to realize that $\Delta\alpha/\alpha$ and its uncertainty are integrated quantities determined by the entire absorption profile. Thus, $\sigma_{\mathrm{v},j}$ should incorporate all the velocity-centroiding information available from a given profile shape. From a spectrum $F(i)$ with 1-σ error array $\sigma_F(i)$, the minimum possible velocity uncertainty contributed by pixel i is given by [5]

$$\frac{\sigma_{\mathrm{v}}(i)}{c} = \frac{\sigma_F(i)}{\lambda(i)\ [\partial F(i)/\partial\lambda(i)]}\,. \tag{3}$$

That is, a more precise velocity measurement is available from those pixels where the flux has a large gradient and/or a small uncertainty. This quantity can be used as an optimal weight, $W(i) \equiv [\sigma_{\mathrm{v}}(i)/c]^{-2}$, to derive the total velocity precision available from all pixels in a (portion of) spectrum,

$$\sigma_{\mathrm{v}} = c\left[\sum_i W(i)\right]^{-1/2}\,. \tag{4}$$

For each transition in an absorber, $\sigma_{\mathrm{v},j}$ is calculated from (3) & (4); the only requirements are the 1-σ error spectrum and the multi-component Voigt profile fit to the transition's absorption profile. The latter allows the derivative in (3) to be calculated without the influence of noise. Once $\sigma_{\mathrm{v},j}$ has been calculated for all transitions, the uncertainty in $\Delta\alpha/\alpha$ follows from (2).

It is important to realize that the uncertainty calculated with the above method represents the absolute minimum possible 1-σ error on $\Delta\alpha/\alpha$; the real error – as derived from a simultaneous χ^2-minimization of all parameters comprising the Voigt profile fits to all transitions – will always be larger than $\delta(\Delta\alpha/\alpha)_{\mathrm{lim}}$ from (2). The main reason for this is that absorption systems usually have several velocity components which have different optical depths in different transitions. Equation (2) assumes that the velocity information integrated over all components in one transition can be combined with the

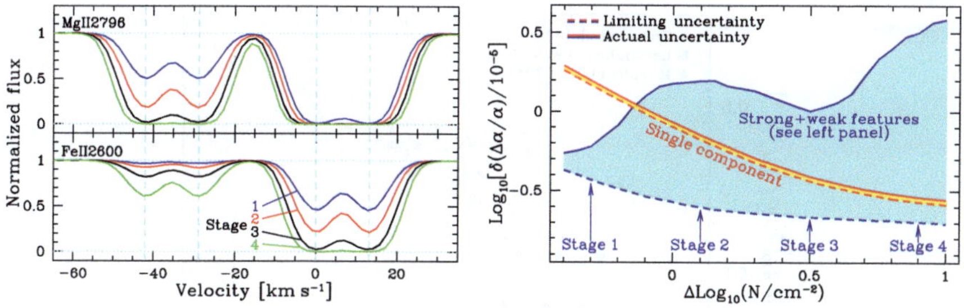

Fig. 1. *Left*: Simulation of two transitions in a multi-component absorber. Labelled are distinct stages of differential saturation in the two main spectral features. *Right*: The difference between the limiting precision, $\delta(\Delta\alpha/\alpha)_{\rm lim}$, and the actual precision (as derived by χ^2-minimization analysis) varies strongly from stage to stage. In a single-component absorber the actual uncertainty tracks $\delta(\Delta\alpha/\alpha)_{\rm lim}$.

same integrated quantity from another transition to yield an uncertainty on $\Delta\alpha/\alpha$. However, in a real determination of $\Delta\alpha/\alpha$, each velocity component (or group of components which define a sharp spectral feature) in one transition is, effectively, compared with only the same component (or group) in another transition. Fig. 1 illustrates this point.

3 Application to Existing Constraints on $\Delta\alpha/\alpha$

We have calculated $\delta(\Delta\alpha/\alpha)_{\rm lim}$ for the absorbers from the three independent data-sets which constitute the strongest current constraints on $\Delta\alpha/\alpha$: (i) The 143 absorbers in our Keck/HIRES sample [2]; (ii) The 23 Mg/Fe II systems from UVES studied by [3]; (iii) The UVES exposures of the $z_{\rm abs} = 1.1508$ absorber towards HE 0515−4414 studied by [4]. For sample (ii), the spectra were kindly provided by B. Aracil who confirmed that they have the same wavelength and flux arrays as those used in [3]. The main difference is the error arrays: our error array is generally a factor ≈ 1.4 smaller than that used by [3] (H. Chand, B. Aracil, 2006, priv. comm.). We have confirmed this by digitizing the absorption profiles plotted in [3]. Thus, $\delta(\Delta\alpha/\alpha)_{\rm lim}$ calculated using our spectra will be *smaller* than the value [3] would derive. The continuum normalization will also be slightly different, but this has negligible effects on the analysis here. For sample (iii), the exposures were reduced using a modified version of the UVES pipeline and their S/N matches very well those quoted by [4]. For samples (ii) & (iii) we use the Voigt profile models published in [3] & [4] to calculate $\delta(\Delta\alpha/\alpha)_{\rm lim}$. For all samples, the atomic data for the transitions (including q-coefficients) were the same as used by the original authors. In practice, when applying (3) and (4) we sub-divide the absorption profile of each transition into 15 km s^{-1} chunks to mitigate the effects illustrated in Fig. 1. This provides a value of $\delta(\Delta\alpha/\alpha)_{{\rm lim},k}$ for each

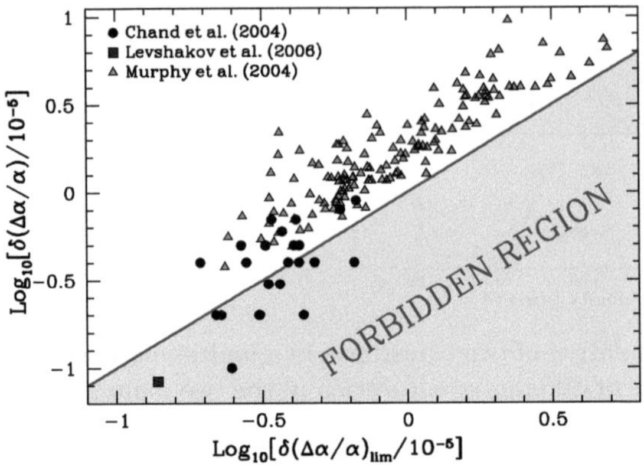

Fig. 2. Quoted errors vs. limiting precision for current samples. The Chand et al. and Levshakov et al. samples fail the basic requirement $\delta(\Delta\alpha/\alpha) > \delta(\Delta\alpha/\alpha)_{\rm lim}$.

chunk k. The final value of $\delta(\Delta\alpha/\alpha)_{\rm lim}$ is simply $\{\sum_k 1/[\delta(\Delta\alpha/\alpha)_{{\rm lim},k}]^2\}^{1/2}$; in all cases this is < 1.4 times the value obtained without sub-divisions.

Figure 2 shows the 1σ error on $\Delta\alpha/\alpha$ quoted by the original authors versus the limiting precision, $\delta(\Delta\alpha/\alpha)_{\rm lim}$. The main results are clear. Firstly, the 1σ errors quoted for the HIRES sample in [2] always exceed $\delta(\Delta\alpha/\alpha)_{\rm lim}$, as expected if the former are robustly estimated. Secondly, for at least 11 of their 23 absorbers, Chand et al. [3] quote errors which are *smaller* than $\delta(\Delta\alpha/\alpha)_{\rm lim}$. Recall that since their error arrays are larger than ours, 11 out of 23 is a conservative estimate. Finally, the very small error quoted by [4] for HE 0515−4414, 0.084×10^{-5}, disagrees significantly with the limiting precision of 0.14×10^{-5}. Thus, the (supposedly) strong current UVES constraints on $\Delta\alpha/\alpha$ fail a basic consistency test which not only challenges the precision claimed by [3] and [4] but which must bring into question the robustness and validity of their analysis and final $\Delta\alpha/\alpha$ values.

4 Conclusion

We have introduced a very simple method of determining the limiting precision on $\Delta\alpha/\alpha$ obtainable from a set of transitions in a QSO absorption system, $\delta(\Delta\alpha/\alpha)_{\rm lim}$. Only the 1-$\sigma$ error spectrum and the model absorption profile (generally constructed from Voigt profiles) are required to calculate $\delta(\Delta\alpha/\alpha)_{\rm lim}$. The method simply equates the total velocity information contained in the absorption profile of a transition with the expected error on $\Delta\alpha/\alpha$ via that transition's sensitivity to α-variation. All absorption systems in our HIRES data-set have quoted uncertainties in $\Delta\alpha/\alpha$ which exceed $\delta(\Delta\alpha/\alpha)_{\rm lim}$, as expected. However, the uncertainties on $\Delta\alpha/\alpha$ currently quoted by [3] and [4] from UVES spectra are, in many cases, *smaller* than $\delta(\Delta\alpha/\alpha)_{\rm lim}$. Clearly, the UVES and HIRES results cannot be reliably compared until more rigorous analyses of the UVES spectra are completed.

References

1. J. K. Webb, V. V. Flambaum, C. W. Churchill, M. J. Drinkwater, J. D. Barrow: Phys. Rev. Lett. **82**, 884 (1999)
2. M. T. Murphy, V. V. Flambaum, J. K. Webb, V. V. Dzuba, J. X. Prochaska, A. M. Wolfe: Lecture Notes Phys. **648**, 131 (2004)
3. H. Chand, R. Srianand, P. Petitjean, B. Aracil: Astron. Astrophys. **417**, 853 (2004)
4. S. A. Levshakov, M. Centurión, P. Molaro, S. D'Odorico, D. Reimers, R. Quast, M. Pollmann: Astron. Astrophys. **449**, 879 (2006)
5. F. Bouchy, F. Pepe, D. Queloz: Astron. Astrophys. **374**, 733 (2001)

On the Variation of the Fine-structure Constant, and Precision Spectroscopy

Chand, H.[1,2], Srianand, R.[2], Petitjean, P.[1,3], and Aracil, B.[4]

[1] Institut d'Astrophysique de Paris – CNRS, 98bis Boulevard Arago, F-75014 Paris, France chand@iap.fr
[2] IUCAA, Post Bag 4, Ganeshkhind, Pune 411 007, India anand@iucaa.ernet.in
[3] LERMA, Observatoire de Paris, 61 Rue de l'Observatoire, F-75014 Paris, France ppetitje@iap.fr
[4] Department of Astronomy, University of Massachusetts, 710 North Pleasant Street, Amherst, MA 01003-9305, USA

Summary. We summarize the constraints on a possible time variation of the fine-structure constant, α, we obtained using UVES/VLT data. In particular, we report new results based on the analysis of 6 Ni II systems. No variation is found to a fractional level of about 0.2×10^{-5} over a redshift range $0.4 \leq z \leq 2.5$. We discuss the importance of precision spectroscopy to improve the constraint on α variation, based on our detailed analysis of a high resolution spectrum ($R \sim 112\,000$) obtained with HARPS.

1 Limits on the Variation of α using UVES/VLT Data

Absorption lines seen in QSO spectra can serve as a laboratory to test the variation over cosmic time of fundamental constants such as the fine-structure constant, α, and the proton-to-electron mass ratio, μ. The recent application of the many-multiplet (MM) method [4] has improved, by an order of magnitude the accuracy of the $\Delta\alpha/\alpha$ measurements based on QSO absorption lines [9]. Analysis of HIRES/Keck data has resulted in the claim of a variation in α, $\Delta\alpha/\alpha = (-0.54 \pm 0.12) \times 10^{-5}$, over a redshift range $0.2 < z < 3.7$ [6]. In order to confirm or refute this result, we have applied the MM method to very high quality (S/N$\sim 60 - 80$, R$\geq 44,000$) UVES/VLT data. In view of the numerous systematic errors involved in the MM method, we have carried out detailed simulations to define proper selection criteria to choose suitable absorption systems in order to perform the best analysis (see [1] for details). The application of these selection criteria to our full sample of 50 Mg II/Fe II systems lead us to restrict our study to 23 Mg II/Fe II systems over a redshift range $0.4 \leq z \leq 2.3$. The weighted mean of the individual measurements from our analysis is $\Delta\alpha/\alpha = (-0.06 \pm 0.06) \times 10^{-5}$ [1],[8]. At present, we are further investigating our procedure for any possible systematics involved in estimating the errors using χ^2 minimization. Indeed, errors could have been underestimated by a factor of about two.

The MM method uses absorption lines from different species (mostly Fe II, Mg II, Si II, Mg I) and makes the implicit assumption of negligible chemical and ionization inhomogeneities between the different species. It is difficult to estimate the exact consequences of this assumption. In order to avoid it, it is possible to use absorption lines from a single species.

We have therefore extended our experiment to obtain a constraint based on the alkali-doublet (AD) method [2] using Si IV as the single species. We applied the AD method to 15 Si IV doublets selected from our UVES/VLT data. The weighted mean of the individual measurements is $\Delta\alpha/\alpha = (+0.15 \pm 0.43) \times 10^{-5}$ over a redshift range of $1.59 \leq z \leq 2.92$. This result corresponds to a factor of three improvement on $\Delta\alpha/\alpha$ measurements based on Si IV doublet, previously published in the literature. The above conclusion of no variation of α has been reached by other independent groups [5], [7] using absorption lines from Fe II only which is more sensitive than using Si IV doublet.

Like Fe II, Ni II is another suitable species to measure $\Delta\alpha/\alpha$ with lines from a single species. Ni II has three lines with very different sensitivity coefficients for α variation. The rest wavelength of Ni IIλ1709 line is insensitive to minor variation of α, thereby providing an anchor to measure the redshift, while the other two lines are very sensitive. We have recently carried out the preliminary analysis of 6 Ni II systems. Our preliminary measurements are shown in the lower right panel of Fig. 1, resulting in the weighted mean value of $\Delta\alpha/\alpha = (-0.13 \pm 0.46) \times 10^{-5}$ over a redshift range of $1.67 \leq z \leq 2.33$.

We summarize our results in Fig. 1 for the MM method (upper panel), AD method (lower left panel) and the new measurements based on Ni II lines (lower right panel). *Combining all our measurements, we find no variations of α to a fractional level of about 0.2×10^{-5}.*

However, while increasing the number of absorbers can lead to an improvement in the constraint on the variation of α, the main limitation however is the calibration accuracy and stability of the spectrograph.

2 $\Delta\alpha/\alpha$ Measurement and Precise Spectroscopy

One concern in our above analysis of $\Delta\alpha/\alpha$ measurement is the accuracy and robustness of the wavelength calibration procedures. Another source of uncertainty comes from the multi-component Voigt-profile decomposition at the typical UVES resolution (R≈50 000). Precision spectroscopy could play an important role in addressing both the issues.

To estimate the improvement in $\Delta\alpha/\alpha$ measurements using precision spectroscopy, we have carried out a detailed analysis of a QSO HE 0515-4414 spectrum obtained using the High Accuracy Radial velocity Planet Searcher (HARPS) mounted on the ESO 3.6 m telescope at the La Silla observatory [3]. The special feature of this spectrograph is its very high stability and high wavelength calibration accuracy. *This experiment uses completely independent instruments and calibration procedures.* We have used the high calibration accuracy of HARPS to test the wavelength calibration accuracy

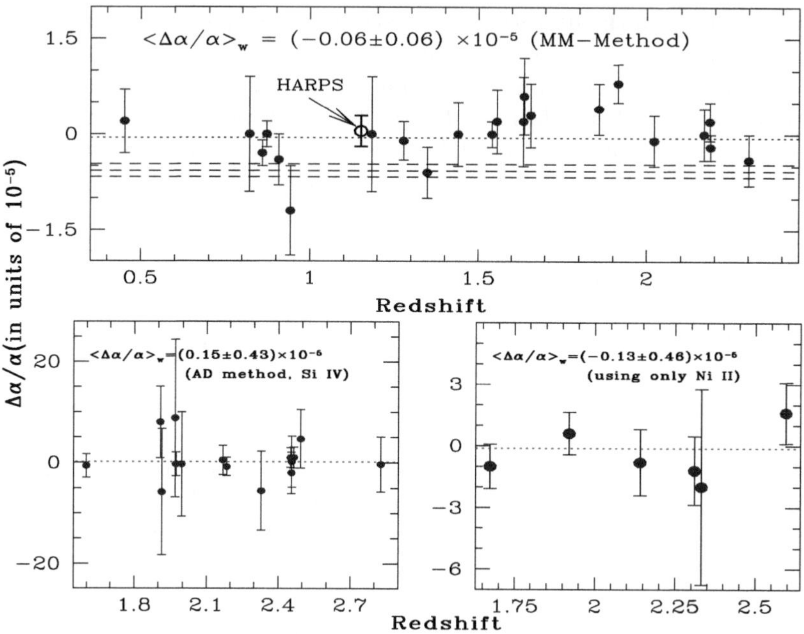

Fig. 1. Results based on MM-method (using Mg II/Fe II systems) are shown in the upper panel [1],[8]. The measurement using HARPS data is shown by an open circle [3]. Dashed lines indicate the result from HIRES/Keck sample [6]. Our results from AD-method using 15 Si IV doublets [2], and our preliminary results based on 6 Ni II absorption systems are shown in, respectively, the lower left and right panels.

of UVES and made an additional measurement. A cross-correlation analysis between the Th-Ar lamp spectra of HARPS and UVES, has been performed to detect any possible shifts between them. We find that the relative shifts between the two spectra are within 2 mÅ. Using Gaussian fits to unblended Th-Ar emission lines, we confirm that the absolute wavelength calibration of HARPS is very robust with rms deviation of 0.87 mÅ. This is about a factor of 4 better than that of UVES ($\sigma = 4.08$ mÅ). Thus the small shifts noted between the HARPS and UVES lamp spectra are well within the typical wavelength calibration accuracy of UVES. We also derive the error on $\Delta\alpha/\alpha$ measurements due to the calibration accuracy. For UVES and HARPS spectra, this is found to be $\sigma = 0.96 \times 10^{-6}$ and $\sigma = 0.19 \times 10^{-6}$, respectively, for a typical system with three well-detached components.

This shows that an instrument such as HARPS with high stability and high-accuracy in wavelength calibration is ideal for the $\Delta\alpha/\alpha$ measurement. Unfortunately it is mounted on a 4-m class telescope and only HE 0515−4414 is bright enough to be observed in a reasonable amount of time. Our analysis also shows that the UVES spectra reduced (or calibrated) with the UVES pipeline and used in the literature to constrain $\Delta\alpha/\alpha$ ([1],[8],[2],[7]) does not suffer from major systematic errors in the wavelength calibration.

To estimate the uncertainty in the multi-component Voigt-profile decomposition at the UVES resolution (R≈50 000), we have obtained an accurate multi-component structure of the DLA-system at $z_{\rm abs} = 1.1508$ toward QSO HE 0515-4414, using the higher resolution HARPS data (R $\approx 112,000$). Our best fit component structure obtained by simultaneously fitting the HARPS data (of higher resolution) and the UVES data (of better S/N ratio) requires, as expected, additional components, as compared to the fit using the UVES data alone [7]. Using this new sub-component decomposition and both, HARPS and UVES data, we performed the $\Delta\alpha/\alpha$ measurement for this system and found $\Delta\alpha/\alpha = (0.05 \pm 0.24) \times 10^{-5}$. This is consistent with earlier measurements for this system using the UVES spectrum alone ($\Delta\alpha/\alpha = [0.01 \pm 0.17] \times 10^{-5}$, [7]).

It can be noted that the precision on $\Delta\alpha/\alpha$ obtained using the HARPS spectrum, which is of higher resolution and lower S/N ratio, is similar to that obtained from the UVES spectrum alone, which is of lower resolution and higher S/N ratio. Therefore, the improvement in the wavelength calibration accuracy by an order of magnitude using HARPS will improve the constraint on $\Delta\alpha/\alpha$, only if high S/N ratio can also be obtained. This will be possible if an instrument similar to HARPS can be mounted on bigger telescopes.

References

1. Chand, H., Srianand, R., Petitjean, P. et al. 2004, A&A, 417,853
2. Chand, H., Petitjean, P., Srianand, R. et al. 2005, A&A, 430,47
3. Chand, H., Srianand, R., Petitjean, P., et al. 2006, A&A,451,45
4. Dzuba, V.A., Flambaum, V. V., & Webb, J. K.,1999, PRL, 82, 888
5. Levshakov, S. A., Centurión, M., Molaro, P. et al. A&A 2005, 434, 827
6. Murphy, M. T., Webb, J. K., Flambaum, V. V.,2003, MNRAS, 345, 609
7. Quast, R., Reimers, D., & Levshakov, S. A. 2004, A&A, 415, L7
8. Srianand, R., Chand, H., Petitjean, P. et al. 2004, PRL, 92121302
9. Webb, J. K., Murphy, M. T., Flambaum, V. V., et al. 2001, PRL, 87, 091301

High-Precision Measurements of $\Delta\alpha/\alpha$ from QSO Absorption Spectra

Sergei A. Levshakov[1], Paolo Molaro[2], Sebastian Lopez[3], Sandro D'Odorico[4], Miriam Centurión[2], Piercarlo Bonifacio[2,5], Irina I. Agafonova[1], and Dieter Reimers[6]

[1] Department of Theoretical Astrophysics, Ioffe Physico-Technical Institute, Politekhnicheskaya Str. 26, 194021 St. Petersburg, Russia
[2] Osservatorio Astronomico di Trieste, Via G. B. Tiepolo 11, 34131 Trieste, Italy
[3] Departamento de Astronomía, Universidad de Chile, Casilla 36-D, Santiago, Chile
[4] European Southern Observatory, Karl-Schwarzschild-Strasse 2, D-85748 Garching bei München, Germany
[5] Observatoire de Paris 61, avenue de l'Observatoire, 75014 Paris, France
[6] Hamburger Sternwarte, Universität Hamburg, Gojenbergsweg 112, D-21029 Hamburg, Germany

Summary. Precise radial velocity measurements ($\delta v/c \sim 10^{-7}$) of Fe II lines in damped Lyα systems from very high quality VLT/UVES spectra of quasars HE 0515–4414 and Q 1101–264 are used to probe cosmological time dependence of the fine structure constant, α. It is found that between two redshifts $z_1 = 1.15$ and $z_2 = 1.84$ the value of $\Delta\alpha/\alpha$ changes at the level of a few ppm: $(\alpha_{z_2} - \alpha_{z_1})/\alpha_0 = 5.43 \pm 2.52$ ppm. Variations of α can be considered as one of the most reliable method to constrain the dark energy equation of state and improvements on the accuracy of the wavelength calibration of QSO spectra are of great importance.

1 Introduction

The late-time acceleration of the universe discovered from the luminosity distance measurements of high-redshift SNe Ia is now regarded as evidence for the existence of dark energy. In many models, the cosmological evolution of dark energy is accompanied by variations in coupling constants as, e.g., the fine-structure constant α [1]. Since theories predict different behavior of dark energy, from slow-rolling to oscillating, to study its dynamics the coupling constants must be measured with highest accuracy at each redshift.

In order to fulfill this requirement we developed a method called '*Single Ion Differential α Measurement*, SIDAM' [2, 3, 4, 5]. Based on the measurements of the relative line position of only one element Fe II, it is free from many systematics which affect other methods, e.g., [6, 7]. It has been shown that SIDAM can provide a sub-ppm precision at a single redshift and that this level of accuracy is mainly caused by systematic errors inherited from the uncertainties of the wavelength scale calibration [5] (ppm stands for parts per million, 10^{-6}).

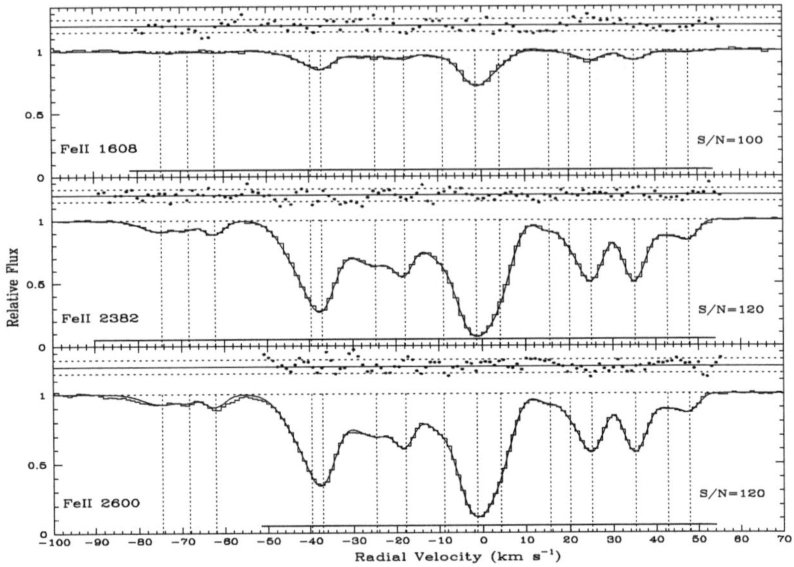

Fig. 1. Histograms are the combined Fe II profiles from the $z = 1.84$ damped Lyα system towards Q 1101–264 [8]. The zero radial velocity is fixed at $z_{\rm abs} = 1.838911$. The synthetic profiles are over-plotted by the smooth curves. The normalized residuals, $(\mathcal{F}_i^{cal} - \mathcal{F}_i^{obs})/\sigma_i$, are shown by dots (horizontal dotted lines restrict the 1σ errors). The dotted vertical lines mark positions of the sub-components. Bold horizontal lines mark pixels used to minimize χ^2. The ranges at $v < -50$ km s^{-1} and at $v \simeq -30$ km s^{-1} in the Fe II $\lambda 2600$ profile are blended with weak telluric lines.

In this presentation we consider our recent results [8] with the SIDAM procedure obtained from the analysis of very high resolution (FWHM $\simeq 3.8$ km s^{-1}, slit width 0.5 arcsec) and high signal-to-noise (S/N = 100-120 per pixel) spectra of Q 1101–264 ($z_{\rm em} = 2.15$, $V = 16.0$). The observations were performed at the VLT Kueyen telescope on 21-23 February, 2006 under the ESO programme No. 076.A-0463. The total exposure time was 15.4 hours.

2 Results

The spectroscopic measurability of $\Delta\alpha/\alpha \equiv (\alpha_z - \alpha_0)/\alpha_0$ at redshift z (α_0 refers to the $z = 0$ value) is based on the fact that the energy of each line transition depends individually on a change in α [9]. It means that the relative change of the frequency ω_0 due to varying α is proportional to the so-called sensitivity coefficient $\mathcal{Q} = q/\omega_0$ [4]. The q-values for the resonance UV transitions in Fe II are taken from [10], and their rest frame wavelengths — from [8, 11].

The value of $\Delta\alpha/\alpha$ can be measured from the relative radial velocity shifts between lines with different sensitivity coefficients. In linear approximation

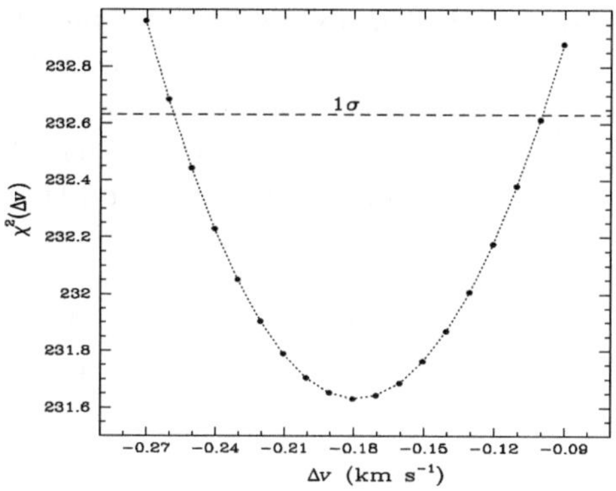

Fig. 2. χ^2 as a function of the velocity difference Δv between the Fe II $\lambda1608$ and $\lambda2382/2600$ lines [8]. The number of data points $M = 305$, the number of degrees of freedom $\nu = 257$. The minimum of the curve at $\Delta v = -0.18$ km s^{-1} gives the most probable value of $\Delta\alpha/\alpha = 5.36$ ppm. The 1σ confidence level is determined by $\Delta\chi^2 = 1$ (dashed line) which gives $\sigma_{\Delta v} = 0.08$ km s^{-1} or $\sigma_{\Delta\alpha/\alpha} = 2.38$ ppm.

($|\Delta\alpha/\alpha| \ll 1$) we have [5]:

$$\frac{\Delta\alpha}{\alpha} = \frac{(v_2 - v_1)}{2\,c\,(\mathcal{Q}_1 - \mathcal{Q}_2)} = \frac{\Delta v}{2\,c\,\Delta\mathcal{Q}}\,, \qquad (1)$$

where index ‘1’ is assigned to the line $\lambda1608$, and index ‘2’ marks one of the other Fe II lines ($\lambda2382$ or $\lambda2600$).

The resulting normalized, vacuum-barycentric, and co-added spectra are shown in Fig. 1. Two independent data reduction procedures (1D and 2D) resulted in almost identical co-added spectra. We found a perfect fit of the Fe II profiles to a 16-component model (shown by the smooth curves in Fig. 1): the normalized χ^2 per degree of freedom equals $\chi^2_\nu = 0.901$ ($\nu = 257$). The computational procedure was the same as in [5]. Since the $\mathcal{Q}$ values for the $\lambda2382$ and $\lambda2600$ lines are equal, their relative velocity shift, $\Delta v_{2600-2382}$, characterizes the goodness of wavelength calibrations, $\sigma_{\rm scale}$, of the corresponding echelle orders. We found that the value of $\Delta v_{2600-2382} = 20$ m s^{-1} is comparable with $\sigma_{\rm scale} \leq 20$ m s^{-1} estimated from the ThAr lines. So, in what follows we consider the $\lambda2382$ and $\lambda2600$ lines as having the same radial velocity, and calculate Δv between this velocity and that of the line $\lambda1608$.

The radial velocity shift between the $\lambda1608$ and $\lambda2382/2600$ lines was derived by comparing synthetic profiles with their observed profiles and minimizing χ^2. To find the most probable value of Δv, we fit the absorption lines with a fixed Δv, changing Δv in the interval from -270 m s^{-1} to -90 m s^{-1} in steps of 10 m s^{-1} (see Fig. 2). For each Δv, the strengths of the subcomponents, their broadening parameters and relative velocity positions were allowed to vary in order to optimize the fit and thus minimize $\chi^2(\Delta v)$. The most probable value of Δv corresponds to the minimum of the curve $\chi^2(\Delta v)$.

In our case it is -180 m s^{-1}. The 1σ confidence interval to this value is given by the condition $\Delta\chi^2 = \chi^2 - \chi^2_{\rm min} = 1$ (the horizontal dashed line in Fig. 2. It is seen from the figure that $\sigma_{\Delta v} = 80$ m s^{-1}, or $\Delta\alpha/\alpha = 5.36 \pm 2.38$ ppm.

We have specifically investigated those systematic effects which could introduce a non-zero difference between the blue and the red lines ($\lambda1608$ and $\lambda2382/2600$, respectively) and thus simulate a variation of $\Delta\alpha/\alpha$ at the ppm level: (1) possible isotopic shifts caused by unknown isotope abundances, (2) effects of the unresolved components, and (3) possible blends. We also compared our previous results [8], where Fe II lines falling in both UVES arms were used, with the present measurements to check up possible shifts caused by different changes of isophote onto the slit during integration and did not reveal any of them. We found that the largest systematic error does not exceed 30 m s^{-1}, i.e. $\sigma_{\Delta\alpha/\alpha,{\rm sys}} \leq 0.89$ ppm [8].

We note that the accuracy of $\sigma_{\Delta\alpha/\alpha} = 2.38$ ppm represents a factor of 1.5 improvement with respect to our previous result $\sigma_{\Delta\alpha/\alpha} = 3.8$ ppm [4] obtained from the archive data of lower resolution ($FWHM \simeq 6$ km s^{-1}) but the same S/N ratio. Thus the higher spectral resolution is shown to significantly contribute to higher accuracy in the $\Delta\alpha/\alpha$ measurements.

The comparison of $\Delta\alpha/\alpha = -0.07 \pm 0.84$ ppm at $z_1 = 1.15$ towards HE 0515–4414 [5] with the measured quantity $\Delta\alpha/\alpha = 5.36 \pm 2.38$ ppm at $z_2 = 1.84$ shows a tentative change of the value of $\Delta\alpha/\alpha$ between these two redshifts: $(\alpha_{z_2} - \alpha_{z_1})/\alpha_0 = 5.43 \pm 2.52$ ppm.

Acknowledgments. SAL acknowledges supports from the RFBR grant No. 06-02-16489, the Federal Agency for Science and Innovations grant NSh 9879.2006.2, and the DFG project RE 353/48-1.

References

1. P. P. Avelino, C. J. A. P. Martins, N. J. Nunes, & K. Olive: Phys. Rev. D **74**, 083508 (2006)
2. S. A. Levshakov: Astrophysical constraints on hypothetical variability of fundamental constants. In: *Astrophysics, Clocks and Fundamental Constants*, ed by S.G. Karshenboim, E. Peik (Springer, Berlin Heidelberg New York 2004) pp 151–166
3. R. Quast, D. Reimers, & S. A. Levshakov: A&A **415**, L7 (2004)
4. S. A. Levshakov, M. Centurión, P. Molaro, & S. D'Odorico: A&A **434**, 827 (2005)
5. S. A. Levshakov, M. Centurión, P. Molaro, et al: A&A **449**, 879 (2006)
6. M. T. Murphy, J. K. Webb, & V. V. Flambaum: MNRAS **345**, 609 (2003)
7. H. Chand, R. Srianand, P. Petitjean, & B. Aracil: A&A **417**, 853 (2004)
8. S. A. Levshakov, P. Molaro, S. Lopez et al: A&A, submit. (2006)
9. J. K. Webb, et al: Phys. Rev. Lett. **82**, 884 (1999)
10. V. A. Dzuba, et al: Phys. Rev. A **66**, 022501 (2002)
11. M. Aldenius, S. Johansson, & M. T. Murphy: MNRAS **370**, 444 (2006)

Probing Fundamental Constant Evolution with Redshifted OH Lines

Nissim Kanekar[1], Jayaram N Chengalur[2], and Tapasi Ghosh[3]

[1] National Radio Astronomy Observatory, Socorro, USA nkanekar@nrao.edu
[2] NCRA-TIFR, Pune, India; ATNF, Epping, Australia chengalur@ncra.tifr.res.in
[3] NAIC-Arecibo, Puerto Rico, USA tghosh@naic.edu

Summary. Radio spectroscopy in the four redshifted 18cm OH lines allows a probe of changes in three fundamental constants, the fine structure constant α, the proton g-factor g_p and the electron-proton mass ratio $\mu \equiv m_e/m_p$. After summarizing the technique, we describe WSRT observations of the conjugate satellite OH lines towards PKS 1413+135, which find weak evidence in support of evolution in the constants, over a lookback time of ~ 2.7 Gyrs.

1 Introduction

Astrophysical studies of redshifted spectral lines provide a powerful probe of putative changes in fundamental constants such as the fine structure constant α, the electron-proton mass ratio $\mu \equiv m_e/m_p$ and the proton gyro-magnetic ratio g_p, over a large fraction of the age of the Universe (e.g. [1, 2, 3, 4]). Most present techniques are based on optical spectroscopy, where wavelength calibration, line blending, intrinsic velocity offsets, isotopic abundances, interloping absorbers, etc, are all possible sources of systematic error (e.g. [1]). Given the importance of changes in fundamental constants for theoretical physics, it is important that independent techniques, with different systematics, be used. In this article, we describe a new technique to probe fundamental constant evolution, based on redshifted radio OH lines, and then discuss recent results from Westerbork Synthesis Radio Telescope (WSRT) observations of these lines in a $z \sim 0.25$ source, PKS 1413+135.

2 Radio OH Lines

The $^2\Pi_{3/2}$ J=3/2 OH rotational ground state is split up into four sub-levels by Λ-doubling and hyperfine splitting; transitions between these sub-levels of the ground state give rise to the four 18cm OH lines. Transitions with $\Delta F = 0$, i.e. with no change in the total angular momentum quantum number F, are referred to as "main" lines, with rest frequencies of 1665.4018 MHz and 1667.3590 MHz, while transitions with $\Delta F = \pm 1$ are called "satellite" lines, with rest frequencies of 1612.2310 MHz and 1720.5299 MHz.

The four 18cm OH lines stem from two very different physical processes, Λ-doubling and hyperfine splitting, and the transition frequencies hence have different dependences on the fundamental constants α, μ and g_p. We have used a perturbative treatment of the OH molecule [5] to obtain these dependences and find that it is possible to measure changes in different combinations of the above constants by comparing the redshifts of the different OH lines ([3]; see also [6] for a similar analysis that only considers changes in α). Specifically, a comparison between the redshifts of the sum and difference of "main" line frequencies is sensitive to changes in $H \equiv g_p \left[\alpha^2/\mu\right]^{0.13}$, while one between the redshifts of the sum and difference of "satellite" frequencies is sensitive to changes in $G \equiv g_p \left[\alpha^2/\mu\right]^{1.85}$ [3]. In addition, one can also compare the redshifts of 1667 MHz OH and HI 21cm absorption in a statistically large absorber sample to obtain an independent probe of any evolution [3, 7]; this traces changes in the quantity $F \equiv g_p \left[\alpha^2/\mu\right]^{1.57}$.

Clearly, OH lines provide an alternative approach to probing fundamental constant evolution. The only drawback to this technique is that these lines are difficult to detect as they are both very weak and occur in regions teeming with terrestrial radio frequency interference (RFI). At present, only five redshifted absorbers have been found in the (strongest) 1665 and 1667 MHz transitions (e.g. [7, 8]), all of which have also been detected in the HI 21cm line. The best present constraint from a comparison between the HI 21cm and "main-line" OH redshifts is $[\Delta F/F] = (0.44 \pm 0.36^{\rm stat} \pm 1^{\rm syst}) \times 10^{-5}$, over $0 < z \lesssim 0.7$, based on four spectral components in two redshifted absorbers [7]. Of course, four measurements are far too few for a reliable estimate of the systematic error arising from relative motions between the HI and OH "clouds" and it is unlikely that the requisite statistically-large OH absorption samples will be obtained until the advent of the Square Kilometer Array, a next generation radio telescope. Despite this, the OH technique still bears promise due to an interesting maser mechanism, discussed in the next section.

3 "Conjugate" Behaviour of the Satellite OH Lines

In rare cases, the 18cm satellite OH lines have the special property of being "conjugate" to each other, with the same shapes, but with one line in absorption and the other in emission. This arises due to competition between two decay routes to the $2\Pi_{3/2}$ (J = 3/2) OH ground state, after the molecules have been pumped from the ground to the higher excited rotational states [9]. In essence, if the last step of the cascade is the 119μ transition $2\Pi_{3/2}$ (J = 5/2) $\rightarrow$ $2\Pi_{3/2}$ (J = 3/2), transitions to the $F = 1$ levels of the ground state are forbidden by the dipole selection rule $\Delta F = 0, \pm 1$, resulting in an over-population of the $F = 2$ levels. This causes an inversion of the 1720 MHz line and an anti-inversion of the 1612 MHz line. On the other hand, if the final step is the 79μ decay $2\Pi_{1/2}$ (J = 1/2) $\rightarrow$ $2\Pi_{3/2}$ (J = 3/2), it results in 1612 MHz inversion and 1720 MHz anti-inversion [9].

Crucially, the conjugate behavior *guarantees that the satellite lines arise from the same gas.* These lines are thus perfectly suited for the purpose of measuring any evolution in α, μ and g_p between the source redshift and today, via a comparison between the sum and difference of satellite redshifts, since systematic velocity offsets between the lines are ruled out by the pumping mechanism. Any measured difference between the redshifts must then arise due to a change in one or more of the above fundamental constants [10].

Only two such conjugate OH systems are known at cosmological distances, at $z \sim 0.247$ towards PKS 1413+135 [10, 11] and at $z \sim 0.765$ towards PMN J0134−0931 [7]. The original WSRT observations of PKS 1413+135 in June 2003 obtained the constraint $[\Delta G/G] = (2.2 \pm 3.8) \times 10^{-5}$, consistent with no changes in α, μ and g_p [10]. Our new WSRT observations of this system are described in the next section.

4 New WSRT Results

The WSRT was used to carry out a deep integration on the satellite OH lines of PKS 1413+135 between May and July 2005 (~ 58 hours). The 1720 MHz and 1612 MHz lines were observed simultaneously, with a velocity resolution of ~ 0.5 km/s, after Hanning smoothing. The RMS noise on the spectra was $\sim 10^{-3}$ per 0.5 km/s channel, in units of optical depth. The sum of the 1612 and 1720 MHz optical depth profiles was found to be consistent with noise, as expected from the conjugate behaviour.

The satellite line profiles were well-fit by a single Gaussian, with the residuals consistent with noise. The peak optical depths and widths of the fitted gaussians were found to be in excellent agreement. However, the peak line redshifts were found to be offset from each other by ~ 0.43 km/s, at the $\sim 1.8\sigma$ level. We also carried out a cross-correlation analysis on the two spectra, to test whether the offset might be the result of the gaussian parameterization. This obtained a similar offset (-0.45 ± 0.27 km/s) between the peak redshifts, with $\sim 1.7\sigma$ significance.

The satellite lines of PKS 1413+135 were also observed with the WSRT in a separate session between October 2004 and January 2005. These data had an error in the frequency settings, causing no spectral baseline to be present on the low velocity side of each line. While the lack of spectral baseline implies that the fits are not very reliable, with the precise results somewhat dependent on the choice of baseline, we note that this run also found a similar offset between the peak 1612 MHz and 1720 MHz redshifts.

The agreement between the shapes (amplitudes and widths) of the satellite lines and the offset between the line redshifts is exactly the signature of an evolution in the fundamental constants that was being sought. A comparison between the line redshifts, using the results derived in [3], gives $[\Delta G/G] = (-2.29 \pm 1.27) \times 10^{-5}$ over $0 < z < 0.25$, where $G \equiv g_p \left[\alpha^2/\mu\right]^{1.85}$. While not as yet statistically significant, this is tantalizing evidence for a change in one of the three constants, α, μ or g_p, with cosmological time. The present result is consistent with either (or both) a lower value of α or a larger value of μ at a lookback time of ~ 2.9 Gyr, assuming that changes in g_p are much smaller than those in α or μ [12]. If μ is assumed constant, we obtain $[\Delta\alpha/\alpha] = (-0.62 \pm 0.34) \times 10^{-5}$, while assuming α constant gives $[\Delta\mu/\mu] = (+1.23 \pm 0.69) \times 10^{-5}$.

Alternatively, the above result can be written as $[\Delta\alpha/\alpha] \times (2 - R) = (-1.23 \pm 0.69) \times 10^{-5}$, defining R by the relation $[\Delta\mu/\mu] = R\,[\Delta\alpha/\alpha]$ and again assuming g_p constant. Given a value of R from a theoretical model, one can immediately obtain allowed values of $[\Delta\alpha/\alpha]$. For example, values of $R \sim -50$ are typical in GUT models (e.g. [12, 13]), giving $[\Delta\alpha/\alpha] \lesssim 10^{-6}$; in such models, changes in G must be dominated by changes in μ (note that any model with $|R| \gtrsim 10$ would give a similar result). If these models are correct, we obtain $[\Delta\mu/\mu] \sim (+1.23 \pm 0.69) \times 10^{-5}$ from $z = 0.247$ to the present epoch, i.e. weak evidence for a larger value of μ at earlier times. This is of similar amplitude, but opposite sign, to the value $[\Delta\mu/\mu] = (-2.0 \pm 0.6) \times 10^{-5}$ obtained by [4], from H_2 absorption lines over $0 < z < 3$.

The only alternative cause for the offset between the satellite line redshifts appears to be a "hidden" spectral component in one of the satellite lines. In principle, this could, at the present S/N, cause a shift in the peak redshift while not affecting the peak amplitude and width. Deeper observations of the satellite OH lines with present instrumentation should be able to both confirm the detection of the shift between peak redshifts and test this possibility.

Acknowledgments
NK thanks Carlos Martins for stimulating discussions and suggestions concerning this work. The WSRT is operated by the ASTRON (Netherlands Foundation for Research in Astronomy) with support from the Netherlands Foundation for Scientific Research NWO.

References

1. M. T. Murphy, J. K. Webb, V. V. Flambaum: MNRAS **345**, 609 (2003)
2. R. Srianand et al.: Phys. Rev. Lett.**92**, 121302 (2004)
3. J. N. Chengalur, N. Kanekar: Phys. Rev. Lett.**91**, 241302 (2003)
4. E. Reinhold et al.: Phys. Rev. Lett.**96**, 151101 (2006)
5. G. C. Dousmanis, T. M. Sanders, C. H. Townes: Phys. Rev. **100**, 1735 (1955)
6. J. Darling: Phys. Rev. Lett. **91** 011301 (2003)
7. N. Kanekar et al.: Phys. Rev. Lett.**95**, 261301 (2005)
8. N. Kanekar, J. N. Chengalur: A&A **381**, L73 (2002)
9. M. Elitzur: *Astronomical Masers*, (Kluwer Academic Publishers 1992)
10. N. Kanekar, J. N. Chengalur, T. Ghosh: Phys. Rev. Lett.**93**, 051302 (2004)
11. J. Darling: ApJ **612**, 58 (2004)
12. P. G. Langacker, G. Segré, M. J. Strassler: Phys. Lett. B **528**, 121 (2002)
13. X. Calmet, H. Fritzsch: Eur. Phys. Jour. C **24**, 639 (2002)

A Molecular Probe of Dark Energy

Rodger I. Thompson[1]

Steward Observatory, University of Arizona, Tucson AZ 85721
rthompson@as.arizona.edu

Summary. Many theories of dark energy invoke rolling scaler fields which in turn predict time variable fundamental constants such as the ratio of electron to proton mass μ. Molecular energy levels are sensitive to μ with the electronic, vibrational and rotational levels affected each in a different manner depending on their quantum numbers. High redshift damped Lyman alpha clouds with molecular hydrogen offer an excellent opportunity to search for this possible variation. New, very accurate, laboratory measurements of the wavelengths of the Lyman and Werner transitions make it now possible to check whether the small variations in wavelength due to a time varying value of μ are present.

1 Introduction

Many theoretical constructs of dark energy invoke rolling scaler fields which predict time variation (rolling) of fundamental constants such as the ratio of the electron to proton mass μ. It has been known for more than 30 years that variations in μ can be tested by observing the electronic, vibration, rotation spectrum of molecular hydrogen in high redshift damped Lyman alpha (DLA) systems (Thompson 1975). Until now, however, the technology of astronomical spectroscopy and the accuracy of laboratory wavelength determinations have not permitted sensitive determinations of μ at high redshift. The advent of large telescopes, sensitive high resolution spectrometers and very accurate laboratory wavelength determination of the Lyman and Werner bands have brought observations to the brink of the accuracy required. Recently Reinhold et al. (2006) have claimed a detection of a variation in μ at $\frac{\Delta\mu}{\mu} = (2.4 \pm 0.6) \times 10^{-5}$ based on observations of molecular hydrogen absorption lines in the quasars Q 0347-383 and Q 0405-443 at absorption redshifts of 3.024897 and 2.5947325 respectively.

2 The Concept

Please refer to the article by Carlos Martins in this volume for the theoretical description of dark energy models that lead to a time variation in μ. This article is dedicated to the process of measuring μ at high redshifts. Thompson

(1975) showed that the rotational and vibrational energies of molecules vary directly with μ and as the $\sqrt{\mu}$ respectively, multiplied by the rotational and vibrational quantum numbers relative to the electronic energy. See also the description in Shu (1991). This means that each line of the Lyman and Werner electronic bands of molecular hydrogen has a unique shift of wavelength for a given shift of μ. This is often expressed as a sensitivity constant $K_i(v_i, J_i)$ for a given line i such that $\lambda_i/\lambda_i^0 = (1+z)(1+K_i\Delta\mu/\mu)$ where z is the redshift, v_i is the vibrational quantum number, and J_i is the rotational quantum number (Varshalovich and Levshakov 1993). A shift in μ then produces predictable shifts in the lines which can not be mimicked by cosmological or velocity redshifts. In general the higher the vibrational quantum number, the larger the shift. At low vibrational quantum numbers some of the shifts are negative rather than positive.

The electronic Lyman and Werner bands of H_2 are observable with ground based telescopes at redshifts of approximately 2 and higher. As shown in Figure 1 the Lyman and Werner bands have regions of overlap. The overlap is very important since it places low vibrational quantum number lines of the Werner system with low or negative shift next to high vibrational quantum number lines of the Lyman bands with large shifts. The absorption lines arise from the lowest J levels of the H_2 electronic and vibrational ground state. Usually only the first 4 or 5 rotational states are observed but transitions to upper level vibrational states of nearly 20 are observed. This produces several hundred lines that may be observed. Most of these, however, are blended or obscured by the Lyman alpha forest.

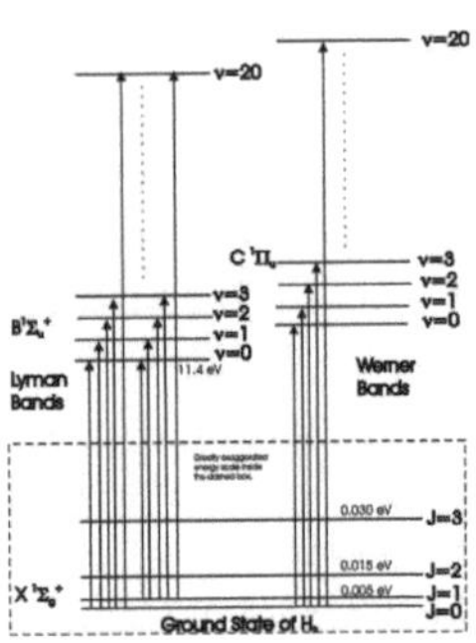

Fig. 1. H_2 energy levels. Note the overlap of the Werner and Lyman bands. Although displaced in the figure to conserve space the zero vibration upper level of the Werner band is very near the vibrational quantum number 14 of the Lyman band upper level.

3 Major Advances

Earlier attempts to measure possible changes in μ, e.g. Foltz et al. (1988), suffered from a lack of telescope aperture and spectroscopic sensitivity. The advent of sensitive spectrometers on large telescopes such at UVES on the VLT brought the sensitivities and resolutions into a range which is on the edge of significance with accuracies in $\Delta\mu$ on the order of $10^{-5}\mu$. Perhaps even more significant have been the improvements in the laboratory determination of the present day rest frame wavelengths of the H_2 lines in the Lyman and Werner bands by Ubachs and Reinhold (2004). The accuracy of wavelengths of most lines are now on the order of 5×10^{-8} of their rest wavelength.

4 Improvements in the Present Analysis

4.1 Master Wavelength Calibration

An exact determination of the wavelength calibration is essential for the determination of possible changes in fundamental constants. The first step is a master calibration lamp image. This step gathers all of the calibration lamp images for a given grating setting. Initial line positions are determined using the photometric source extraction procedure SExtractor (Bertin and Arnouts 1996). The position of prechosen isolated lines are then refined with a Gaussian fit and then the images are shifted and median combined to form a cosmic ray free image. A severe editing is done on the thorium and argon line list to only retain very high quality unblended lines. A master wavelength solution is then performed on the master image. Shifts of this solution are determined from a cross correlation of the master calibration image with the appropriate calibration lamp image associated with the spectrum.

4.2 "Sizing" the Continuum

Most of the H_2 absorption lines are blended with or lie very close to lines from the Lyman alpha forest and other absorption lines. Establishing an accurate continuum is essential to determining an accurate wavelength solution, particularly for those lines whose centers of absorption may have been shifted by nearby lines. In the same way as a canvas is sized or prepared before a painting the continuum must be properly prepared before doing the H_2 wavelength analysis. In this case the continuum is considered to be any part of the spectrum that is not an H_2 line rather than the true continuum. The continuum is sized by fitting each H_2 line with a Voigt profile to find the extent of the line. The continuum in that region is then determined by a least squares quadratic interpolation of the 3 point smoothed spectrum on either side of the line. Figure 2 shows an example of a sized continuum.

4.3 Self Calibrating Spectrum

The usual procedure to test for a change in μ is to determine the wavelength of each H_2 line and then plot any deviation from the expected line position versus the shifts predicted from the sensitivity coefficients. The overlap of the

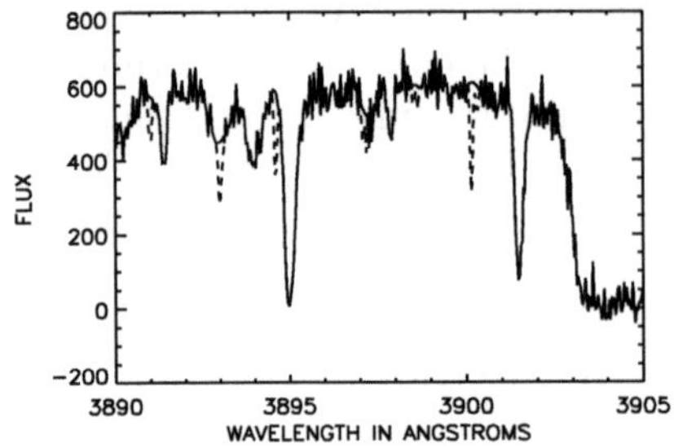

Fig. 2. The sized continuum is shown by the solid line and the removed H_2 lines are indicated by the dashed line.

Lyman and Werner bands offers another opportunity for analysis. The overlap mixes lines with low expected shifts with line that have a high expected shift. The low shift lines can be used to refine the wavelength solution and test whether the lines with expected high shifts have indeed shifted. In many cases high and low expected shift lines lie within a small fraction of an angstrom of each other. This is somewhat analogous to placing an iodine cell in the beam as is done in planetary detection observations.

5 Future Prospects

High resolution spectrometers on very large telescopes offer and opportunity to greatly increase the signal to noise and spectral resolution over present observations (See the contribution by L. Pasquini on CODEX on the ELT in this volume). A more immediate prospect in the Potsdam Echelle Polarimeteric and Spectroscopic Instrument (PEPSI) on the Large Binocular Telescope (LBT). PEPSI has a spectral resolution of 300,000 and uses both primaries of the LBT for an effective collecting area of a 12m telescope. As an example, given the expected sensitivity of PEPSI, the redshift 4.224 H_2 system in PSS J 1443+2724 can be observed at R = 300,000 at a signal to noise of 50 in 30 hrs.

References

1. E. Bertin & S. Arnouts: Astron. & Astroph. **117**, 393, (1996)
2. C.B. Foltz et al.: ApJ., **324**, 267, (1988)
3. E. Reinhold et al.: Phys. Rev. Let. **96**, 151101, (2006)
4. F. Shu: *The Physics of Astrophysics: Volume 1 Radiation*, 1st ed. (University Science Books, Mill Valley California 1991) pp 322–325
5. R. I. Thompson: Astrophysical Letters **16**, 3, (1975)
6. W. Ubachs & E. Reinhold: Phys. Rev. Let. **92**, 1-1302, (2004)
7. D.A. Varshalovich and S.A. Levshakov: JETP Lett, **58**, 237, (1993)

Part IV

Beyond Photon Noise

Establishing Wavelength Standards in the near Infra-red: Th-Ar

Florian Kerber[1], Gillian Nave[2], Craig. J. Sansonetti[2], Gaspare Lo Curto[3], Paul Bristow[1], and Michael R. Rosa[4]

[1] European Southern Observatory, Karl-Schwarzschild-Str.2, 85748, Garching, Germany fkerber@eso.org, bristowp@eso.org
[2] National Institute of Standards and Technology, Gaithersburg, MD 20899, USA gillian.nave@nist.gov, craig.sansonetti@nist.gov
[3] European Southern Observatory, Chile glocurto@eso.org
[4] Space Telescope European Co-ordinating Facility, Karl-Schwarzschild-Str.2, 85748, Garching, Germany mrosa@stecf.org

Summary. We describe the joint efforts by the European Southern Observatory (ESO), the Space Telescope European Co-ordinating Facility (ST-ECF), and the United States National Institute of Standards and Technology (NIST) to establish the Th-Ar hollow cathode lamp as a standard for the calibration of VLT (Very Large Telescope) spectrographs in the near infrared (IR). In the near IR only a limited number of wavelength standards are available. Th-Ar hollow cathode lamps provide a rich spectrum in the ultraviolet (UV)-visible region and have been used in astronomy for a long time. We report new measurements using the 2-m UV/visible/IR Fourier transform spectrometer (FTS) at NIST that establish more than 2000 lines as wavelength standards in the range 900 nm to 4500 nm. This line list is used as input for a physical model that provides the wavelength calibration for the Cryogenic High-Resolution IR Echelle Spectrometer (CRIRES), ESO's new high-resolution (R≈100 000) IR spectrograph at the VLT. Based on these data and additional measurements investigating other properties relevant for operations we conclude that Th-Ar lamps hold the promise of becoming a standard source for wavelength calibration in near IR astronomy.

1 Wavelength Calibration in the Near Infra-Red and the Th-Ar hollow cathode lamp

Traditionally, astronomical spectroscopy in the near infra-red (IR) has relied on atmospheric features of the night sky for wavelength calibration (Rousselot et al. 2000). The main limitation of the method is in the number of lines available in a given wavelength range and at a given resolution. Experience with ESO's Cryogenic High-Resolution IR Echelle Spectrometer (CRIRES) during commissioning shows that at high resolution the intensity of the OH night lines will be too low to be used for calibration purposes in many cases. A detailed description of CRIRES is given in Käufl et al. (2004).

On the other hand, the use of calibration sources such as hollow cathode lamps (HCLs) or gas cells is limited in the near-IR, mostly because these sources have not been extensively measured in the near-IR and hence only a limited number of wavelength standards are available.

The Th spectrum was studied from 278 nm to about 1000 nm at high resolution more than 20 years ago by Palmer & Engleman (1983) and it is a well established standard for wavelength calibration of astronomical spectrographs in the UV and visible. Two valuable studies of the Th-Ar spectrum in the near-IR have recently been published, but neither is directly applicable to the operation of CRIRES:

- Hinkle et al. (2001) produced an atlas of the Th-Ar spectrum covering selected regions in the range 1000 nm to 2500 nm. They established wavelength standards using the McMath 1-m laboratory Fourier Transform Spectrometer (FTS) at the US National Solar Observatory. For technical reasons their list of about 500 lines contains gaps in wavelength coverage.
- More recently, a fundamental analysis of the Th-Ar spectrum was provided by Engleman, Hinkle & Wallace (2003). Their list contains more than 5000 lines derived from observations of a high current Th-Ar HCL with the McMath FTS. Such a lamp produces a very rich Th spectrum, but it is very different from low current commercially available lamps and is not well-suited for operation at an astronomical facility. Although the spectrum from the high current lamp is significantly different from commercial Th-Ar lamps, the line list is highly valuable for identification of the lines in low current spectra.

2 Laboratory Work and Results

Spectra of the Th-Ar lamps operated at 20 mA were recorded on the NIST 2-m FTS. The FTS (Nave, Sansonetti, & Griesmann 1997) was fitted with a CaF_2 beamsplitter, silver coated mirrors, and InSb detectors. The optimum alignment of the spectrometer depends slightly on wavelength. Hence interferograms for two different wavelength regions were recorded – the first optimized for wavelengths between 800 nm and 2000 nm and the second for wavelengths greater than 2000 nm. A resolution of 0.01 cm^{-1} (0.001 nm at 1 μm) was used for the short wavelength region and a resolution of 0.005 cm^{-1} (0.0125 nm at 5 μm) for the long wavelength region.

To obtain a good signal-to-noise ratio many interferograms were co-added for each spectrum, corresponding to acquisition times of up to 20 h. Radiometric calibration was achieved using a calibrated tungsten ribbon lamp.

Analysis of the spectra is ongoing, but we expect to measure a total of about 2500 lines in the range 900 nm to 4800 nm. We anticipate that the final wavenumber values for strong narrow lines will be accurate to 0.001 cm^{-1}, but the deviation for weak and broad lines may be substantially greater. The final line list will contain wavenumber, wavelength, line width, intensity, and identification for about 2500 lines. A detailed description of the use of HCLs, the laboratory work and the results is given by Kerber et al. (2007).

3 Additional Measurements

To better understand the properties of Th-Ar lamps and to prepare for their routine operation as calibration sources for an astronomical spectrograph, we conducted additional measurements using a commercial FTS at ESO and the High Accuracy Radial velocity Planet Searcher (HARPS) (Mayor et al. 2003).

We investigated the spectral output of the lamp as a function of operating current. We found a highly linear increase in the total detected light as the power dissipated in the lamp was increased. When we analysed for argon and thorium lines separately, however, we found a more complex behaviour. For every spectrum, we calculated the ratio of the intensity of each observed line to the intensity of the same line in the 10 mA spectrum. We then averaged the ratios for all lines of a particular species at each current. The results, which represent the average behaviour of the line intensities as a function of current, are displayed in Fig. 1. There is a very pronounced distinction between the behaviour of the metal and the gas lines. Qualitatively this behaviour can be understood because at the higher lamp current more ions are available in the plasma and there is a corresponding increase in sputtering. A first test using about 250 lines shows that this method can be used as a diagnostic tool to distinguish Th and Ar lines with a high degree (> 90 %) of confidence.

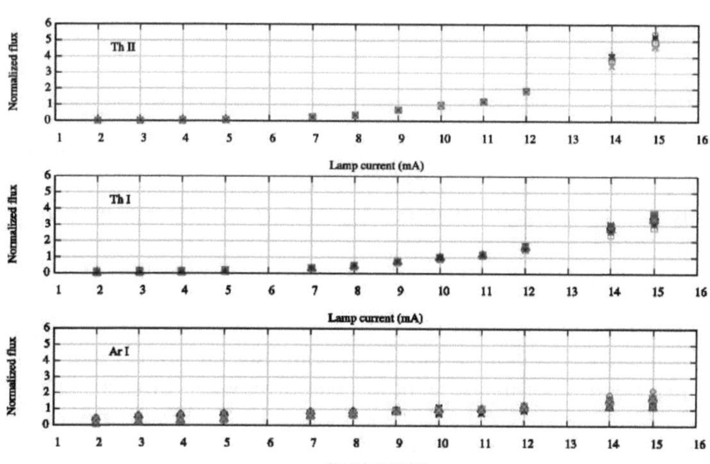

Fig. 1. Variation of line intensities as a function of operating current. All values are normalised to the intensity observed at 10 mA.

Based on these findings one can devise ways to optimise the use of the lamps for CRIRES calibration. For example, the lamp could be run at higher currents for observations at wavelengths where the intensity and spectral density of lines is intrinsically lower.

4 Summary

We have described our efforts to establish Th-Ar HCLs as a standard calibration source for near-IR spectroscopy. To this end we have:

- measured wavelengths for more than 2000 lines in the range 900 nm to 4500 nm in the spectrum of a commercial Th-Ar hollow cathode lamp.
- made additional measurements in order to gain insight into the properties of Th-Ar lamps, most notably the variation of the spectral output as a function of current.
- described the distinct behaviour of the intensities of the gas and metal lines with respect to current which can be also used as a diagnostic tool for line identification.

We conclude that Th-Ar HCLs hold the promise of becoming an excellent standard source for wavelength calibration for near IR astronomy. With this development the situation in the near IR will become very similar to the UV-visible region, and it will be possible to support high accuracy absolute wavelength calibration without having to rely on atmospheric features.

Acknowledgements: We gratefully acknowledge financial support from the European Southern Observatory (ESO) and the European Space Agency (ESA). We thank H.U. Käufl and the members of the CRIRES team.

References

1. Engleman Jr., R., Hinkle, K. H., & Wallace, L.: Journal of Quantitative Spectroscopy & Radiative Transfer, **78**, 1, (2003)
2. Hinkle, K. H., Joyce, R. R., Hedden, A., Wallace, L., & Engleman Jr., R.: PASP **113**, 548, (2001)
3. Käufl, H.-U., Ballester, P., Biereichel, P., et al. In: *Proceedings of the SPIE*, 5492, 1218, (2004)
4. Kerber, F., Nave, G., Sansonetti, C. et al. : In: *The Future of Spectrophotometric and Polarimetric Standardization*, ASP Conf. Ser. in press, (2007)
5. Mayor, M., Pepe, F., Queloz, D. et al: The Messenger **114**, 20, (2003)
6. Nave, G., Sansonetti, C. J., & Griesmann, U. In: *Fourier Transform Spectroscopy: Methods and Applications. OSA Technical Digest Series*, 3, 38
7. Palmer, B. A., & Engleman Jr., R.: Los Alamos National Laboratory Report, LA-9615, (1983)
8. Rousselot, P., Lidman, C., Cuby, J.-G., Moreels, G., & Monnet, G.: A&A **354**, 1134, (2000)

Atomic Data for Astrophysics - Parameters, Precision, Priorities

Sveneric Johansson

Atomic Astrophysics, Lund Observatory, Lund University, Lund, Sweden
Sveneric.Johansson@astro.lu.se

1 Introduction

The needs for high-accuracy atomic data in astrophysics have increased tremendously over the last 15 years with the development of astronomical spectroscopy. High-resolution instruments in space (e.g. HST/STIS, FUSE) and on ground (e.g. KECK, VLT) have raised the demands on laboratory data. During the same period of time many laboratories for atomic spectroscopy have closed down or have seen a drastic reduction in manpower. This development will hinder the extraction of detailed information from astronomical spectra. Better laboratory measurements do not necessarily contribute to the progress of physics but are conditional for progress in many areas of astrophysics.

In this paper we will discuss the balance and coupling between parameters, priorities and precision - i.e. which atomic parameters are vital for analyses of astrophysical spectra, which elements, parameters and transitions should be given the highest priority and what precision is needed in the various cases? We will elucidate these issues by giving some examples of recent and ongoing work.

2 Parameters

There are two fundamental parameters - wavelengths and oscillator strengths (gf-values) - involved in *abundance analyses* based on line spectra of astrophysical plasmas, e.g. stellar atmospheres, nebulae, the interstellar medium etc. In some sense the internal properties of the energy levels are also of fundamental importance when using the lines for abundance work. We will also as a parameter include line structure, which means hyperfine structure (hfs) and isotope shift (IS). These effects cause line broadening and asymmetries in stellar spectra - asymmetries that should not be taken care of by adjusting stellar parameters. A valuable reference source for recent atomic data is the triennal report from Comm. 14 of IAU [1].

3 Energy Levels

The *identity* of a spectral line is given by the upper and lower levels of the corresponding transition. The names and identities of these energy levels are often ignored in numerical astrophysics, but it is of importance in the interpretation of nebular spectra. In LTE analyses of stellar abundances only the excitation energy is needed, but for studies of excitation mechanisms in nebular regions the atomic physics behind the observed lines plays a significant role. Therefore, it would be desirable to see adequate level notations in the line lists published in astronomical journals. The ideal case would be to see a transformation to the established spectroscopic notation system used among laboratory spectroscopists, which would facilitate the communication between users and producers.

The *accuracy* of energy levels is important for calculations of precise wavelengths, Ritz wavelengths, especially for forbidden lines. These cannot be measured directly in the laboratory but are derived from accurate energy level values [2].

The *priority* when deciding which elements that need further studies in the laboratory and how much the present knowledge of the level system has to be extended is set by the astrophysical relevance. In some spectra, e.g. Fe II, more than 1000 energy levels are known, and that is definitely not enough for ultraviolet stellar spectroscopy and, which is shown below, not enough even for optical spectroscopy.

4 Wavelengths

Even if high-resolution instruments in astronomy have improved the wavelength *accuracy* in astrophysical spectra it is still possible to measure wavelengths at a much higher accuracy in the laboratory. This statement is true for the low ionization stages (I-III), where Fourier transform spectroscopy (FTS) is used. Below 1500 Å, where FTS is not available, the wavelength accuracy is still not acceptable for analyses of some HST data. For higher ionization stages there is still a lack of accurate wavelengths, especially for forbidden optical lines.

Accurate wavelengths are needed for internal calibration of both laboratory and stellar spectra. The establishment of standard wavelengths of unprecedented quality may be produced in the near future by making use of the frequency comb technique. Building up energy level diagrams from precise wavelengths will for some elements generate Ritz wavelengths at shorter wavelengths (below 1500 Å) and for forbidden lines. The former can then be used as secondary standards, whereas the latter are applied in nebular physics.

A rather new field demanding wavelengths with very high accuracy or rather *precision*, is the study of the variation of the fine structure constant through the time of the Universe (see papers by Murphy, Chand, Levshakov in these proceedings). To provide precise laboratory wavelengths that are independent of source and calibration we have at Lund recorded spectra of several elements simultaneously [3, 2]. Other astrophysical projects requiring accurate wavelength data are discussed by Dravins [4] at this meeting, who studies the dynamics in stellar atmospheres by measuring very small wavelength shifts (the bisectors) in photospheric lines.

Some specific astrophysical problems require accurate data for lines that are not a priori of interest in stellar analyses. We will give an example of such a problem that has deserved high *priority* in the laboratory. The abundance of oxygen is in many objects best determined by analysing the parity- forbidden [OI] line at 6300 Å. It was shown in [5] that this line is often blended by a Ni I line with an internal structure caused by IS (see Fig.1). The Ni I line has now been measured in the laboratory [6] - both the log gf value and the IS - and 4 of the 5 atomic parameters needed for modelling the stellar feature have experimental values. The gf-value for the [OI] line is calculated.

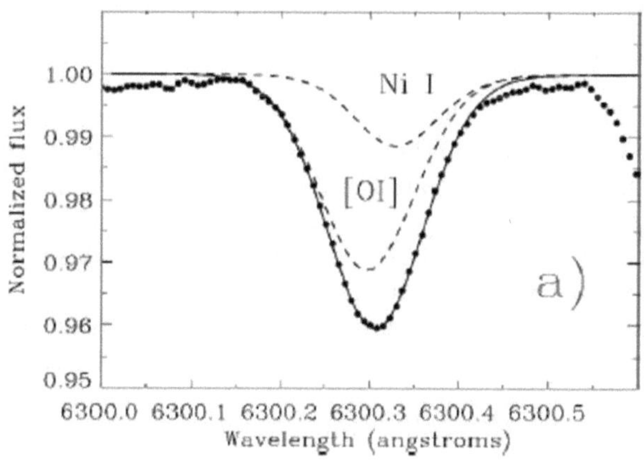

Fig. 1. The [OI] line at 6300Å blended with a Ni I line having an internal isotope structure (Figure from [5])

5 Oscillator Strengths

The radiative transition probability or oscillator strength (f-value) is the most critical parameter in abundance work, and f-values are therefore the most reguired data from the astronomical community. There are, however, different problems involved in the production of theoretical and experimental f-values. Firstly, one can calculate the f-value for any transition in any ion in any element, but we don't know the accuracy of the calculated value. A poor match between observed and calculated lines in a stellar analysis comparing spectra from a real target and its model can always be interpreted as poor atomic data. Secondly, because of time-consuming experiments, we cannot measure f-values for all transitions, as it generally needs measurements of both radiative lifetime (τ) and branching fractions (BF). With the techniques available we cannot measure lifetimes for all excited levels, and it is presently difficult to measure the branching ratio between lines in the infrared and vacuum-ultraviolet regions.

We can distinguish between four types of transitions of importance for different fields of astrophysics: i) Permitted lines, from levels with $\tau \sim$ 1-100 ns, for abundance analyses in stars and the ISM; ii) Intercombination lines, i.e. spin-forbidden LS-transitions, from levels with $\tau \sim 1\mu$s-1ms, for diagnostics of nebular densities iii) forbidden transitions between metastable states, violating electric dipole radiation rules, with $\tau \sim$1ms-10s, for diagnostics of low-density plasmas, and iv) permitted emission lines from "pseudo-metastable" states, with $\tau \sim 1\mu$s-1ms for the upper level, for analysis of laser lines in low-density nebular regions.

5.1 Permitted Lines

It is a delicate task to make the *priorities* of which f-values to measure both regarding choice of spectrum and element as well as choice of lines for a particular spectrum. We have initiated and performed a specific international effort, the FERRUM Project [7], to measure f-values for the iron group elements, especially Fe II. The aim of the project is to get: A set of Fe II lines in the optical region with 1) excitation energies from 3 to 13 eV 2) accurate wavelengths (1-3mÅ) and 3) experimental f-values accurate to 10-25%. These data will provide critical abundance analyses and test the assumptions of stellar model atmospheres.

A new collaboration (Castelli, Hubrig, Johansson) on high-resolution UVES spectra of a few peculiar stars will greatly test and improve the application of the FERRUM Project. In the star HR6000 (classified as a Fe-P-Mn star) the optical spectrum in the 4800-6000 Å region contains a huge number of 4d-4f transitions of Fe II, with a lower E.P. of slightly above 10 eV. These lines were first identified in CP stars by Johansson & Cowley [8] based on the laboratory work on Fe II [9]. We have now at Lund derived experimental f-values for some of the 4d-4f transitions based on complicated lifetime measurements. These f-values can be used to derive the iron abundance from 10 eV lines and compare with the abundance obtained from 3 eV and 8 eV lines in the same wavelength region, all with experimental f-values.

In the same wavelength region of HR6000 there are a great number of unidentified lines, which are also present in laboratory iron spectra. The lines are most probably due to 4d-4f transitions of Fe II, but to higher parent terms. This means that the lower E.P. is close to 13 eV and the upper level just below the ionization limit. By combining the stellar and laboratory spectra it will be possible to extend the analysis of Fe II. By verifying the population of Fe II levels at 12.8 eV in this star and assuming thermal population we should remember that the number of known energy levels below 12.8 eV is of the order of 500. Considering this fact it is amazing that the continumm looks so well defined and that the spectrum is not a mess of undefined absorption.

5.2 Forbidden Lines

Optical and ultraviolet spectra of objects having nebular regions in the environment show numerous forbidden lines, i.e. transitions disobeying the selection rules for electric dipole radiation. Since these lines are hardly produced in spectra of laboratory sources the f-values (or rather A-values) used in astrophysical analyses are from theoretical calculations. Since the lines connect levels at very low excitation the calculated A-values are probably rather reliable. However, this cannot be verified until measurements have been made. A series of measurements of radiative lifetimes of metastable states of Fe II and Ti II have been performed within the FERRUM Project at the storage ring, CRYRING, in Stockholm (see e.g. Hartman et al [10]).

5.3 Laser Lines

As discussed by Johansson & Letokhov (these proceedings) stimulated emission of radiation has been proposed for some Fe II transitions appearing in gas condensations close to the central star(s) of η Carinae. As shown in Fig.2, the HLyα line pumps via a short-lived state (2) Fe^+ ions into "pseudo-metastable" states (3) of Fe II, having a radiative lifetime of a few ms. This level is a bottle-neck in the transition scheme, and an inverted population is created. However, the state can decay slowly into short-lived states (4), which opens the possibility for stimulated emission. States (4) connect to the original state (1) and thereby close the circuit [11]. The strange thing with the "pseudo-metastable" state 3 is that it has such a long lifetime $\tau \sim$1ms in spite of possible decay channels through LS- permitted transitions. However, the "pseudo-metastable" states belong to the ground configuration of Fe II, where the levels are metastable with lifetimes of the order of 1 s. The complex atomic structure of Fe II with many valence electrons shows the strange feature of having the highest 4s-states located *above* the lowest 4p-states causing decays from 4s *down* to 4p.

The gf-values from (1) and A-values from (2) and (3) in the laser scheme in Fig. 2 are quite uncertain as many of the levels involved are mixed, making the theoretical calculations very complicated. We plan to measure the 1$\rightarrow$2 transitions using synchrotron radiation, and the lifetimes of states (3) using some kind of iron trap.

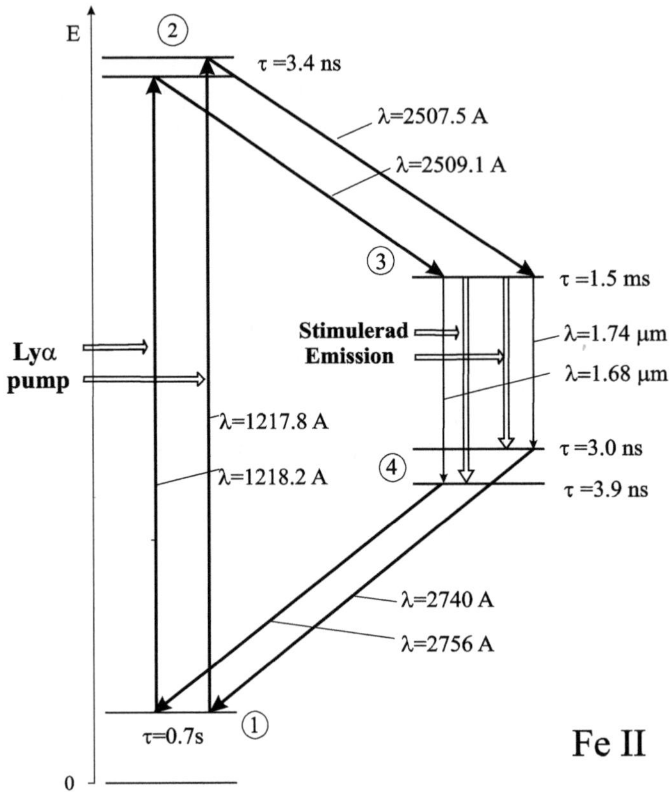

Fig. 2. Level scheme of Fe II showing the conditions for obtaining inverted population in the 3→4 transition at 1.7 μm from the "pseudo-metastable" state (3). The Lyα pumped channel 1→2 is followed by strong spontaneous emission in the $\lambda\lambda$2509,2507 lines in spectra of η Carinae. The fast decay of state (4) to (1) closes the radiative circuit.

References

1. http://www.astro.lu.se/Research/astrophys/iau/
2. M. Aldenius, S. Johansson: these proceedings
3. M. Aldenius, S. Johansson, M.T. Murphy: MNRAS **370**, 44 (2006)
4. D. Dravins: these proceedings
5. C. Allende Prieto, D.L. Lambert, M. Asplund: ApJ **556**, L63 (2001)
6. S. Johansson, U. Litz'en, H. Lundberg, Z. Zhang: ApJ **584**, L107 (2003)
7. S. Johansson, A. Derkatch, M. Donnelly et al: Phys. Scripta **T100**, 71 (2002)
8. S. Johansson, C.R. Cowley: A&A **139**, 243 (1984)
9. S. Johansson: Phys. Scripta **18**, 217 (1978)
10. H. Hartman, A. Derktach, M. P. Donnelly et al: A&A **397** 1143 (2003)
11. S. Johansson, V. Letokhov: A&A **428**, 497 (2004)

Optimal Extraction of Echelle Spectra

Nikolai Piskunov

Dept. of Astronomy and Space Physics, Uppsala University, Box 515, S-75120 Uppsala, Sweden, piskunov@astro.uu.se

1 Introduction

The extraction of the echelle spectra registered with a CCD detector represents a big challenge because of three reasons: (1) the pixel sampling is often close or worse then optimal, (2) spectral orders are curved and tilted with respect to the CCD rows (or columns) and (3) every pixel contains additional noise coming from various sources as illustrated in Figure 1. The main goal of an optimal extraction is to recover as much of the science signal while minimizing the contribution of the noise. Here we present the Slit Function Decomposition algorithm which replaces the summation in a sliding window with a reconstruction of the slit illumination profile. The reconstruction is formulated as an inverse problem solved by iterations and it is robust against most of the systematic problems including cosmic rays and cosmetic defects.

2 Main Assumptions

For the derivation of the algorithm below we view the whole spectrograph as an optical system which constructs a sequence of overlapping monochromatic images of the entrance slit convolved with the PSF. This is a single most important assumption which implies that the (useful) signal in every pixel can be described as a sum of all contributing monochromatic images of slit scaled by the spectrum in the corresponding wavelengths. The second assumption we make is that the monochromatic slit images are parallel to the columns of the CCD. This assumption can be relaxed using 2D model of the PSF.

3 Equations

In this section we will quickly derive the main equations for reconstructing the sLit illumination function L and the sPectrum P from a 2D CCD image of a spectral order. Following the main assumptions formulated above the signal S at any point x, y of the focal plane for an infinite spectral resolution can be represented as:

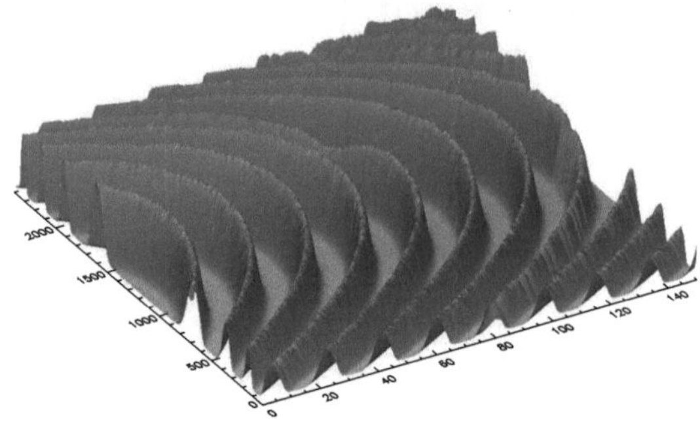

Fig. 1. A fragment of the focal plane with multiple spectral orders produced by an echelle spectrometer with a prism cross-disperser.

$$S(x, y) = P(x) \times L(y - y_c) \tag{1}$$

where the $y_c(x)$ is a 1D function that traces the order location. The intensity $S(x, y)$ is sampled by the CCD pixels and the corresponding pixel signal $S_{x,y}$is given by convolution:

$$S_{iX,iY} = P_{ix} \times \int_{iY} L(y - y_c)dy \tag{2}$$

where iX and iY are now pixel indices. In practice we only hope to reconstruct the slit function L on a discrete grid which has a smaller step size δy than the CCD pixel as illustrated in Figure 2. The integration is then replaced by summation:

$$S_{iX,iY} = P_{ix} \sum_{iy \in iY} \omega^{iy}_{iX,iY} L_{iy}. \tag{3}$$

Lower and upper case notation are used for distinguishing between the whole CCD pixel and the L-grid. The weights ω tell which fraction of the step iy falls into pixel iY. The equation 3 represents a model of a signal content for a CCD pixel based on 1D slit function. Such model does not include any of the typical distorsion (e.g. cosmic ray hits, CCD cosmetic defects etc.) and, therefore, those problems do not affect the solution. 1D slit function model works fine if the slit images are vertical. If this is not case we must use a more general 2D approximation to the PSF and re-write to equation 3 as:

$$S_{iX,iY} = \sum_{jX=iX-1}^{iX+1} P_{jX} \sum_{ix,iy \in \{jX,iY\}} \omega^{ix,iy}_{jX,iY} L_{ix,iy} \tag{4}$$

The summation over adjacent PSFs (jX) is needed because slit images corresponding to different wavelengths and different parts of PSF can now contribute to the same CCD pixel. On the other hand if the number of such

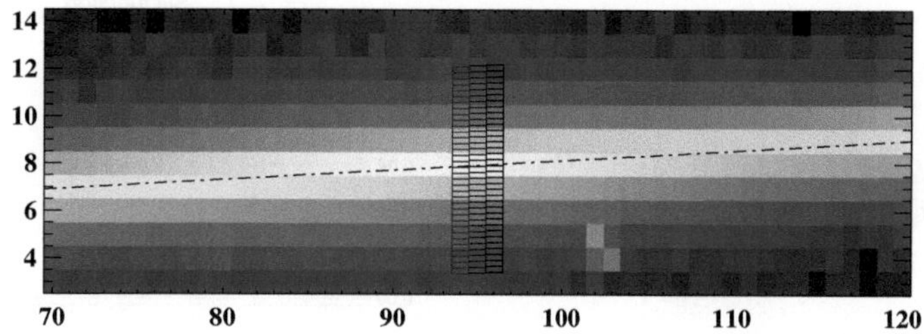

Fig. 2. A fragment of a spectral order. Dash-dotted line traces the order position. Three rectangular grills show consecutive monochromatic images of the slit. Vertical spacing between horizontal lines shows the (over-)sampling of the slit function. One can see the drift of the slit location from one column to the next.

contributors is larger than 3 it indicates the the PSF is heavily undersampled in the spectral direction and perhaps there is no point to apply 2D model anyway.

We proceed now with derivation of the equations for P and L using 1D model of the PSF. In order to do that we need to compare the model with the data:

$$\sum_{iX,iY} \left[S_{iX,iY} - E_{iX,iY}\right]^2 + \Lambda \cdot \Re(L) = \min \tag{5}$$

where $E_{iX,iY}$ is the actual number of photons registered by pixel iX, iY. The second term is the regularization for the slit function. Its purpose is to make the L-grid independent of the pixel size and the tilt of the spectrum relative to the CCD rows, that is if the data does not provide enough information to constrain the slit function on a given grid regularization takes over and makes sure the slit function is smooth. The usual form of regularization fulfilling this goal is $\Re \equiv 1/\delta y^2 \sum_{iy}(L_{iy+1} - L_{iy})^2$ and the appropriate balance between the data fitting and the regularization is achieved by selecting the Lagrangian multiplier Λ. Setting the derivative over L_{iy} and P_{iX} equal to zero we get the necessary equations:

$$\sum_i \left(\mathrm{A}_{ij} + \Lambda \cdot \mathrm{B}_{ij}\right) L_i = \mathrm{R}_j \tag{6}$$

$$P_i = \mathrm{C}_i/\mathrm{D}_i \tag{7}$$

Detailed expressions for matrices and vectors involved can be found in Piskunov & Valenti (2002). System 6 has band-diagonal structure simplifying the solution.

The solution of system 6 & 7 is obtained by iterations. The slit function must be area normalized closing the system of equations. The convergence is quick and robust. Figure 3 shows an example of decomposition.

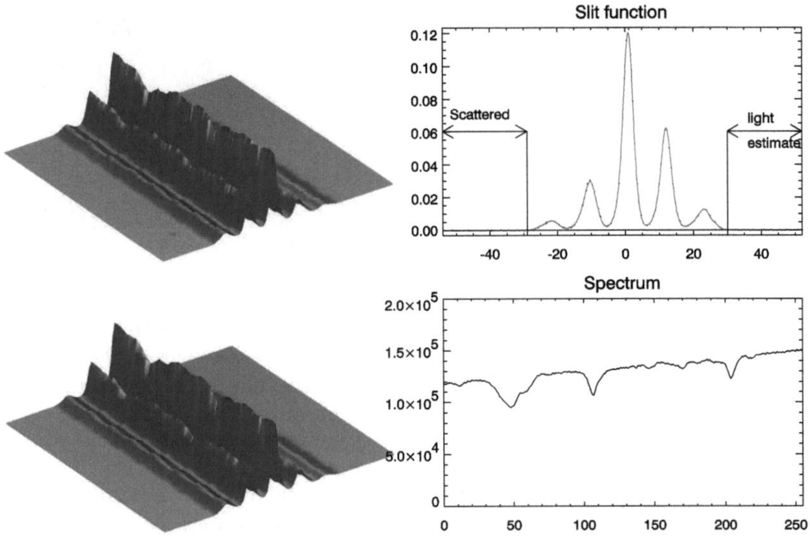

Fig. 3. A segment of a stellar spectrum (HD 217522) obtained with an image slicer (upper left) and the corresponding decomposition into an oversampled spatial profile (upper right) and the spectrum (lower right). The reconstructed model image (Eq. 3) is shown in the lower left.

4 Limitations, Generalization and Free Parameters

The algorithm proved to be robust due to the fact that it uses no assumption about the shape of the slit function. It was successfully used with many instruments and it works fine as long as the spectral order signal dominates the total number of counts in the region used for decomposition. It fails in a region with very faint spectrum containing multiple strong cosmic spikes.

We accommodate for the change of the slit function by splitting a spectral order in multiple regions (typically few hundred column wide) and performing decomposition on each part separately. This allows for slow variation of L along the order while slit function normalization insures the continuity of P. More complicated case of tilted slit are treated by using 2D model (eq. 4).

The decomposition involves 3 free parameters: the ratio between the CCD pixel size and δy (oversampling), Λ and the width of the extraction region. Too small oversampling will not be able to reproduce the shape of L while larger oversampling makes L insufficiently defined by the data and requires higher regularization.

Reference

1. N. Piskunov, J.A. Valenti: A&A **385**, 1095 (2002)

Hydrodynamical Model Atmospheres and 3D Spectral Synthesis

Hans-Günter Ludwig[1] and Matthias Steffen[2]

[1] CIFIST, GEPI, Observatoire de Paris-Meudon, 92195 Meudon Cedex, France
Hans.Ludwig@obspm.fr
[2] Astrophysikalisches Institut Potsdam, 14482 Potsdam, Germany
msteffen@aip.de

1 Radiation-hydrodynamics Modeling – Overview

In this paper we discuss three issues in the context of three-dimensional (3D) hydrodynamical model atmospheres for late-type stars, related to spectral line shifts, radiative transfer in metal-poor 3D models, and the solar oxygen abundance. To establish the context we start by giving a brief overview about the model construction, taking the radiation-hydrodynamics code CO^{5}BOLD (COnservative COde for the COmputation of COmpressible COnvection in a BOx of L Dimensions with L=2,3; [3]) and the related spectral synthesis package Linfor3D as examples.

Based on a Godunov-type finite volume approach, CO^{5}BOLD provides the time-dependent solution for a one-component compressible radiating fluid in an external gravity field on a fixed, non-staggered 3D Cartesian grid (allowing variable spacing). Operator splitting separates Eulerian hydrodynamics, optional tensor viscosity, and radiation transport. Directional splitting decomposes the 3D hydrodynamics problem into 1D sub-steps which are treated by an approximate Riemann solver of Roe type, modified to work with an arbitrary equation of state, and to properly handle an external gravity field. This scheme is very robust and well adapted to handle transonic flows and shocks in a highly stratified medium. By design, the code guarantees the numerical conservation of mass, momentum, and energy. For any prescribed chemical composition, CO^{5}BOLD uses a tabulated equation of state taking into account partial ionization of H I, He I, and He II, as well as the formation and dissociation of H_2 molecules.

The role of radiation in the hydrodynamical simulations is to describe the energy balance due to radiative heating and cooling. The radiative energy exchange is computed from the solution of the non-local transfer equation on a system of a large number of rays traversing the computational volume under different azimuthal and polar angles. Realistic stellar opacities are used, optionally based on ATLAS or MARCS opacity data. The frequency dependence of the radiation field is treated in a multi-group approximation – the so-called *opacity binning method* (OBM; [9, 7, 10]) – where frequencies are sorted into a small number of bins (typically 4 ... 6) according to the ratio

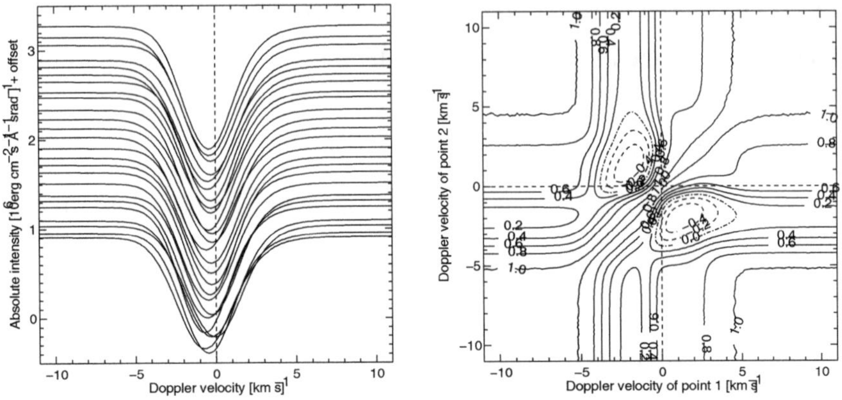

Fig. 1. Left panel: Time series of line profiles of a Fe I line at 6082 Å in a 3D solar CO^{5}BOLD model. The wavelength is given as corresponding Doppler velocity with respect to the line's laboratory wavelength. The line profiles have been offset proportional to time running from top to bottom. The time interval between profiles is the same, fluctuations in the continuum brightness cause the non-equidistant appearance. **Right panel:** Contour plot of correlation coefficients of the intensity between two wavelength points in the profiles shown in the left panel.

of monochromatic to Rosseland optical depth. So far, strict LTE is assumed, thus scattering cannot be treated. Radiation pressure is ignored.

The code Linfor3D accepts CO^{5}BOLD models as background structures on which spectral synthesis calculations at high wavelength resolution – usually focusing on one particular spectral line – can be performed. When calculating the emergent spectrum, Linfor3D takes into consideration the full 3D flow geometry including Doppler shifts caused by the macroscopic hydrodynamical velocities. It represents the effects of thermal and pressure broadening in standard fashion, but leaves out the ad-hoc broadening mechanisms of micro- and macro-turbulence introduced in 1D atmospheric models. Similar to CO^{5}BOLD,strict LTE is assumed in Linfor3D. Resulting spectral line profiles provide detailed information about intrinsic line shapes, and convective line shifts with respect to a line's laboratory wavelength.

2 High Precision Line Shifts from 3D Models?

A CO^{5}BOLD simulation constitutes a statistical realization of the atmospheric flow field in the stellar surface layers. If one is not interested in studying time-variable phenomena but only in the mean state of the atmosphere, the fluctuations present a noise source which limits the precision to which flow and related spectroscopic properties can be determined. This is similar to the observational situation where the intrinsic variability of a star limits the precision to which its radial velocity, e.g. in planet searches, can be measured. The left panel of Fig. 1 illustrates the temporal variability of a Fe I line calculated from a hydrodynamical solar model. Each line profile is a horizontal average over the surface of the computational box. Shown are 25 instants in time which are sufficiently separated that they can be considered statistically uncorrelated. We ask: what is the precision due to the statistics (ignoring systematic effects) to which we can determine the line shift?

It is straight forward to show (see [8]) that the expectation value of the disk-integrated line profile corresponds to the expectation value of the profile of the local hydrodynamical model. Hence, the statistical uncertainties of the profile emerging from the model directly correspond to the uncertainties of the predicted disk-integrated profile. From Fig. 1 it is obvious that the the statistical fluctuations are not just pixel-to-pixel random noise like in the case of photometric Poisson noise. The line profiles change their overall shape, i.e. different wavelength points show a considerable degree of correlation. The linear correlation coefficient between intensities at two wavelength points 1 and 2 is given by

$$C\left[I_1, I_2\right] \equiv \frac{\langle \Delta I_i \, \Delta I_j \rangle}{\sigma_{I_1} \sigma_{I_2}} = \frac{\langle I_1 I_2 \rangle - \langle I_1 \rangle \langle I_2 \rangle}{\sigma_{I_1} \sigma_{I_2}} . \tag{1}$$

I denotes the intensity, $\langle . \rangle$ the temporal average. $\Delta I_i \equiv I_i - \langle I_i \rangle$ is the intensity deviation from the mean. The right panel of Fig. 1 shows the correlation matrix of the example line depicted.

In order to quantify the line shift λ_s we need a model of the procedure by which it is measured, which in turn emerges from the definition of λ_s. Here, we assume that the measuring procedure of λ_s can be described by a function Λ of potentially all available (assumed discrete) intensities I_i: $\lambda_s = \Lambda\left(I_i\right)$. In order to make algebraic headway we simplify and linearize Λ around the expectation value of the line profile described by the values $\langle I_i \rangle$. To leading order in ΔI we obtain for the variance of the line shift the standard expression of the error propagation for correlated variables

$$\sigma^2_{\lambda_s} \approx \sum_{i,j} \frac{\partial \Lambda}{\partial I_i} \frac{\partial \Lambda}{\partial I_j} \langle \Delta I_i \, \Delta I_j \rangle = \sum_{i,j} \frac{\partial \Lambda}{\partial I_i} \frac{\partial \Lambda}{\partial I_j} \sigma_{I_i} \sigma_{I_j} \, C\left[I_i, I_j\right] . \tag{2}$$

The summation is performed over all pixels which are relevant for the measurement of λ_s. Equation (2) emphasizes the role of the covariance matrix of the intensities $\langle \Delta I_i \, \Delta I_j \rangle$ – or equivalently the standard deviations of the intensities and their correlation matrix – plays for the magnitude of the uncertainty of the line shift. In the present context we discussed line shifts but relation (2) of course also holds for other measures like, e.g., the equivalent width of a line. The statistical quantities in relation (2) can be estimated from the time series provided by the hydrodynamical model. Asymptotically, for a given line one will arrive at a fixed value for the correlation matrix $C\left[I_i, I_j\right]$. If one wants to improve the accuracy of the line shift one has to beat down the uncertainties in the intensities σ_{I_i}. This can be achieved by longer time series or larger horizontal extent of the hydrodynamical model.

Our example Fe I line shows a RMS temporal scatter of its position of $0.16\,\mathrm{km\,s^{-1}}$. The value was obtained by directly (and somewhat heuristically) measuring the location of the line core without formalizing the process by explicitely constructing a measurement function Λ. The statistical independence of the 25 individual snapshots implies an uncertainty of about $30\,\mathrm{m\,s^{-1}}$ for the line shift. While the specific value depends on the chosen line and selection of snapshots we think that it gives an indication of the precision one is typically working with in todays hydrodynamical standard models. Higher precision is possible but computationally also more costly. Of course, at some point real uncertainties will be dominated by systematic shortcomings of a model.

3 3D Radiative Transfer in Metal-poor Atmospheres

As mentioned earlier, the radiative transfer in the 3D models is commonly approximated by the opacity binning method (OBM) assuming strict LTE. While the approach is working fine in atmospheres of about solar metallicity, metal-poor atmospheres pose a challenge to the OBM. At first glance, this may come as surprise because the dramatic decrease of the number of spectral lines relevant for the radiative energy exchange should simplify the radiative transfer. However, the actual situation is quite different. First, scattering in the continuum becomes important for the thermal structure of metal-poor atmospheres. In the OBM, scattering is treated as true absorption so that one must expect some effects on the resulting temperature structure. Second, experience has shown that the OBM does not work as accurately in metal-poor atmospheres as in atmospheres of solar metallicity. It turned out that this deficit is not related to the treatment of the line blocking but already shows up for the radiative transport in the continuum.

Figure 2 shows an example of an atmosphere of a metal-poor giant. Plotted are temperature profiles of 1D ATLAS6 (see [6]) model atmospheres in radiative-convective equilibrium. The only difference among the models is the way in which the radiative transfer was treated. In three cases labeled "scattering", "no scattering", "scattering as true absorption" a high wavelength resolution was employed, and scattering was treated exactly, scattering opacity was neglected, or treated as true absorption, respectively. The by far dominating scattering opacity under the considered conditions is Rayleigh scattering by hydrogen atoms. As evident from Fig. 2, the temperature structure is noticeably influenced by scattering. The OBM used in the 3D models was also implemented in the 1D atmosphere code and a resulting radiative-convective equilibrium calculated. Comparison with the exact radiative transfer solution shows a close correspondence from the deep layers up to lower optical depth of $\log\tau \approx -3$. However, while useful in practice this is only

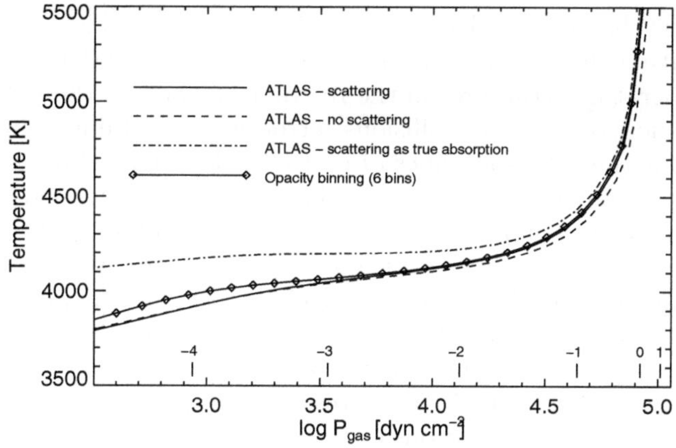

Fig. 2. Comparison of 1D model atmospheres ($T_{\rm eff}$ = 5000 K, log(g) = 2.94, [M/H] = −2) in radiative-convective equilibrium based on different treatments of the radiative transfer. (for details see text)

fortuitous. The OBM based temperature structure should in fact follow the "scattering as true absorption" case since in the OBM scattering opacity is treated as true absorption. At present, the reason for the rather poor performance of the OBM is unclear. Identifying its cause, improving the OBM, and including scattering effects in 3D geometry are challenges to be met in near-future developments of 3D model atmospheres for metal-poor stars.

4 3D Models and the Solar Oxygen Abundance

Recent spectroscopic abundance determinations by [1], based on a 3D hydrodynamic model atmosphere, led to a much debated downward revision of the solar C, N, and O abundances. Their result for the oxygen abundance is $\log \epsilon_{\rm O} = 8.66 \pm 0.05$ (on the scale $\log \epsilon_{\rm H} = 12$), causing a dramatic deterioration of the agreement between the thermal structure derived from helioseismic inversions and theoretical solar models, respectively. Motivated by this problem, we (see [2]) are currently using a 3D CO^{5}BOLD simulation with 5-bin frequency-dependent radiative transfer based on MARCS opacities to see whether the results by [1] can be confirmed. This independent redetermination of the solar oxygen abundance is based on 2 forbidden and 7 permitted O I lines, using a number of different observations, including both disk-center ("intensity") and full-disk ("flux") spectra. In addition to 25 snapshots from the simulation, we also derive abundances from different 1D atmospheres for comparison. Special care is taken to provide realistic error estimates.

The following preliminary conclusions can be drawn at this point: (i) "intensity" and "flux" spectra give practically the same result. (ii) the oxygen abundance derived from the 3D CO^{5}BOLD simulation is only slightly lower (by –0.04 dex) than that derived from the 1D empirical model by [4] (hereafter HM), indicating that the 3D mean model and the 1D HM model have very similar temperature structures in the relevant layers. (iii) the unknown cross sections for neutral particle collisions introduce uncertainties in the NLTE corrections for the O I triplet lines of up to 0.1 dex; depending on the weight of these lines, the resulting error in the mean oxygen abundance derived from our set of lines is about 0.05 dex. (iv) Our preliminary best estimate for the solar oxygen abundance is $\log \epsilon_{\rm O} = \mathbf{8.72 \pm 0.06}$, which is close to the value recommended by [5], $\log \epsilon_{\rm O} = 8.736 \pm 0.078$. A remaining problem of our analysis is that the two forbidden [O I] lines give significantly different abundances, which cannot be explained by NLTE-effects or deficiencies of the model atmosphere. We hope to resolve this problem by analyzing the observed center-to-limb variations of these two line profiles.

5 Remarks on Precision Spectroscopy and 3D Models

Hydrodynamical model atmospheres are on their way of becoming a standard tool for the analysis of stellar spectra. Their ability of making detailed predictions about the shape of spectral lines in convective atmospheres can only be fully exploited if observed spectra of sufficient resolution are available. Ideally, spectrographs should be able to provide a spectral resolutions above 10^5 – something that we would like instrument builders to keep in mind.

High-fidelity abundance work benefits from the theoretical knowledge of the precise intrinsic line shape. However, in practice one is nonetheless often confronted with ambiguities, e.g. in the case of blends, which remain unresolved by considering disk-integrated line profiles only. The center-to-limb variation of a line shape can provide crucial further constraints. Combining interferometry with high-resolution spectroscopy (like in the UVES-I project presented by A. Quirrenbach, this volume) can open-up this source of information for stellar work.

References

1. M. Asplund, N. Grevesse, A.J. , Sauval, C. Allende Prieto, D. Kiselman, A&A **417**, 751 (2004)
2. E. Caffau, H.-G. Ludwig, M. Steffen, T.R. Ayres, R. Cayrel, P. Bonifacio, B. Plez, B. Freytag: A&A, (2007), in preparation
3. B. Freytag, M. Steffen, B. Dorch, AN **323**, 213 (2002)
4. H. Holweger, E. Müller, *Solar Phys.* **39**, 19 (1974)
5. H. Holweger, in: *Solar and galactic composition*, ed by R.F. Wimmer-Schweingruber, AIP Conf. Proc. 598, p. 23 (2001)
6. R.F. Kurucz, ApJ **40**, 1 (1979)
7. H.-G. Ludwig, S. Jordan, M. Steffen. A&A **284**, 105 (1994)
8. H.-G. Ludwig, A&A **445**, 661 (2006)
9. Å. Nordlund, A&A **107**, 1 (1982)
10. A. Vögler, J.HM.J. Bruls, M. Schüssler, A&A **421**, 741 (2004)

Intrinsic Lineshifts in Astronomical Spectra

Dainis Dravins

Lund Observatory, Box 43, SE-22100 Lund, Sweden
dainis@astro.lu.se

Summary. Spectral-line displacements away from the wavelengths naively expected from the Doppler shift due to radial motion may originate as convective shifts (correlated velocity and brightness patterns), as gravitational redshifts, or be induced by wave motions. Convective shifts are important tools for testing 3-dimensional stellar hydrodynamics; analogous shifts may be expected even in intergalactic absorption lines (convection driven by AGNs in clusters of galaxies).

1 Wavelength Shift vs. Radial Velocity

The *absolute shift* of any spectral line is its displacement from the wavelength naively expected from the Doppler shift caused by radial motion. Its study requires knowledge of observed and laboratory wavelengths, and also the radial motion of the object. In the past, such studies were possible only for the Sun, but now also some other stars are accessible thanks to *astrometric* determinations of stellar radial motion [7]. However, *differential lineshifts* between different spectral lines in the same object can be studied without knowing its radial velocity. Further, 'wavelength shift' is not *one* quantity but – since all lines are asymmetric to some extent – rather a function of intensity position in the line, conveniently shown by the line bisector (median).

2 Convective Lineshifts in Stellar Photospheres

Hydrodynamic simulations of stellar photospheres can be tested by whether they succeed in reproducing observed lineshapes, asymmetries, and shifts [2, 3, 11]. The latter originate from correlated velocity and brightness patterns: typically, greater photon contributions from the bright and rising (thus blueshifted) convective elements dominate in the average, causing a 'convective blueshift'. Lines also become asymmetric, their bisector (median) often taking on a shape like the letter 'C' [3]. The success of such models suggests that many physical processes probably are beginning to get understood. On the other hand, similar spectral-line shapes may sometimes be predicted by models with different parameter combinations, showing the need for further observational constraints. New diagnostic tools are needed to clarify effects of non-LTE, or of effects from 3-dimensional radiative transfer.

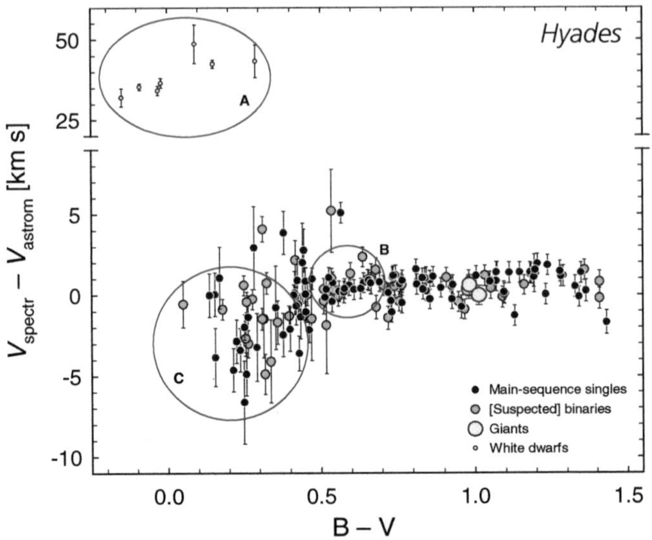

Fig. 1. Lineshifts in the Hyades seen as differences between spectroscopic 'velocities' and astrometric values [8]. (A): White-dwarf gravitational redshifts place them far off main-sequence stars. (B): An increased blueshift in stars somewhat hotter than the Sun is expected due to more vigorous surface convection. (C): Spectra of hotter stars are blueshifted, probably due to rising atmospheric shock-waves [9].

3 'Ultimate' Information Content of Stellar Spectra

Hydrodynamic models can predict subtle properties for various idealized lines, not yet feasible to falsify or verify through observations. One limit is set by 'astrophysical' noise: no comparison with observations can result if modeled lines do not exist in real spectra, or are too blended for meaningful measurement. While the synthesis of blended multi-line regions is not impossible in principle, it is currently unfeasible since it would require detailed knowledge of all blending lines, including their precise laboratory wavelengths, data that are not yet available.

This leads to a question of the 'ultimate' information content of stellar spectra: what limits are set by imperfect instrumentation, by photon noise (i.e., telescope size), by lack of laboratory data, or by pure astrophysics? Individual lines are often too noisy to give a credible line asymmetry or shift; however by averaging groups of similar lines one may expect noise from weak blends to average out, revealing the intrinsic line-shapes.

An example of an informative signature that only emerges in averages over many lines, is in Figure 2. The F4 V type star θ Scl (not much unlike Procyon at F5 IV-V) displays a sudden 'blueward hook' of its [average] bisector quite close to the continuum. This effect was suggested by Procyon spectra already long ago [6], and later given an explanation from hydrodynamic modeling

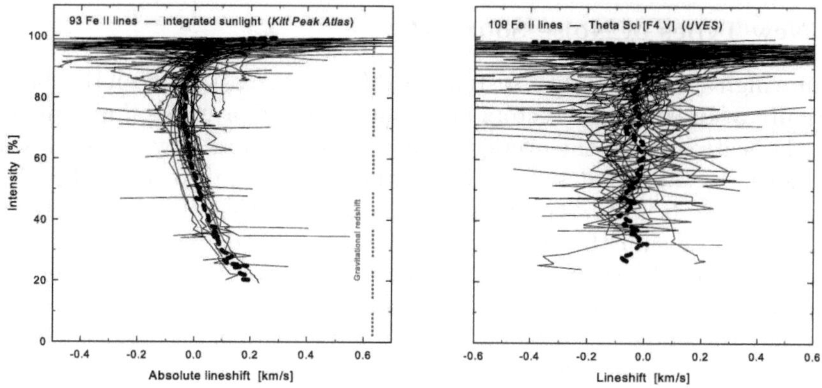

Fig. 2. Left: Bisectors of Fe II lines in integrated sunlight, placed on an absolute wavelength scale. Each thin curve is the bisector of one spectral line; the thick curve is their average. Right: Fe II lines in the F4 V star θ Scl, observed within the ESO/VLT UVES Paranal project [4]. The sudden 'blueward hook' of its average bisector close to the continuum is believed to be real.

[1]. The blueward hook originates from extended Lorentzian wings in line components from upflowing (blueshifted) granules. There lines are stronger (due to steeper temperature gradients), saturate, and develop damping wings. These extended blueshifted wings dominate the resulting bisector close to the continuum, where the more Gaussian-shaped intergranular components give only small redshifted contributions. This signature thus reveals different degrees of line saturation across stellar surface inhomogeneities.

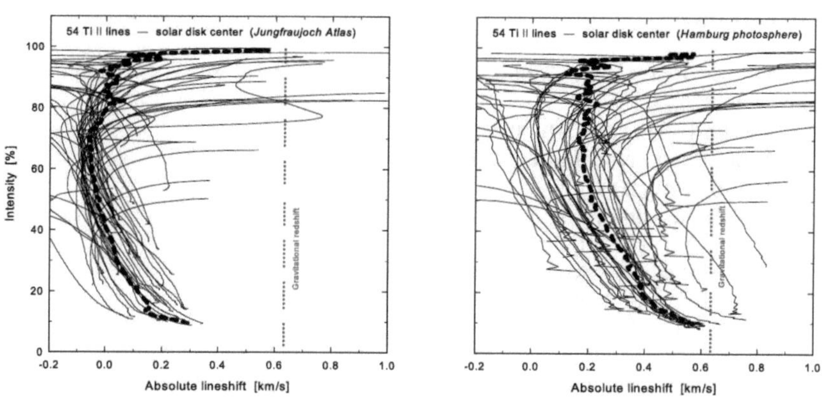

Fig. 3. Bisectors of 54 Ti II lines in the solar disk-center spectrum from the Jungfraujoch Atlas (left; recorded with a grating spectrometer); and (right) from the Kitt Peak Fourier transform spectrometer, as normalized in Hamburg [10]. The data sets differ in average wavelength shift, and in scatter about their average.

3.1 New Types of Noise Sources?

Despite high spectral quality, systematic differences exist between the datasets of Figure 3. Individual bisectors are similar but the average shifts differ, as does the scatter about the average. Different zero-points of wavelength scales are known, but different spreads about the average appear not to be. Possibly, these arise from low-spatial-frequency noise in the FTS. Its spectrum is synthesized as a sum of sinusoidals, where the amplitude of each is measured sequentially by the scanning FTS. Phase errors for the longest sinusoidal components will displace whole groups of spectral lines without necessarily deforming the shapes of each of them (like a big wave that is moving whatever smaller-scale structures float on top of it).

4 Wavelength Shifts from Intergalactic Convection

Analogous lineshifts may affect intergalactic absorption lines in quasar spectra. Active galactic nuclei near cluster cores emit energetic cosmic rays, heating the surrounding plasma, and driving convective motions across clusters of galaxies [5, 12, 13]. Corresponding 3-D hydrodynamic models do reproduce observed structures of the intergalactic gas, and should be possible to use as input for spectral-line synthesis. Differential wavelength shifts among lines of different strength, ionization level, etc., could then diagnose intergalactic hydrodynamics, even for the distant Universe. Such shifts must be modeled and segregated from other possible shifts – perhaps due to cosmologically changing 'constants' – if such other shifts are to be reliably deduced.

References

1. C. Allende Prieto, M. Asplund, R. J. García López, & D. L. Lambert: ApJ **567**, 544 (2002)
2. M. Asplund: ARA&A **43**, 481 (2005)
3. M. Asplund, Å. Nordlund, R. Trampedach, C. Allende Prieto, & R. F. Stein: A&A **359**, 729 (2000)
4. S. Bagnulo, E. Jehin, C. Ledoux, R. Cabanac, C. Melo, & R. Gilmozzi: ESO Messenger No. 114, 10 (2003)
5. B. D. G. Chandran: ApJ **632**, 809 (2005)
6. D. Dravins: A&A **228**, 218 (1990)
7. D. Dravins, L. Lindegren, & S. Madsen: A&A **348**, 1040 (1999)
8. S. Madsen, D. Dravins, & L. Lindegren: A&A **381**, 446 (2002)
9. S. Madsen, D. Dravins, H.-G. Ludwig, & L. Lindegren: A&A **411**, 581 (2003)
10. H. Neckel: Sol. Phys. **184**, 421 (1999)
11. Å. Nordlund, D. Dravins: A&A, **228**, 155 (1990)
12. C. S. Reynolds, B. McKernan, A. C. Fabian, J. M. Stone, & J. C. Vernaleo: MNRAS **357**, 242 (2005)
13. J. C. Vernaleo, C. S. Reynolds: ApJ **645**, 83 (2006)

Study of Line Bisectors and its Relation with Precise Radial Velocities in the Search for Extrasolar Planets

A. F. Martínez Fiorenzano

Fundación Galileo Galilei - INAF, Calle Alvarez de Abreu 70, 38700 Santa Crúz de La Palma (Tenerife), España. fiorenzano@tng.iac.es

Summary. In the search for exoplanets, we prepared a suitable software to use the same spectra for radial velocity (RV) determinations (with the spectrum of the I_2 cell imprinted on) to measure variations of the stellar line profiles. This novel approach, can be useful in high precision RV surveys based on I_2 cell data. The software has been used on data acquired within our survey in the search for exoplanets around wide binary stars, ongoing at the Telescopio Nazionale Galileo (TNG), where spurious estimates of the RVs due to activity or contamination by light from the companions were revealed. This technique can also be considered to correct the measured RVs, in order to search for planets around active stars.

1 Introduction

The Radial Velocity (RV) technique is the most successful to reveal the presence of substellar objects around stars. However, the intrinsic noise (jitter due to activity) is a natural limitation in the search for exoplanets.
The analysis of spectral line asymmetries through line bisectors (LB) provide a strong tool to understand the origin of the observed RV variations. The constancy in shape and orientation of LB ensures that the observed RV variations correspond to a center of mass motion.
Stellar activity, due to interactions between magnetic fields and turbulent plasma, produces active regions where strong magnetic fields create temporary features, causing variations of spectral line profiles: plages alter the granulation pattern and spots changes the rotation profile. Both may be studied through LB. Variations of line profiles are responsible of spurious RV variations.

2 Analysis

To determine LB from the stellar spectra, the spectrum of a fast rotator star helps to remove the I_2 lines from the target spectrum. A mask made from a solar catalog is employed to compute a Cross Correlation Function (CCF) with chunks of spectra to get average absorption profiles, these are added and normalized to get one average absorption to compute the LB for every spectrum. To study quantitatively the line asymmetries, we compute the Bisector Velocity Span (BVS) which is the difference of velocities given by LB in two zones of the profile: Top and Bottom (see [2] for details).

3 Results

The plots of BVS vs. RV show cases of correlation due to light contaminating the target spectrum; anti correlation due to stellar activity or a lack of correlation with constant LB in stars with known planets (Fig. 1). Given a linear correlation it is possible to apply corrections to the observed RVs in the search for signals of the RV curves (see [2] for details).
It is the first time, to our knowledge, that the same spectra employed for RVs measurements is analyzed after removal of I_2. The measured variations shows spreads fully consistent with internal errors. We can distinguish contaminated spectra and discard them from the RV curve [1]. We identify a few cases of RV variation due to stellar activity.

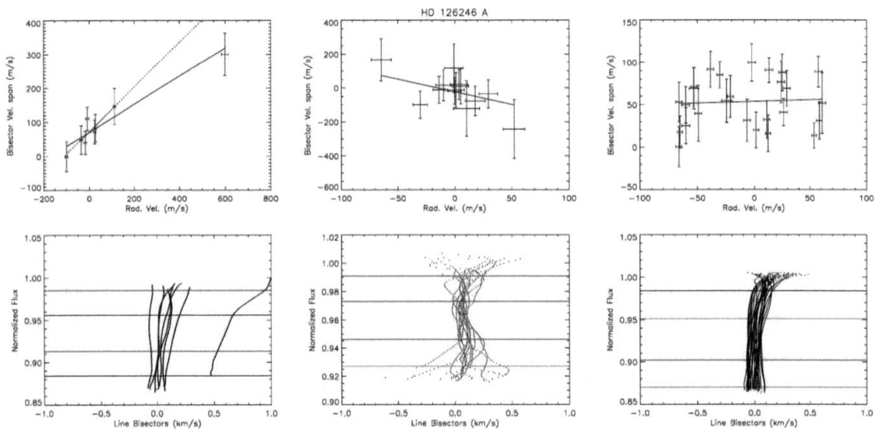

a b c

Fig. 1. BVS vs. RV plots (upper panels) and line bisectors (lower panels). a) Case of correlation due to light contamination HD8071B. b) Anti correlation due to stellar activity HD126246A. c) Lack of correlation for the known star with planet ρCrB.

References

1. A. F. Martínez Fiorenzano et al. 2005, A&A, 442, 775
2. A. F. Martínez Fiorenzano 2006, PhD Thesis, astro-ph/0603788

A Pan-Spectral Method of Abundance Determination

A. Sapar, A. Aret, L. Sapar, and R. Poolamäe

Tartu Observatory, 61602 Tõravere, Estonia `sapar@aai.ee`

Summary. We propose a new method for determination of element abundances in stellar atmospheres aimed for the automatic processing of high-quality stellar spectra. The pan-spectral method is based on weighted cumulative line-widths $Q_\lambda = \int_{\lambda_0}^{\lambda} \left| \frac{dR_\lambda}{dZ} \right| (1 - R_\lambda) d\lambda$, where R_λ is residual flux and Z is abundance of studied element. Difference in quantities Q_λ found from synthetic and observed spectra gives a correction to the initial abundance. Final abundances are then found by rapidly converging iterations. Calculations can be made for many elements simultaneously and do not demand supercomputers.

1 Description of the Method

Essential developments in observational high-precision and high-resolution spectroscopy, and fast-growing computing facilities stimulate to seek new automatic methods of analysis of stellar spectra. We recently started to develop a pan-spectral (or broad-band) method for determination of abundances of chemical elements and their isotopes in stellar atmospheres and to compose a software necessary for its application.

The method is based on weighted cumulative line-widths Q_λ defined as

$$Q_\lambda = \int_{\lambda_0}^{\lambda} \left| \frac{dR_\lambda}{dZ} \right| (1 - R_\lambda) d\lambda \ , \tag{1}$$

where R_λ is the residual flux (intensity) and $Z = \log(N_{elem}/N_{tot})$ is the abundance of studied element or isotope.

Integral (1) can be easily reduced to the equivalent width of a spectral line by omitting the derivative $|dR_\lambda/dZ|$ and integrating only over wavelengths of this line. We call the quantity Q_λ cumulative because integral is taken over all lines starting from some initial wavelength to the current wavelength. And we call it weighted because the derivative of residual flux R_λ with respect to abundance Z can be treated as a weight function. This derivative automatically excludes spectral regions insensitive to changes of the abundance of studied element and gives a large contribution in the most sensitive regions, i. e. in the centers of non-saturated lines and in the steep wings of strong lines of this element.

Weighted cumulative widths Q_λ for both observed and synthetic spectra are calculated using the derivative of R_λ found from the synthetic spectra. For small abundance changes the derivative can be replaced by differences, thus

$$\frac{dR_\lambda}{dZ} = \frac{R_\lambda^+ - R_\lambda^-}{Z^+ - Z^-} , \tag{2}$$

where flux R_λ^+ corresponds to abundance $Z^+ = Z + \delta Z$ and flux R_λ^- corresponds to abundance $Z^- = Z - \delta Z$. Correction to the abundance of studied element Z can then be found by formula

$$\Delta Z = \frac{(Z^+ - Z^-)(Q_\lambda^{obs} - Q_\lambda^{syn})}{Q_\lambda^+ - Q_\lambda^-} , \tag{3}$$

where Q_λ^{obs} is found from the observed spectrum and Q_λ^{syn} from the synthetic spectrum with initial (guessed) abundance Z. Quantities Q_λ^+ and Q_λ^- correspond to synthetic spectra with abundances Z^+ and Z^-, respectively.

Abundance of element can be found iteratively using the corrections ΔZ, the number of iterations needed depends on difference between initial and best-fit abundance values. Abundance determined by such a procedure is a best fit for all lines of the element, diminishing thus an influence of possible atomic data errors. We would like to point out that here the contribution of blended lines is duly taken into account. Abundances can be found simultaneously for many elements. Our calculations show that the method works well for line-rich elements, but for line-poor elements further refinements are necessary.

The method can be used also for determination of the effective temperature and gravity, using as the weight function the derivative of residual flux with respect to T_{eff} or $\log g$, respectively.

2 Testing the Method

2.1 Observational Data

We applied the weighted cumulative width method to determine element abundances in the atmosphere of the chemically peculiar star HD 175640. High resolution spectral atlas of this star, covering the 3 040 – 10 000 Å region, was recently published by Castelli and Hubrig [1]. The spectrum with resolution 90 000 – 110 000 and signal-to-noise ratio 200 – 400 was obtained at ESO 8 m UT2 telescope with the UV-Visual Echelle Spectrograph (UVES). Castelli and Hubrig [1] also computed a synthetic spectrum for the whole observed interval with the SYNTHE code [4] and determined abundances for 48 ions of 39 elements by the visual best fit of synthetic and observational spectra of selected lines. The model atmosphere accepted was an ATLAS12 model [2] with parameters $T_{eff} = 12\,000$ K, $\log g = 3.95$, $\xi = 0$ km s^{-1}. Improved and extended Kurucz atomic line list used in computations is available at F. Castelli website (`http://wwwuser.oat.ts.astro.it/castelli/stars.html`).

We had to ignore most of the IR spectrum due to telluric lines and also some spectral regions with weak overlap or gaps between adjoined echelle spectral orders.

2.2 Synthetic Spectrum

In computations of synthetic spectra we based on the Kurucz ATLAS9 [3] model atmospheres and on the line lists and initial element abundances provided by Castelli and Hubrig [1]. Computations of the synthetic spectra were carried out with program SMART composed by our team during last decade [5]. SMART is a compact and simple software and does not demand large computer facilities. We used the same spectral resolution (500 000) as Castelli and Hubrig in computations with SYNTHE [4]. Gauss broadening and the noise cutoff parameters were found to achieve best fit of observed and synthetic spectra. Obtained synthetic spectrum agrees reasonably well with the observed one.

2.3 Application of the Pan-spectral Method

To estimate the domain of applicability of the method, we obtained abundances of 12 line-rich elements (Mn, Ti, Cr, Fe, Y, Mg, Si, Ca, O, S, Yb, Ni) in the atmosphere of HD 175640. Regions of Balmer and Paschen continua were studied separately because abundances of studied metals turned out to be larger in outer layers of atmosphere (UV region) than in inner layers (visual region).

We started with abundances derived by Castelli and Hubrig [1] and iteratively found corrections to these values. Abundances of Si, O and S from UV spectrum have not been found since there were only few weak lines of these elements in UV region. Final abundances were found with only two iterations since necessary corrections were small. Weighted cumulative widths Q_λ for chromium computed with initial abundance are presented in Fig. 1. Corresponding corrections to Cr abundance are +0.16 dex in UV and -0.14 dex in visual region.

Obtained abundances of the studied elements and comparison with results of Castelli and Hubrig [1] are presented in Table 1. Our results differ somewhat from the abundances obtained by Castelli and Hubrig. However, the differences in obtained element abundances may partly be caused by use of different spectral synthesis programs, not only by difference of used methods. This is particularly true for S and Si. Also the difference of used model atmospheres should be mentioned.

We conclude that the proposed method is promising for analysis of chemical composition and other main parameters of stellar atmospheres.

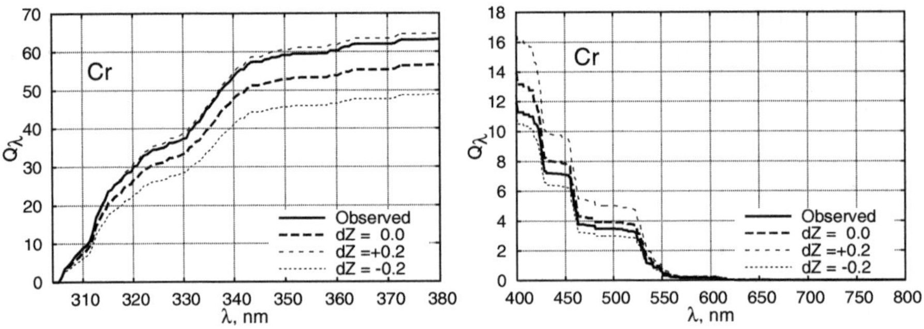

Fig. 1. Weighted cumulative width Q_λ for Cr.

Table 1. Element abundances $[N/N_\odot]$ for HD175640

	Castelli & Hubrig (2004)						Present study	
	UV			Vis			UV	Vis
	ion I	ion II	ion III	ion I	ion II	ion III		
Mn	+2.47	+2.40	–	+2.46	–	–	+2.75	+2.57
Ti	–	+1.43	–	–	+1.30	–	+1.62	+1.31
Cr	+1.19	+1.03	–	+1.13	+0.96	–	+1.24	+0.96
Fe	-0.36	–	–	-0.21	-0.30	–	-0.23	-0.33
Y	–	+3.38	–	–	+3.01	–	+3.26	+3.04
Mg	–			+0.18	-0.25	–	-0.19	-0.09
Si	–			–	-0.23	-0.09	–	+0.21
Ca	–	-0.15	–	+0.42	+0.06	–	+0.10	+0.24
O	–			+0.03	–	–	–	+0.10
S	–			–	-0.41	–	–	+0.03
Yb	–	+3.14	+3.66	–	+2.76	–	+3.64	+2.79
Ni	–	-0.22	–	–	-0.35	–	-0.12	-0.43

We are grateful to Estonian Science Foundation for grant ETF 6105.

References

1. F. Castelli, S. Hubrig: A& A **425**, 263 (2004)
2. R. Kurucz: Model Atmospheres for Individual Stars with Arbitrary Abundances. In: *The Third Conference on Faint Blue Stars*, ed by A. G. D. Philip, J. Liebert, R. A. Saffer (L. Davis Press, Schenectady 1997) p 33
3. R. Kurucz: CD-ROM, No. 13 (1993)
4. R. Kurucz: CD-ROM, No. 18 (1993)
5. A. Sapar, R. Poolamäe: "SMART": A Compact and Handy FORTRAN Code for the Physics of Stellar Atmospheres. In: *ASP Conf. Ser. 288: Stellar Atmosphere Modeling*, ed by I. Hubeny, D. Mihalas, K. Werner (Astronomical Society of the Pacific, San Francisco 2003) p 95

Spectroscopic Binary Mass Determination Using Relativity

Shay Zucker[1] and Tal Alexander[2]

[1] Dept. of Geophysics & Planetary Sciences, Raymond and Beverly Sackler Faculty of Exact Sciences, Tel Aviv University, Tel Aviv 69978, Israel
shayz@post.tau.ac.il

[2] Faculty of Physics, Weizmann Institute of Science, PO Box 26, Rehovot 76100, Israel Tal.Alexander@weizmann.ac.il

Summary. High-precision radial-velocity techniques, which enabled the detection of extrasolar planets, are now sensitive to relativistic effects in the data of spectroscopic binary stars (SBs). These effects can be used to derive the absolute masses of the components of eclipsing single-lined SBs and double-lined SBs from Doppler measurements alone. High-precision stellar spectroscopy can thus substantially increase the number of measured stellar masses, thereby improving the mass-radius and mass-magnitude calibrations.

1 Introduction

Measuring precise radial-velocity (RV), with long-term precisions of a few meters per second, is one of the most important applications of high-precision spectroscopy. The most notable scientific achievement of precise RV measurements has been the detection of planets orbiting solar-type stars [1, 2]. We suggest another application of high-precision RVs, namely, the detection of relativistic effects in the Doppler shifts of close spectroscopic binary stars (SBs).

The typical velocities of components of close binary stars can be as high as $150\,\mathrm{km\,s^{-1}}$, $\beta \equiv v/c \sim \mathcal{O}(10^{-4})$. The classical Doppler shift formula predicts a relative wavelength shift $\Delta\lambda/\lambda$ of the order β. The next order corrections are of order $\beta^2 \sim \mathcal{O}(10^{-8})$, which translate to $\mathcal{O}(1\,\mathrm{m\,s^{-1}})$. Terms of order β^3 are beyond foreseen technical capabilities. We thus limit our analysis to $\mathcal{O}(\beta^2)$ effects – the transverse Doppler shift (time dilation) and the gravitational redshift.

2 Single-Lined Spectroscopic Binary

The Keplerian RV curve of a single-lined spectroscopic binary (SB1) can be presented as:

$$V_{R1} = K_1 \cos\omega \cos\nu - K_1 \sin\omega \sin\nu + eK_1 \cos\omega + V_{R0} \,, \tag{1}$$

where K_1 is the primary star RV semi-amplitude, ω is the argument of periastron, e is the eccentricity, V_{R0} is the center-of-mass RV, and ν is the time-dependent true anomaly. The customary procedure to solve an SB1 is to fit this orbital model to the observed RV data. This fit is achieved through some optimization algorithm that scans the (P, T, e) space (period, periastron time and eccentricity). For each trial set of values for these three parameters the algorithm produces the corresponding $\nu(t)$ and then solves analytically for $K_{C1}(= K_1 \cos\omega)$, $K_{S1}(= -K_1 \sin\omega)$, and V_{R0}, which appear linearly in the expression for V_{R1}.

We have recently shown [3] that including the β^2 effects changes (1) by simply modifying the linear elements K_{S1}, K_{C1} and V_{R0}. The modified elements are given by:

$$K'_{S1} = -K_1 \left(1 + \frac{V_{R0}}{c}\right) \sin\omega \tag{2a}$$

$$K'_{C1} = K_1 \left[\left(1 + \frac{V_{R0}}{c}\right) \cos\omega + \frac{e}{\sin^2 i} \frac{2K_1 + K_2}{c}\right] \tag{2b}$$

$$V'_{R0} = V_{R0} + \frac{1 - e^2}{\sin^2 i} \frac{K_1}{c} \left(\frac{3}{2} K_1 + K_2\right) + \frac{V_0^2}{2c} \tag{2c}$$

The modified linear elements are more difficult to interpret now. The three quantities K'_{S1}, K'_{C1}, and V'_{R0} depend on the six elements K_1, K_2, ω, $\sin i$, V_{R0}, and V_0 (the absolute value of the three-dimensional center-of-mass velocity). Thus, equations (2) are underdetermined and we cannot infer the six elements above, unless some additional independent information is available, or further assumptions are introduced.

Such independent information may be available through precise photometry of eccentric eclipsing binaries. There, the shape of the eclipse as well as the phase differences between primary and secondary eclipses can be used to estimate ω and $\sin i$. In this case, we may derive K_2 – the RV amplitude of the secondary star. By obtaining K_2 we effectively turn the binary into a double-lined spectroscopic binary (SB2), in which both K_1 and K_2 are measured. Together with the known inclination, we then obtain full knowledge of the component masses.

Curiously, another result of including the relativistic terms is that an eccentric binary should always display an apparent RV signature, even in the extreme case where the inclination is exactly zero and the orbit is observed face on. Nevertheless, such small values of the inclination are extremely rare and this possibility is not realistic.

3 Double-Lined Spectroscopic Binary

In the case of a Keplerian SB2, there are two sets of measured RVs. The two sets of RVs share the same fundamental orbital elements (P, T, e, ω, and

V_{R0}), and the only difference between them is their amplitudes K_1 and $-K_2$. The common procedure is to scan the space of the four parameters (P, T, e, ω) and then solve analytically for the three linear elements K_1, K_2, and V_{R0}. However, when we incorporate the relativistic corrections above, we see that we now have two RV curves with two different sets of derived amplitudes, arguments of periastra, and center-of-mass velocities. The two RV curves still share the same period, periastron time and eccentricity.

The four equations corresponding to the two equations (2a) and (2b) have four unknowns: K_1, K_2, ω and $\sin i$. V_{R0} appears in these four equations always divided by the speed of light. Thus we can safely use its approximate derived value, from either set of measured velocities , since the discrepancy will be only $\mathcal{O}(\beta^3)$. Solution of this set of equations will yield a more accurate estimate of the first three unknowns, but more importantly, it will yield an estimate of $\sin i$. Retaining only leading order terms, we can arrive at the following solution:

$$\sin^2 i = \frac{3e}{\omega_2' - \omega_1'} \frac{K_{S2}' + K_{S1}'}{c} \tag{3}$$

where ω_1' and ω_2' are the apparent arguments of periastra of the two components. The fact that the two apparent arguments of periastra differ is a pure relativistic effect. Note that classically, there is no way to estimate $\sin i$ from pure RV data alone. Usually, $\sin i$ is obtained only from the analysis of SBs that are eclipsing or astrometric binaries.

4 Discussion

We present here an approach to extract more information from RV data of a spectroscopic binary, based on the inclusion of relativistic effects. To test the applicability of this new approach we have presented in [3] a simulated test case where we examine the SB2 12 Boo [4]. For each trial inclination, we generated 1000 random realizations of the system, where we added three simulated HARPS measurements to the existing data. We then applied our new approach to the simulated data and estimated the inclination. Figure 1 shows the results of the simulations and demonstrates that with reasonable efforts, $\sin i$ can be measured satisfactorily. An additional advantage of this test case is that the inclination of 12 Boo has already been measured by interferometry and is known to be 108° [5]. Thus if this test is performed, the derived $\sin i$ can be compared to the known value. Note that many practical issues still need to be solved in real life observations, such as the correct way to deal with the light-travel time effect and the RV bias caused by the effect of tidal distortion.

In any case, it is essential that the data quality be high enough to be sensitive to variations of the required order, namely one meter per second or less. Currently, the best precision is obtained by HARPS, where the RV

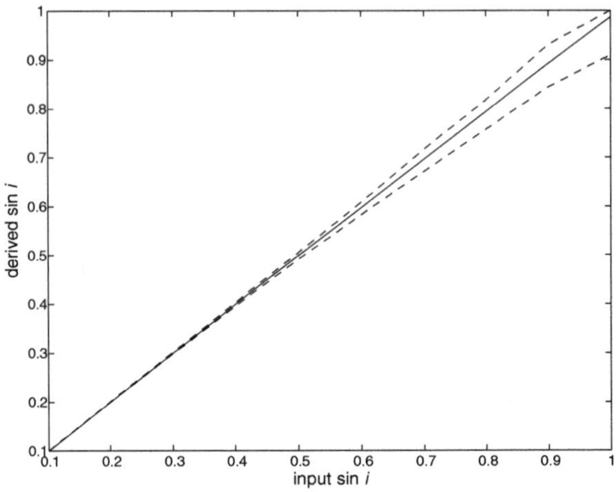

Fig. 1. The results of simulations of 12 Boo including relativistic effects, assuming RV precision of $1\,\mathrm{m\,s^{-1}}$. The plot shows percentiles of the derived $\sin i$ against the input simulated value. Shown are the first and third quartile (*dashed line*) and the median (*solid line*)

error can be as low as $1\,\mathrm{m\,s^{-1}}$ in certain cases and maybe even less [6]. In the future, much better precisions can be hoped for, on instruments designed for the "extremely large telescopes" (e.g., [7]).

In summary, we have shown that it should be possible to detect relativistic effects in close spectroscopic binaries, and that this possibility may have more practical implications than merely validating relativity. It can also serve as a path to derive more information about the binary orbits, namely the inclinations and the absolute masses, and thus help to calibrate the mass–radius and mass–magnitude relations. No other method exists yet to derive this information purely from RV measurements.

References

1. M. Mayor, D. Queloz: Nature **378**, 355 (1995)
2. G.W. Marcy, R.P. Butler: ApJ **464**, L147 (1996)
3. S. Zucker, T. Alexander: ApJ **654**, L83 (2007)
4. J. Tomkin, F.C. Fekel: AJ, **131**, 2652 (2006)
5. A.F. Boden, G. Torres, C.A. Hummel: ApJ, **627**, 464 (2005)
6. C. Lovis, M. Mayor, F. Bouchy et al: A&A, **437**, 1121 (2005)
7. L. Pasquini, S. Cristiani, H. Dekker et al: CODEX: measuring the acceleration of the universe and beyond. In: *IAU Symp. 232, The Scientific Requirements for Extremely Large Telescopes*, ed by P. Whitelock, B. Leibundgut, M. Dennefeld (Cambridge Univ. Press, Cambridge 2006) pp 193–197

Part V

Asteroseismology/Oscillations

Asteroseismology Across the HR Diagram

Mário J. P. F. G. Monteiro[1,2]

[1] Centro de Astrofísica da Universidade do Porto, Rua das Estrelas, 4150-762 Porto, Portugal (mjm@astro.up.pt)

[2] Departamento de Matemática Aplicada, Faculdade de Ciências da Universidade do Porto, Portugal

High precision spectroscopy provides essential information necessary to fully exploit the opportunity of probing the internal structure of stars using Asteroseismology. In this work we discuss how *Asteroseismology* combined with *High Precision Spectroscopy* can establish a detailed view on stellar structure and evolution of stars across the HR diagramme.

1 Introduction

Stars are fundamental units of the visible Universe, making stellar structure and evolution a conner stone of Astrophysics. However, the modelling of stellar structure and evolution still has quite a few fundamental questions to address and resolve. Among these we have convection, atomic diffusion and non-standard effects(like rotation and additional mixing processes).

In order to reproduce with models the observables for a star there is a collection of parameters that must be choosen. The mathematical solution is then adjusted with the parameters to a limited set of available boundary conditions provided by the observations. An improvement or test of the modelling is only possible if the number of parameters is smaller than the observables. This is usually not the case, so complementary observational information is required to effectively constrain the models and the physics used to calculate these.

To achieve this we must combine all available observational techniques that can provide information on specific stars. Besides the "classical" observables provided by photometry, spectroscopy and astrometry we now have the possibility to use interferometry (e.g. [21, 11]) and *Asteroseismology.* In particular the detection of periodic variations (oscillations) in most stars – the field known as Asteroseismology – has provide us with a powerful tool to analyse the inner structure of stars (for a detailed review see [8, 9] and references therein). The potential is great as almost all stars across the HR diagram have been found to display oscillations (some of these are identified in Fig. 1).

In this work we will discuss briefly how these stellar oscillations are use to provide additional observational constraints on the models. We will concentrate on some examples of Asteroseismology with the potential for providing

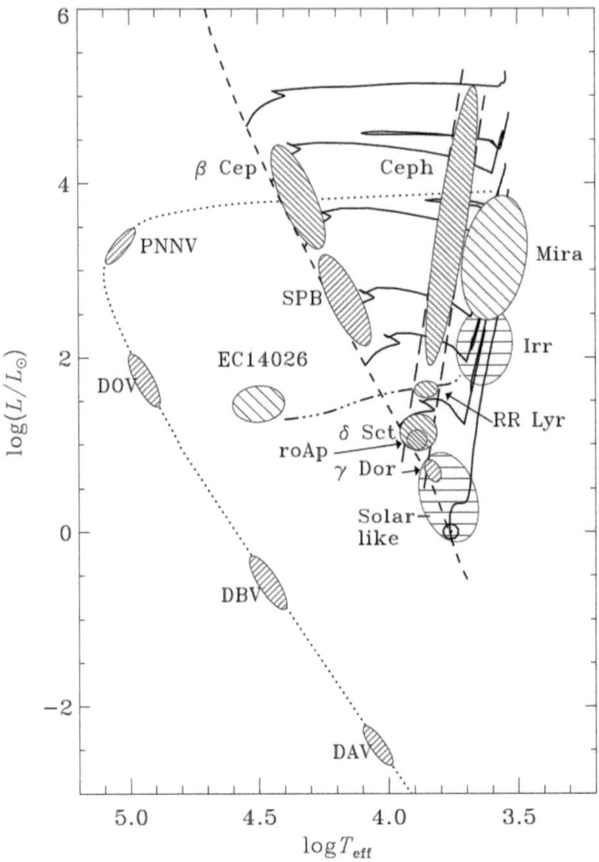

Fig. 1. Asteroseismic HR Diagram from [7] illustrating some of the known pulsators. The study of solar-like pulsators have benefit greatly from the tools developed for Helioseismology – seismic study of the Sun – which are now being applied to analyse ground and space observations of other stars.

tests that can help to understand some of the fundamental problems in stellar structure and evolution.

Considering the topic of this volume we also give special attention to the benefit of having precise spectroscopic information to produce conclusive seismic tests on specific pulsators in the HR diagram. Spectroscopy can be used in two ways; either by providing precise frequency determinations (and amplitudes) or by allowing a precise determination of global stellar characteristics. In either case the quality of the seismic inference is greatly improved by the quality of the spectroscopy measurements done as already shown by some of the currently available facilities (see these proceedings).

2 Uncertainties in Stellar Structure and Evolution

From observation the global stellar characteristics are the quantities more "easily" available (in particular luminosity and effective temperature), and eventually, the surface metallicity and gravity. These can be reproduced using stellar models by adjusting global parameters like the age, the initial Helium abundance, and if not known - the mass. But the key aspect is that the models have an extensive set of options (mixing length parameter, relative abundances of specific heavy elements, extent of overshoot, additional mixing, the atmosphere, diffusion, to name a few) to build a particular solution that fully complies with the observed parameters. These are all equally valid solutions unless further observational constraints can be found to discriminate and eliminate some.

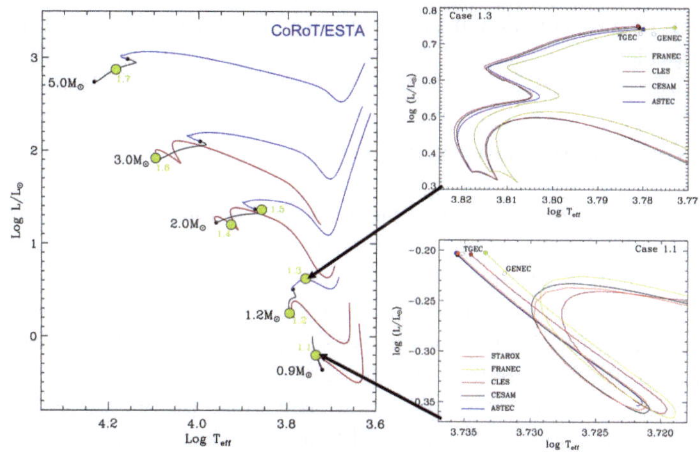

Fig. 2. Uncertainties/differences in the modelling of stellar structure and evolution evaluated under the CoRoT/ESTA effort using several evolution codes [19]. The interpretation of the observations is strongly dependent on the modelling leading to different conclusions for specific regions of the HR diagram where particular aspects of the physics may have an important role.

The freedom to adjust several possible solutions is contained in (1) the prescriptions we select, (2) the uncertainties of our best prescriptions for calculating the internal structure of a star (see Fig. 2) and (3) the observational uncertainties of the stellar measurements. There is such a wide range of competing uncertainties that it is very difficult (some would say impossible) to be conclusive when reproducing the photometric and astrometric observables.

In item (1) we also need to remind that there are dynamical effects on evolution usually not included, arising from the complex interaction between convection [24], rotation (e.g. [23]) and magnetic fields (including winds). Contrary to standard current procedures, the effects of these components of the physics on the evolution of the stars may have to be included in the modelling, in particular for specific stars in the HR diagram. But a detailed (fully validated) inclusion of those effects in our models is still missing in most cases.

Regarding item (2) we may try to establish what are the uncertainties on the modelling and in particular in uncertainties introduced by numerical difficulties in adequately implement the physics on the models. Some of these effects have been quantified but some aspects still need further improvement (see [19] for details).

High precision spectroscopy can help with item (3) as the range of allowed solutions can be significantly reduced if better, more precise stellar characteristics can be measured. One example that requires improvement is the determination of the abundances in stars. The models are extremely sensitive to these and no other observational data can replace the information content of a direct measurement of individual abundances of all relevant chemical species.

3 Asteroseismic Constraints

Some known pulsators show the richness of oscillation data that we have found in our own Sun (e.g. [5]). With such a multitude of data it has become clear that much more stringent constraints could be placed on the basic physical ingredients used to model the structure and evolution of these stars. If measured, oscillations frequencies increase greatly the available observational constraints on our models (e.g. [6, 10, 12]).

This makes Asteroseismology one of the most promising techniques for solving some of the long standing problems in Stellar Astrophysics. The underlying goal is to be able to understand how some assumptions in the modelling affect the observed properties, and so to use the observed seismic behaviour to constrain those assumptions. For a detailed discussion on how the frequencies can complement the photometric, spectroscopic and astrometric observations, providing a great improvement on tests of stellar models, see [6], [7] or [17] and references therein.

Different approaches within Asteroseismology have been developed to infer specific constraints on the interior of different type of pulsators. Here we only mention some cases, in order to illustrate briefly the type of information we can obtain from seismology which have a direct consequence on the modelling of stellar structure and evolution.

A direct comparison of observed and model frequencies is usually the first step taken in using seismic data. This is however a very difficult task as the potential for using an individual frequency depends strongly on knowing what type of mode it corresponds to. This is in most cases very difficult, but high precision spectroscopy can be a step forward by providing the data necessary for mode identification (see Aerts, these proceedings). Some recent examples where it was attempted include the case of β Cephei stars [16] or young pre-main sequence stars [22] and references therein.

The near surface structure of most stars is poorly know and our Sun has shown that it is inadequately represented in our models. Consequently a direct comparison of frequencies is in most cases dominated by that aspect invalidating a test of the internal structure. To overcome this difficulty several approaches have been developed based on the assumption that some adequate combinations of the frequencies can better test the interior structure without being very sensitive to what happens near the surface. This is the approach being developed for studying pulsators with several identified frequencies of oscillation. The Sun has shown us that the detail we may achieve is very high even if the data we expect for other stars is not as precise and abundant as what we have for our star. The development of these tools is being actively extended (as an example see [18, 3, 15, 9]) for studying other pulsators. There is a wide variety of applications showing that frequencies can add an extremely important observational constraint allowing us to probe in detail the characteriostics of the stars being observed. The detailed analysis by Creevey et al. [10] illustrates the sensitivity of the parameters (quantities defining the models) to the observables (measured quantities).

An interesting example is the study of sub-giant stars which display solar-like oscillations. As the star approaches the bottom of the red-giant region it becomes very difficult to differentiate stellar age and stellar mass using only luminosity and effective temperature. If we add to the global parameters the oscillation frequencies the determination of stellar mass and age can be greatly improved allowing a more precise study of how the Helium core changes in this phase of evolution (e.g. [4]).

Another set of very promising applications focus on the pre-main sequence evolution [20, 14, 1]. Much can be gained from using seismology to study this complex phase of evolution in low and intermediate mass stars.

4 Final Remarks

With the new observational opportunities provided by the space missions in Asteroseismology, which will complement the coverage of the HR Diagram with ground based observations, we may expect to be able to address some of the unsolved problems in the modelling of stellar structure and evolution.

The space missions in Asteroseismology: MOST launched in 2000 [25], CoRoT launched in 2006 [2], and Kepler to be launched in 2008, will provide seismic data for several stars of different masses and in different phases of evolution. All phases, from the pre-main sequence up to the post-main sequence evolution can be covered, complementing the already available data from the ground on quite a few families of pulsating stars. Additional efforts are being also prepared for ground based observations of which SONG is an example ([13] and these proceedings).

The much need increase in precision provided by a new generation of spectroscopic instruments (see these proceedings) on measuring stellar characteristics and oscillations are also expected to have a strong impact on Asteroseismology over the next decade.

References

1. E. Alecian, M.-J. Goupil Y. Lebreton, et al.: A&A in press (2007)
2. A. Baglin, M. Auvergne, P. Barge, et al: in *First Eddington Workshop*, (Eds) F. Favata, I.W. Roxburgh and D. Galadi, ESA SP **485**, 17 (2003)
3. M. Bazot, S. Vauclair, F. Bouchy, et al.: A&A **440**, 615 (2005)
4. T.R. Bedding, R.P. Butler, F. Carrier, et al.: ApJ **647**, 558 (2006)
5. T.R. Bedding, H. Kjeldsen: in *Highlights of Recent Progress in the Seismology of the Sun and Sun-Like Stars*, IAU JD17, 22 (2007)
6. T.M. Brown, J. Christensen-Dalsgaard, B. Weibel-Mihalas, R.L. Gilliland: ApJ **427**, 1013 (1994)
7. J. Christensen-Dalsgaard: ApSS **261**, 1 (1999)
8. J. Christensen-Dalsgaard: in *Helio- and Asteroseismology: Towards a Golden Future*, (ed) D. Danesy, ESA-SP **559**, 1 (2004)
9. J. Christensen-Dalsgaard: in *A New Era of Helio- and Asterosiesmology*, (Eds) B. Fleck and K. Fletche, ESA SP **624** in press (2007)
10. O.L. Creevey, M.J.P.F.G. Monteiro, T.S. Metcalfe, et al.: ApJ inpress (2007)
11. E. Di Folco, F. Thévenin, P. Kervella, et al: A&A **426**, 601
12. P. Eggenberger, F. Carrier, F. Bouchy: NewA **10**, 195 (2005)
13. F. Grundahl, H. Kjeldsen, S. Frandsen, et al: MnSAI **77**, 458 (2006)
14. J.P. Marques, M.J.P.F.G. Monteiro, J. Fernandes: MNRAS **371**, 293 (2006)
15. A. Mazumdar, S. Basu, B.L. Collier, P. Demarque: MNRAS **372**, 949 (2006)
16. A. Mazumdar, M. Briquet, M. Desmet, C. Aerts: A&A **459**, 589 (2006)
17. M.J.P.F.G. Monteiro: in *Observed HR Diagrams and Stellar Evolution*, (Eds) T. Lejeune and J. Fernandes, A.S.P. Conf. Ser., Vol **274**, 77 (2002)
18. M.J.P.F.G. Monteiro, J. Christensen-Dalsgaard, M.J. Thompson: in *Stellar Structure and Habitable Planet Finding*, (Eds) F. Favata, I.W. Roxburgh, D. Galadì-Enríquez, ESA SP **485**, 291 (2002)
19. M.J.P.F.G. Monteiro, Y. Lebreton, J. Montalbán et al.: in *The CoRoT Book*, (Eds) F. Favata, A. Baglin & J. Lochard, ESA SP (in press)
20. F. Palla, I. Baraffe: A&A **432**, L57 (2005)
21. F.J. Pijpers, T.C. Teixeira, P.J. Garcia P.J., et al.: A&A **406**, L15 (2003)
22. V. Ripepi, S. Bernabei, M. Marconi et al.: A&A **449**, 335 (2006)
23. I.W. Roxburgh: A&A **454**, 883 (2006)
24. C.W. Straka, P. Demarque, D. Guenther, et al: ApJ **636**, 1078 (2006)
25. G. Walker, J. Matthews, R. Kusching, et al.: PASP **115**, 1023 (2003)

High-Precision Spectroscopy of Pulsating Stars

C. Aerts[1,2], S. Hekker[1,3], M. Desmet[1], F. Carrier[1], W. Zima[1], M. Briquet[1], and J. De Ridder[1]

[1] Institute of Astronomy, Catholic University of Leuven, Celestijnenlaan 200D, B-3001 Leuven, Belgium conny@ster.kuleuven.be
[2] Department of Astrophysics, Radboud University Nijmegen, P.O. Box 9010, 6500 GL Nijmegen, The Netherlands
[3] Leiden Observatory, Leiden University, P.O. Box 9513, 2333 RA Leiden, The Netherlands

Summary. We review methodologies currently available to interprete time series of high-resolution high-S/N spectroscopic data of pulsating stars in terms of the kind of (non-radial) modes that are excited. We illustrate the drastic improvement of the detection treshold of line-profile variability thanks to the advancement of the instrumentation over the past two decades. This has led to the opportunity to interprete line-profile variations with amplitudes of order m/s, which is a factor 1000 lower than the earliest line-profile time series studies allowed for.

1 Line-Profile Variations due to Pulsations

In the research domain of asteroseismology, one tries to probe the internal structure parameter of stars from their observed pulsation properties. Prerequisites to succeed in that are accurate pulsation frequency values and an unambiguous identification of the spherical wavenumbers (ℓ, m) of at least two, but preferrably many more, non-radial pulsation modes. Asteroseismology has been applied successfully across the whole HR diagram. For a recent extensive overview of its successes so far, we refer to Kurtz (2006).

The introduction of high-resolution spectrographs with sensitive detectors in the 1980s had a large impact on the field of pulsation mode identification. Spectroscopic data indeed offer a very detailed picture of the pulsation velocity field. Mode identification requires that moderate to large telescopes be available during a long time span. It remains a challenge to obtain time-resolved spectra with a high resolving power and with a high signal-to-noise ratio (> 300) covering the overall beat period of the multiperiodic pulsation. The required temporal resolution must be such that the ratio of the integration time and the pulsation period(s) remains below a few percent. The latter condition is necessary in order to avoid smearing out of the effects of the pulsations in the line profiles during the cycle. This requirement is difficult to meet for rapid pulsators with periods of order minutes. This is the reason why spectroscopic mode identification has been achieved mainly for massive main sequence stars with spectral type from O to F and with pulsation modes excited by the κ mechanism with periods above one hour.

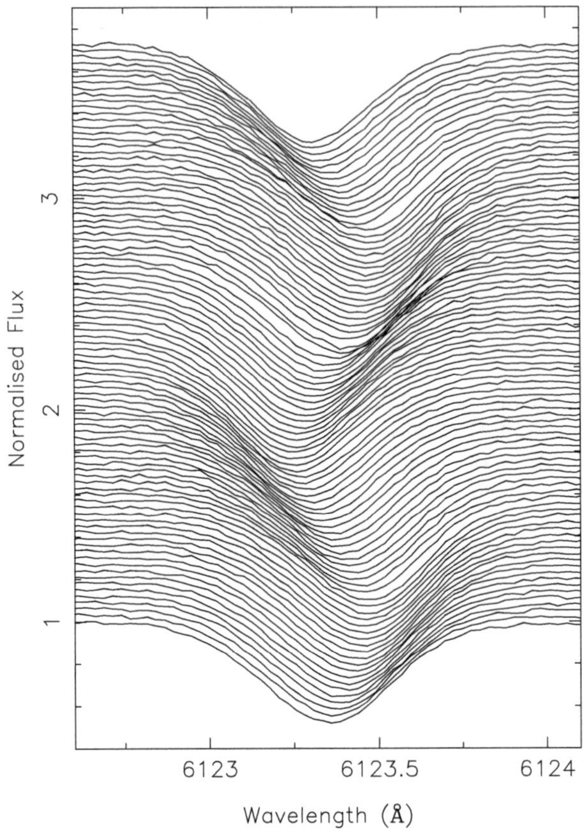

Fig. 1. Observed line-profile variations of the δ Sct star ρ Puppis obtained in 1995 with the Coudé Auxiliary Telescope of the European Southern Observatory in Chile. Data taken from Mathias et al. (1997).

An example of a nice data set is shown in Fig. 1. The application to stars with stochastic excitation, to magnetic pulsators and to compact pulsators has only recently been attempted and needs further improvements.

2 Methodology for Line-Profile Analysis

The pulsation velocity due to a spheroidal pulsation mode with infinite lifetime in the approximation of a non-rotating star equals

$$\boldsymbol{v}_{\rm puls} = (v_r, v_\theta, v_\varphi) = N_\ell^m v_{\rm p} \left(1, K\frac{\partial}{\partial\theta}, \frac{K}{\sin\theta}\frac{\partial}{\partial\varphi}\right) Y_\ell^m(\theta,\varphi)\exp\left({\rm i}\omega t\right), \quad (1)$$

when a system of spherical coordinates (r, θ, φ) with origin at the centre of the star and with polar axis along the rotation axis is used. In this expression, N_ℓ^m

is a normalisation factor for the $Y_\ell^m(\theta, \varphi)$, v_p is proportional to the pulsation amplitude, ω is the cyclic pulsation frequency, and K is the ratio of the horizontal to the vertical velocity amplitude: $K = GM/(\omega^2 R^3)$. This velocity acts together with rotational broadening and with intrinsic line broadening due to pressure and temperature effects. Moreover, we only detect the velocity component projected onto the line-of-sight. This leads us to the conclusion that line-profile variations due to a single pulsation mode are determined by six unknown parameters among which two are integer numbers (ℓ, m). For each additional mode, three unknowns are added in the linear approximation ignoring non-linear coupling between the modes. The above Eq. (1) is far too simple when Coriolis, centrifugal or Lorentz forces come into play.

It is clear from these arguments that the derivation of the full details of the pulsational velocity field from observed line-profile variations cannot be a simple task. Nevertheless, the richness of the information in these type of data is such an asset compared to photometric data (which essentially only allow estimation of ℓ) that spectroscopic mode identification has become an entire subfield by itself within asteroseismology.

Fairly recent overviews of the methodologies for spectroscopic mode identification are available in Telting & Schrijvers (1997), Aerts & Eyer (2000), and Mantegazza (2000). Rather than repeating what is available in these papers, we point out two newer versions of the methods available since then. One is the numerical implementation of the so-called moment method (Briquet & Aerts 2003). In this work, the authors have generalised previous versions of this technique to multiperiodic pulsators. This technique is based on the time-variations of the lowest-order moments of a line profile and works well for slow rotators. It has meanwhile been applied to several stars whose pulsational broadening dominates over rotational and intrinsic broadening.

Zima (2006), on the other hand, generalised the Telting and Schrijvers (1997) method, which is based on the amplitude and phase variations of the modes across the line profile. Zima (2006) included a statistical significance criterion for the mode identification and subsequently applied the method to the rapidly-rotating δ Sct star FG Vir (Zima et al. 2006). A recent combined application of both methods to the pulsations of the β Cep star ν Eri, observed during a multisite campaign, was made by De Ridder et al. (2004), while De Cat et al. (2005) applied both techniques to the observed line-profile variations of several slowly pulsating B stars.

In all of the abovementioned examples, one carefully selected isolated and unblended spectral line was used for the analysis. This is a good strategy whenever pulsation amplitudes above a few km/s are encountered and when the required temporal resolution is achieved. For lower-amplitude pulsators, however, or when time-resolved spectroscopy leads to a too low signal-to-noise ratio, one may also apply the methodology discussed above to a time series of cross-correlation functions (CCFs) derived from different lines in the spectrum. This induces complications, however, because one takes a weighted average over a much more extended line-forming region in the stellar atmosphere compared to the case where one uses only one line. The pulsation amplitude and phase may have slightly different behaviour across this whole region, such that one models their average value in that case. This works fine on the condition that no nodal surfaces of the pulsation eigenfunction occur in the line-forming region. An adaptation of the moment method in such a case was already made by Mathias & Aerts (1996), who applied it to the low-amplitude ($< 5\,\mathrm{km/s}$) δ Sct star 20 CVn. More recently, Aerts et al. (2004) and De Cat et al. (2006) also used CCFs to analyse extensive spectroscopic data of a sample of γ Dor stars, with pulsation amplitudes typically below $2\,\mathrm{km/s}$.

3 Towards Lower Amplitudes

The first and so far only application of the line-profile methodology to solar-like oscillations with finite lifetimes was made by Hekker et al. (2006). They used extensive time series of CCFs of four pulsating red giants derived from CORALIE spectra obtained with the 1.2m Euler telescope in an attempt to identify the pulsation modes. Their results are based on a simulation study in which they assessed the effect of the finite mode lifetime on the line-profile diagnostics, resulting in the conclusion that the phase across the CCF can no longer be used for mode identification. Due to this, and the very low amplitudes of $\sim$m/s, only an estimate of (ℓ, m) of the dominant mode could be obtained. This led to the surprising result that these stars seem to have non-radial modes, while it was assumed so far in the modelling that the frequencies were due to radial modes. The application to solar-like pulsators is of relevance for exoplanet search as well, since a Keplerian radial-velocity shift is then superimposed on the pulsation velocity. It remains to be studied how well one can separate between those two effects at such low amplitudes.

References

1. C. Aerts, J. Cuypers, P. De Cat, et al: A&A **415**, 1079 (2004)
2. C. Aerts, L. Eyer: ASPC **210**, 113 (2000)
3. M. Briquet, C. Aerts, A&A **398**, 687 (2003)
4. P. De Cat, M. Briquet, M., J. Daszynska, et al: A&A **432**, 1013 (2005)
5. P. De Cat, L. Eyer, J. Cuypers, et al: A&A **449**, 281 (2006)
6. J. De Ridder, J. Telting, L.A. Balona, et al: MNRAS **351**, 324 (2004)
7. S. Hekker, C. Aerts, J. De Ridder, F. Carrier: A&A **458**, 391 (2006)
8. D.W. Kurtz: CoAst **147**, 6 (2006)
9. L. Mantegazza: ASPC **210**, 138 (2000)
10. J.H. Telting, C. Schrijvers: A&A **317**, 723 (1997)
11. W. Zima: A&A **455**, 227 (2006)
12. W. Zima, D. Wright, J. Bentley, et al: A&A **455**, 235 (2006)

Mapping Atmospheric Motions in Classical and Type II Cepheids

Monika Jurkovic[1] and József Vinkó[1]

Department of Optics & Quantum Electronics, University of Szeged,Hungary
mojur@titan.physx.u-szeged.hu, vinko@physx.u-szeged.hu

Summary. Type II Cepheids populate the same region of the instability strip as Classical Cepheids, but they belong to older stellar population. Their different stellar structure may result in different pulsational characteristics. We study the relative motion of the photospheric and chromospheric layers of Type II Cepheids by comparing radial velocities of metal lines and $H\alpha$. It is shown that in the $P = 1-3$ days regime the two types are markedly different, but their atmospheric motions become quite similar at longer periods.

1 Introduction

Classical Cepheids (CCs) are young, massive, Population I supergiants that populate the middle and upper part of the instability strip. Type II Cepheids locate in the same part of the Hertzsprung–Russell diagram, but they belong to much older stellar population. It is reflected mainly from their galactic position and kinematics. They are thought to be low-mass stars that cross the instability strip evolving from the horizontal branch toward the AGB.

Early spectroscopic observations of Type II Cepheids suggested that their pulsation may be different from that of Pop I Cepheids. Sometimes they exhibit $H\alpha$ emission at certain phases, which is not observed in CCs. The atmospheric motions of the two types, e.g. the velocity gradients may also be different. Comparing the pulsational characteristics of the two types may help to understand the physical and evolutionary state of Type II Cepheids.

2 Observation and Data Reduction

We have started a long-term project for spectroscopic monitoring of Type II Cepheids at David Dunlap Observatory, Canada with the 74" telescope. The first results of this project were summarized in [1]. In this paper we analyze the spectra obtained between August 1997 and September 1998. During this period, both the Cassegrain spectrograph (resolving power $\sim 11,000$) and the echelle spectrograph (resolving power ~ 40000 at $H\alpha$) were applied, although the echelle could be used only for the brighter CCs. 19 of the program stars that were observed at more than 4 different phases were analyzed further.

The spectra were reduced in *IRAF* using the standard procedures. The radial velocities were determined by cross-correlating the Cepheid spectra with those of standard velocity stars. We have used the IAU list of standard velocity stars (from the Astronomical Almanach), but their velocities were updated from the CfA list [2].

We have used the radial velocities of the $H\alpha$ line to get information for the movement of the upper atmospheric layers. The $H\alpha$ velocities were derived from cross-correlation between 6550 and 6580 Å. For the photosphere, we have computed the average radial velocities of several metallic lines. From the Cassegrain spectra the cross-correlation was computed in the region from 6600 to 6705 Å. Unfortunately, this region was not covered by the echelle spectra, so the orders between $5980 - 6225$ Å and $6320 - 6435$ Å were cross-correlated with the spectra of radial velocity standards. Care was taken to avoid telluric-contaminated regions.

Following the procedure given in [1], we have constructed the radial velocity differences as

$$\Delta v = v_{H\alpha} - v_{phot}, \tag{1}$$

where v_{phot} refers to the average cross-correlation radial velocities of the metallic lines mentioned above. It is known that Δv depends strongly on pulsational phase [1],[3],[4],[5]. As in these papers, we have chosen the maximum of the observed Δv (Δv_{max}) as an indicator of the atmospheric motion. Low Δv_{max} ($\sim$ a few kms^{-1}) means that the photosphere and the upper atmosphere move together during the pulsational cycle. On the other hand, high Δv_{max} (> 30 kms^{-1}) means that the photosphere and the chromosphere move in the opposite direction at certain phases. It is usually the rapid expansion phase ($\phi \sim 0.7 - 1.0$) when Δv reaches the maximum, i.e. when the photosphere starts to expand while the upper layers still collapse.

3 Results

Figure 1 left panel shows the Δv_{max} parameter as a function of the pulsational period. Our observational sample has been supplemented by the data from [1], [3], [4], [5] and [6]. Because Δv_{max} may be strongly phase-dependent, we have selected only those stars that had at least 4 observed velocities. The whole sample consists of 97 stars, 60 of them are Type II Cepheids (circles in Fig. 1) and 37 are CCs (filled squares).

For short-period Cepheids ($P < 3$ days) only the Type II stars show high Δv_{max}. However, in the case of longer period stars, the distribution of the Δv_{max} is similar: longer period Cepheids tend to have higher Δv_{max}, regardless of their types.

In Fig. 1 right panel the maximum velocity difference versus the radial velocity amplitude is plotted for Cepheids with known velocity amplitude. Again, the two types show no significant difference on this diagram: both

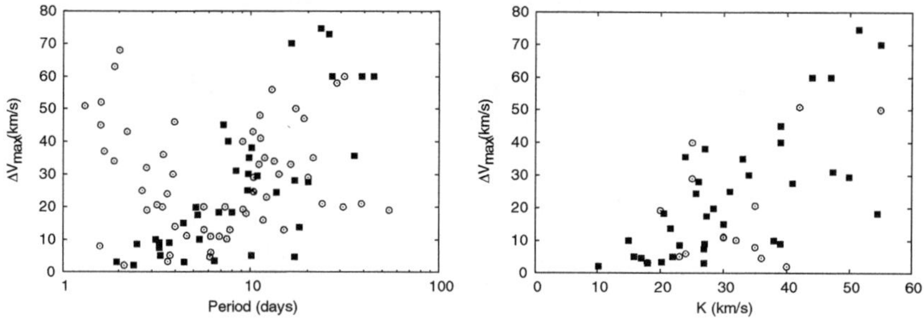

Fig. 1. Maximum radial velocity difference versus period (left panel) and velocity amplitude (right panel) (*filled squares: Classical Cepheids, circles: Type II Cepheids*).

Classical and Type II Cepheids that have higher Δv_{max} also have larger velocity amplitude.

4 Simple Model Computation

A simple variant of the one-zone model [7] has been used to illustrate the dependence of the maximum velocity difference on the period and the amplitude of the pulsation. We have assumed that the radius of the photosphere (R_p) performs a sinusoidal oscillation:

$$R_p = R_p(0) + Asin(\omega t). \tag{2}$$

The motion of the outer layer (with outer radius R and thickness z) has been followed by integrating the equation of motion:

$$\frac{d^2\xi}{dt^2} = \frac{GM}{R_0^3}\left(\xi^2\frac{P}{P_0} - \xi^{-2}\right), \tag{3}$$

where M is the stellar mass, $\xi = R/R_0$, P is the pressure and the 0 index refers to the equilibrium values. The variation of the pressure is computed from the adiabatic equation of state.

Two types of model sequences have been calculated: one is for models with masses and radii of CCs and the other one is for Type II Cepheids. Within each model sequence, the period and the velocity amplitude was varied and the maximum velocity difference was determined after integrating (3).

Surprisingly, this very over-simplified model shows similar [1] velocity difference - period or amplitude dependence as the real stars in Fig. 1, at least for CCs. The results of model computations for the CC sequence are plotted in Fig. 2. It is visible that the computed maximum velocity differences have a local maximum at $P = 10$ days, and they generally increase with increasing velocity amplitude, similarly to the observations in Fig. 1. On the other hand,

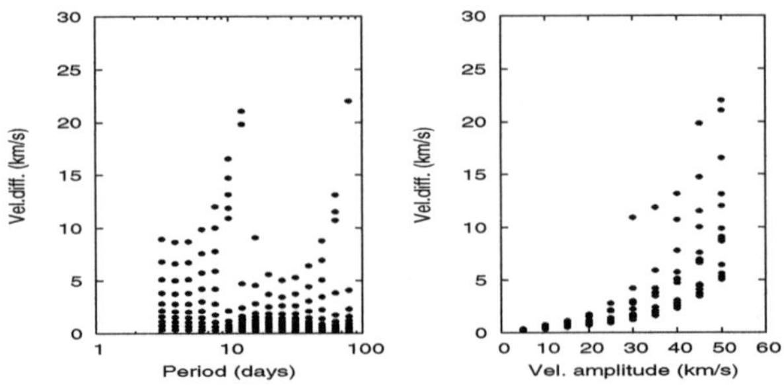

Fig. 2. Model calculation results for Classical Cepheids

the Type II models turned out to be more unstable and sensitive to the initial parameters.

5 Conclusions

Based on our measurements and data taken from the literature we concluded that in the $1 < P < 3$ days period regime Type II Cepheids show Δv_{max} differences up to 70 kms^{-1}. At periods greater than 3 days Classical and Type II Cepheids have a similar Δv_{max} distribution. Increasing velocity amplitude generally increases Δv_{max} for both Cepheid types.
Acknowledgments This research was supported by Hungarian OTKA Grants T042509, TS049872.

References

1. J. Vinkó; N. Remage Evans; L. L. Kiss; L. Szabados: MNRAS **296**, 824-838 (1998)
2. R. P. Stefanik; D. W. Latham: CfA Digital Speedometers, as reported at IAU General Assembly XXI, Buenos Aries (1991)
3. E.G. Schmidt; K. M. Lee; D. Johnston; P. R. Newman; S. A. Snedden: AJ **126**, 906-917 (2003)
4. E.G. Schmidt; S. Langan; K. M. Lee; D. Johnston; P. R. Newman; S. A. Snedden: AJ **126**, 2495-2501 (2003)
5. E.G. Schmidt; D. Johnston; K. M. Lee; S. Langan; P. R. Newman; S. A. Snedden: AJ **128**, 2988-2996 (2004)
6. O. K. L. Petterson; P. L. Cottrell; M. D. Albrow; A. Fokin: MNRAS **362**, 1167-1182 (2005)
7. R. F. Stellingwerf: A&A **21**, 91-96 (1972)

Iron Abundances of Southern Double-mode Cepheids from High-resolution Echelle Spectroscopy

K. Sziládi[1], J. Vinkó[1], L. Szabados[2], M. Kun[2] and E. Poretti[3]

[1] Department of Optics & Quantum Electronics, University of Szeged, POB 406, Szeged 6701, Hungary szkati@titan.physx.u-szeged.hu
[2] Konkoly Observatory of the Hungarian Academy of Sciences, POB 67, Budapest 1525, Hungary
[3] Osservatorio Astronomico di Brera, Via Bianchi 46, 23807 Merate, Italy

Summary. We determined [Fe/H] for 17 double-mode Cepheids of the southern hemisphere using high-resolution echelle spectra. For five of the program stars these are the first iron abundances in the literature. The available data suggest that the period ratio for galactic beat Cepheids is mainly determined by their metallicity. Therefore, the observed period ratios do not reflect large differences in their mass, effective temperature or luminosity.

1 Introduction

Double-mode Cepheids or beat Cepheids are radially pulsating stars, with two periods. They are useful to test the theory of pulsation by comparing observed and computed period ratios. It is possible to constrain evolutionary models by comparing masses derived from evolutionary and pulsational models. The crucial parameter is metallicity. Metallicity is an input parameter for the models, which must be determined from observations. Andrievsky et al. ([2, 3]) determined the chemical composition of four double-mode Cepheids (TU Cas, BQ Ser, VX Pup, EW Sct), and revealed an interesting correlation between the period ratio of the excited modes and metallicity. D'Cruz et al. ([4]) derived [Fe/H] for seven of our program stars by comparing observed and theoretical period ratios.

2 Observations and Data Reduction

High resolution echelle spectra of 17 southern beat Cepheids were taken with the FEROS spectrograph at the ESO 2.2m telescope on La Silla in May and June 2004. The spectral range covered was 3900−9200 Å with no gaps. The resolving power was 48000 and the S/N ratio $\sim$ 200−350 was reached. All spectra were reduced using IRAF, in two independent ways. First, we used the Fiber Optic Echelle data reduction task (imred.echelle.dofoe). In the second case, after bias subtraction and flat-field division, we transformed each echelle order to a one-dimensional spectrum and reduced them individually. The results from these two methods are in good agreement with each other. The final spectra were also double-checked with the results of an independent reduction using the ESO-MIDAS pipeline. Again, very good agreement has been found.

For the chemical analysis we used the SPECTRUM[4] code by Richard Gray. We calculated the abundance from neutral and ionized iron lines using the code BLACKWEL, an auxiliary program of SPECTRUM. This program solves the radiative transfer equation in LTE and gives the logarithmic abundance for each line. We have assembled a line list containing 77 weak unblended Fe I and 18 Fe II lines. Their equivalent widths were measured by fitting Voigt-profiles to the continuum-normalized spectra.

3 Results

As a reference, the solar iron abundance was derived for each night using daylight solar spectra. Using the well known solar parameters, T_{eff}=5777 K, $\log g$=4.437, v_t=1.5 km s^{-1}, we see that the abundances are independent of the lower level of transition and equivalent widths. We got 7.49 ± 0.19 for the average iron abundance, which is similar to the known literature value (7.50).

Similarly, we calculated the iron abundance for 17 stars. Figure 1 shows an example with the derived values for BQ Ser. We used the measured parameters as input data for ATLAS9 to generate a spectrum in order to check with an independent method that our parameters are correct (Fig. 2). Fig. 3 illustrates the sensitivity of the results to the input parameters. When we use incorrect effective temperature, the iron abundances depend on the excitation potential, which is physically untenable. The gravity has been calculated from the ionization balance of neutral and ionized iron lines. Microturbulent velocity was determined from the condition that there the individual iron abundances should not depend on the line equivalent widths.

At the present the iron abundance determination was completed for 17 beat Cepheids. The spectroscopic parameters are given in Table 1.

Fig. 4 shows the comparison of our [Fe/H] data with data taken from literature. Our data disagrees with those from Andrievsky et al. ([3]), but agrees much better with their results from 2002 ([1]). The best agreement is with data from D'Cruz et al. ([4]), but those are from pulsation models, independent of spectroscopy.

Our homogeneous data improve the observed relation between the period ratio and metallicity. Figure 4 (right panel) shows this relation from our data. This is the main result of our program.

---

[4]www.phys.appstate.edu/spectrum/spectrum.html

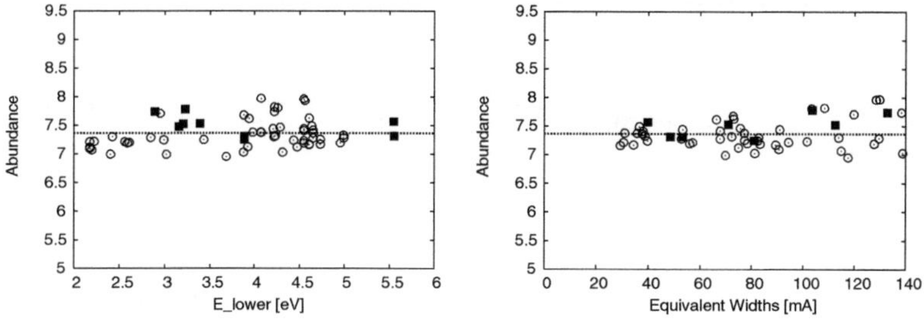

Fig. 1. Derived Fe abundance for BQ Ser from Fe I *(circle)* and Fe II *(square)* lines. Parameters: T_{eff}=6000 K, log$g$=2.0 v_t=3.7 km s^{-1}. *Left:* Abundance vs. energy of lower level of transition (in eV). *Right:* Abundance vs. equivalent width (in mÅ)

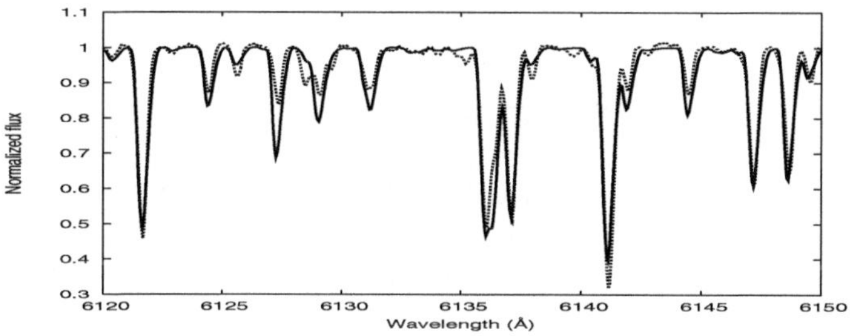

Fig. 2. A comparison of ATLAS9 program generated spectrum with measured spectrum.

Table 1. Parameters for the beat Cepheids. The columns contain the following data: star, T_{eff}, log$g$, microturbulent velocity, Fe abundance and its uncertainty.

Star	T_{eff} (K)	log$g$	v_t (km s^{-1})	A_{Fe}	σ	Star	T_{eff} (K)	log$g$	v_t (km s^{-1})	A_{Fe}	σ
Y Car	6500	2.2	4.1	7.44	0.16	V458 Sct	6250	2.26	4.2	7.58	0.17
EY Car	5875	1.25	4.2	7.32	0.22	BQ Ser	5917	1.97	3.77	7.37	0.25
GZ Car	6125	2.02	5.0	7.42	0.21	U TrA	6000	2.18	4.8	7.34	0.16
UZ Cen	5875	1.82	4.0	7.30	0.18	AP Vel	5875	2.0	3.9	7.43	0.19
BK Cen	6000	2.12	4.0	7.56	0.19	AX Vel	6375	2.34	5.1	7.40	0.18
V1048 Cen	6250	2.0	4.2	7.31	0.23	GSC 8607 0608	5750	1.5	3.7	7.38	0.22
VX Pup	6500	2.51	4.4	7.29	0.18	GSC 8691 1294	6125	2.25	3.65	7.47	0.22
EW Sct	6000	1.8	3.73	7.48	0.20	TU Cas	6250	2.0	4.0	7.33	0.24
V367 Sct	6125	1.67	4.5	7.56	0.22						

Acknowledgements. This work was supported by Hungarian OTKA Grants No. TS 049872, T 042509 and T 046207. We are thankful to ESO for spectroscopic observations in Chile (La Silla). The NASA ADS Abstract Service was used to access data and references.

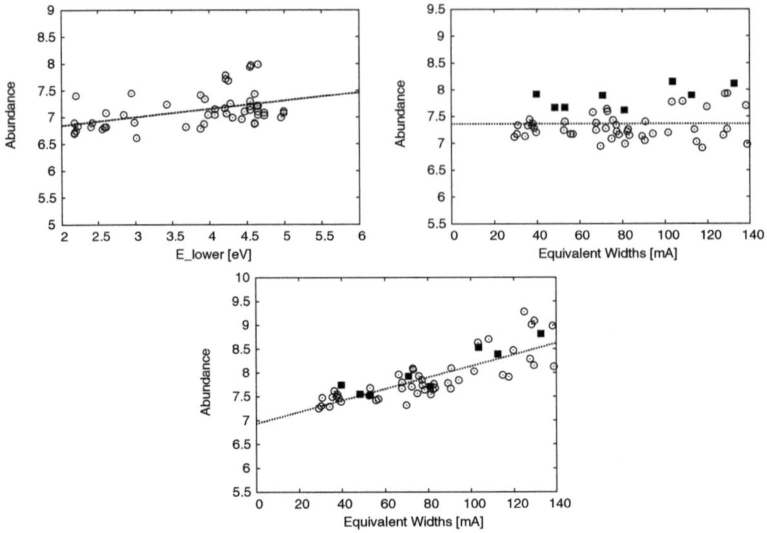

Fig. 3. The sensitivity of the derived Fe abundances of BQ Ser to the stellar parameters (the symbols have the same meaning as in Fig. 1). *Top left:* wrong effective temperature (T_{eff}=5500 K), *top right:* wrong gravity (log$g$=3.0) and *bottom:* wrong microturbulent velocity (v_t=1.5 km s^{-1}).

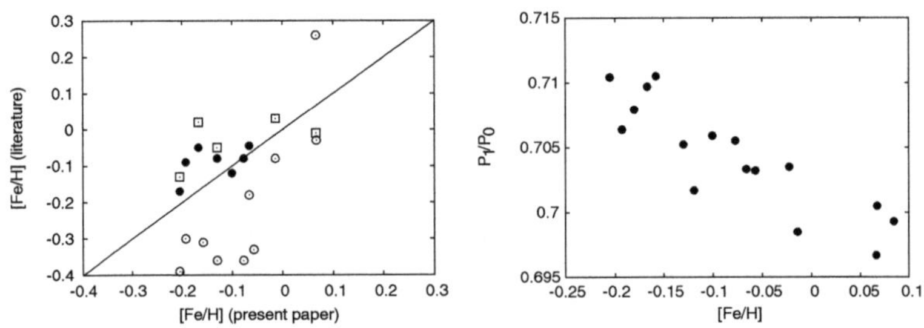

Fig. 4. *Left:* A comparison of our results with data from literature. Andrievsky et al. (2002) *(open square)*, Andrievsky et al. (1994) *(open circle)*, D'Cruz et al. (2000) *(filled circle)*. *Right:* The period ratio – metallicity relation based on our results.

References

1. S. M. Andrievsky, V. V. Kovtyukh, R. E. Luck et al: A&A [381], 32 (2002)
2. S. M. Andrievsky, V. V. Kovtyukh, E. N. Makarenko et al: MNRAS, [265], 257 (1993)
3. S. M. Andrievsky, V. V. Kovtyukh, I. A. Usenko et al: A&AS, [108], 433 (1994)
4. N. L. D'Cruz, S. M. Morgan, E. Böhm-Vitense: AJ, [120], 990 (2000)

Part VI

Planets

Radial Velocity Planet Detection using a Gas Absorption Cell

William D. Cochran[1], Artie P. Hatzes[2], Michael Endl[1], Diane B. Paulson[3] and Robert A. Wittenmyer[1]

[1] McDonald Observatory, The University of Texas at Austin, Austin Texas USA
wdc@astro.as.utexas.edu, mike@astro.as.utexas.edu, robw@astro.as.utexas.edu

[2] Thüringer Landessternwarte, Tautenburg, Germany artie@tls-tautenburg.de

[3] NASA Goddard Space Flight Center, Greenbelt Maryland USA
Diane.B.Paulson@gsfc.nasa.gov

Summary. The use of a gas absorption cell as the velocity metric has become a standard method of achieving the very high velocity precision necessary for detection of low-mass planetary companions to nearby stars. At present, I_2 is the most commonly used gas. This gas absorption cell technique has a number of significant advantages: 1) The I_2 spectrum is an absorption spectrum superimposed directly on the stellar spectrum. Thus, the reference spectrum follows exactly the same illumination pattern as the stellar spectrum. 2) The extremely narrow I_2 absorption lines permit modeling of small-scale variations in the spectrograph instrumental profile. 3) It is easy and inexpensive to retrofit an existing spectrograph with an I_2 cell. The I_2 cell technique has been exceptionally successful in the detection of extra-solar planets.

1 Introduction

Classical techniques for measurement of stellar radial velocities with respect to a separate comparison source spectrum rarely achieve standard errors better than a few tenths of a kilometer per second. The principal limiting factor, as discussed by [1], is the difference in illumination of the spectrograph optics by the stellar and wavelength comparison sources. The stellar beam must traverse the terrestrial atmosphere and the telescope optics. The comparison source spectrum is generally taken at a different time (either before or after the stellar spectrum), has a different beam illumination pattern, and usually follows a significantly different path through the spectrograph. Most modern spectrographs do not have either the mechanical or thermal stability to give an intrinsic velocity precision at the $1\,\mathrm{m\,s^{-1}}$ level. In addition, the telescope pupil which feeds the spectrograph can be a function of time and of the direction in which the telescope is looking.

In order to achieve very high radial velocity precision, one can either stabilize and control the entire spectrograph to the necessary level of precision (the approach of HARPS), or one can use a gas absorption cell to measure, model, and remove these effects.

2 Gas Absorption Cells as the Velocity Metric

The use of a gas reference absorption spectrum is the least expensive and most easily implemented option for an existing multi-user spectrograph. The gas cell can easily be regulated in temperature and pressure, giving a constant absorption spectrum. The stellar beam traverses the gas cell before entering the spectrograph, so any mechanical or thermal variability of the spectrograph will affect both equally. The system is immune to atmospheric transparency variations. The gas cell absorption spectrum provides a simultaneous monitor of the spectrograph instrumental profile. The optional use of an optical fiber scrambler will ensure that the spectrograph sees the same pupil for every observation. An exposure meter will ensure optimal exposure levels, and will automatically measure the mid-exposure time very precisely.

The use of a gas absorption cell as a velocity metric is not a new idea. It has long been an established technique in the solar astronomy community [2, 3, 4]. It was then rediscovered by the stellar astronomers [5, 6, 7, 8, 9, 10] for application to planet hunting. The first planet-detection use of a gas absorption cell was the pioneering work of Campbell and Walker using the Canada-France-Hawaii Telescope. This program used the 3-0 band R-branch absorption lines of HF at 8670-8770Å as the velocity metric. The HF absorption cell was extremely difficult to use. HF is a very corrosive acid. The cell was made from monel, and the windows were sapphire. The HF gas was maintained in equilibrium with HF liquid. Thus, very precise temperature control was required in order to maintain constant gas pressure. Small pressure variations were one of the limiting factors since HF exhibits significant pressure shifts.

Use of a molecular iodine gas cell gets around the challenges presented by HF. I_2 provides an acceptable compromise of spectral complexity, throughput, and bandpass for visible light. There are other gas mixtures that offer promise for use in the near-IR. A gas cell is inexpensive and very easy to make. A typical modern gas absorption cell is shown in Figure 1. It is simple to retrofit just about any existing high-resolution spectrograph with an I_2 cell. The entire project can cost less than 1000 euros.

3 Details of the Gas Cell Technique

The basic methods for precise measurement of stellar radial velocities using I_2 cells were explained by [8, 9, 10]. The gas absorption spectrum serves as the fundamental velocity metric. After constructing a cell, we obtain a very high-resolution (R $\sim$ 500,000) high signal/noise spectrum of the cell with an FTS spectrometer. This spectrum must be obtained at the temperature at which the cell will be operated at the telescope.

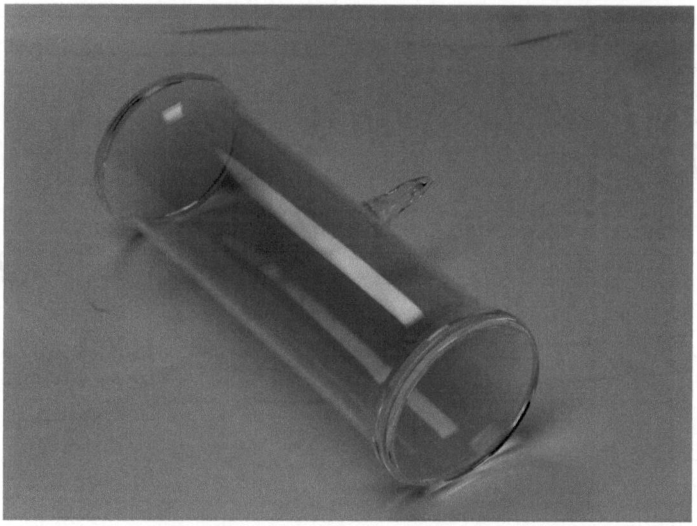

Fig. 1. A typical gas absorption cell after construction and before installation of the heater, temperature sensor, and insulation. This cell is about 15 cm long and 5 cm in diameter. The exterior of the windows are anti-reflection coated.

3.1 Modeling the Observed Spectrum

The observing procedure consists of taking a "template" spectrum of each target star. This is a high S/N spectrum taken without the I_2 cell. We must then deconvolve the spectrograph instrumental profile (IP) function from this template spectrum. To measure the IP, we must obtain a spectrum of the I_2 cell, but with the spectrograph optics illuminated in exactly the same manner as for stellar observations. Since most spectrograph flat-field lamps fail to meet this requirement, it is common to measure the IP by taking spectra of a hot, rapidly rotating star with and without the I_2 cell. Routine program observations of the target star are taken with the I_2 cell in place. The velocity computation then consists of computing the model spectrum, I_m, which best matches the observed spectrum:

$$I_m = k[T_{I_2}(\lambda)I_S(\lambda + \delta\lambda)] * IP \qquad (1)$$

where I_S is the intrinsic stellar spectrum, T_{I_2} is the transmission function of the I_2 cell, k is a normalization factor, $\delta\lambda$ is the wavelength (Doppler) shift, and IP is the spectrograph instrumental profile function.

The modeling process is shown in Figure 2. The observed spectrum (dots) of a star taken through the I_2 cell shows absorption lines from both the stellar photosphere and from the gas cell. The "pure" stellar template spectrum (taken without the I_2 cell) is shown in the dashed line, and the high resolution FTS I_2 spectrum is shown in the dotted line. The model spectrum

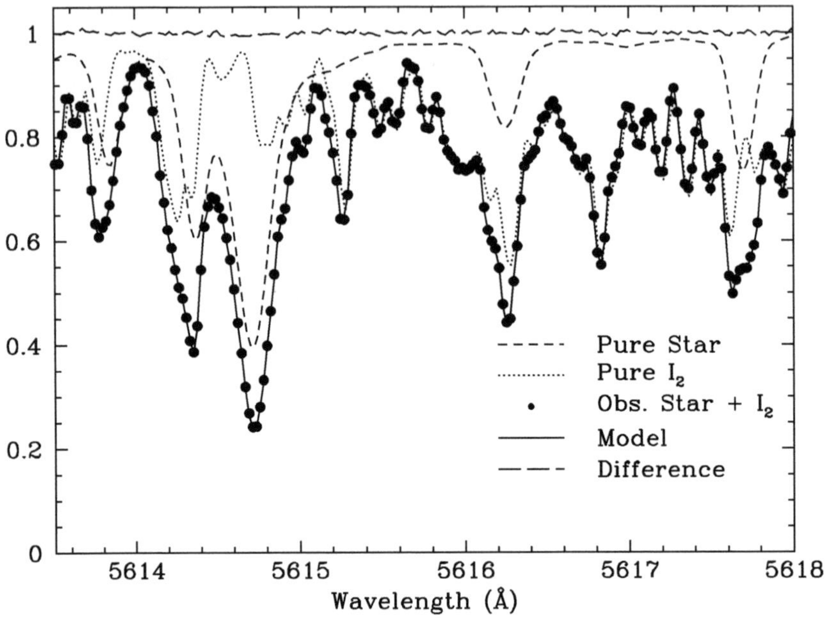

Fig. 2. Modeling the observed spectrum. The pure star spectrum is Doppler shifted by the proper amount and multiplied by the pure I_2 spectrum. This product is then convolved with the instrumental profile to give the model spectrum. The difference between the observed and model spectra is shown on the top.

computed according to equation 1 is shown in the solid line. The difference of the observed and model spectra is shown as the long-dashed line at the top.

Of course, there are a lot of subtleties in actually computing the model. The real key is proper modeling of the IP, and there are a variety of ways to do this. It is common to parameterize the IP as a sum of several Gaussian components, but other functions may be used as well. One may convolve this with a slit function, if that is appropriate for the observing situation. Alternatively, one can allow the amplitude of the IP at each point to be a free parameter. Figure 3 illustrates the IP parameterization process for one typical echelle order from the McDonald Observatory 2.7m cs23. The spectrum has been divided into 20 different chunks. Each chunk has been parameterized by a superposition of five Gaussian functions, and the sum is convolved with a top-hat slit function, giving the resulting curve. What works for one instrument may be completely inappropriate for a different spectrograph. The IP will often vary over short regions on the detector, so one must divide each echelle order of the spectrum into a number of independent "chunks" and then treat each chunk separately.

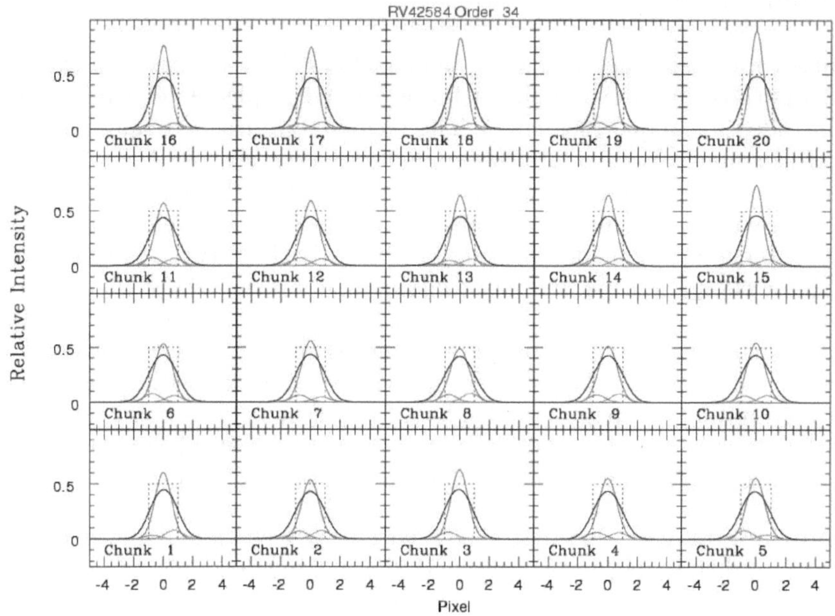

Fig. 3. An example of the instrumental profile determination for one order of a McDonald Observatory 2.7m cs23 echelle spectrum. The gray curves are the individual Gaussian functions, and the dotted line is the slit function. The heavy curve is the final IP.

3.2 Velocity Precision

The intrinsic radial velocity content of a spectrum depends on the mean absolute value of the slope of the stellar spectrum. As has been shown by [10, 11],

$$\frac{\delta v}{c} = \frac{\delta I}{\lambda \partial I / \partial \lambda} \tag{2}$$

Thus, the precision with which you can measure δv depends on the precision with which you can measure the spectrum I and the sharpness of the spectral lines $\partial I/\partial \lambda$. Thus, one wants to use a spectral resolution sufficient to resolve the stellar photospheric lines, but higher resolving power does not help any (and may actually hurt by decreasing the S/N ratio). If one puts in the appropriate values for a bandpass of about 1000Å of spectrum of a G or K dwarf, a S/N of 250 will yield about $1\,\mathrm{m\,s^{-1}}$ precision. In addition to photon noise, other sources of noise include intrinsic stellar noise (which is often called "jitter"), intrinsic spectrograph limits, and algorithmic imperfections.

4 Conclusions

Gas absorption cells offer a means to achieve 1 m s^{-1} RV precision on existing general-purpose multi-user high resolution spectrographs. A cell can be built and installed for about 1000 euros. The technique can be extended into the near IR for application to late M dwarfs and brown dwarfs. A new generation of extremely large telescopes is being planned. Although time on these telescopes will be very precious, some amount of observing will certainly be devoted to extrasolar planet searches. To use a HARPS-type instrument on a 20 meter class telescope, one must either scale up the size of the spectrograph optics in proportion to the mirror diameter, or feed the spectrograph with a much smaller image diameter using adaptive optics. However, it is almost certain that new large telescopes will be equipped with general purpose, high-resolution spectrographs, operating either in the visible or in the near-IR. The use of a gas absorption cell offers the most affordable method to achieve high RV precision on the largest aperture telescopes.

Acknowledgments: This material is based upon work supported by the United States National Aeronautics and Space Administration under Grants NNG04G141G and NNG05G107G issued through the Terrestrial Planet Finder Foundation Science program.

References

1. Griffin, R., and Griffin, R., *MNRAS*, **162**, 243 (1973).
2. Beckers, J. M., *Nature*, **260**, 227 (1976).
3. Koch, A., and Wöhl, H., *A&A*, **134**, 134 (1984).
4. Libbrecht, K. G., "A Search for Radial Velocity Oscillations in Procyon," in *I.A.U. Symposium 132: The Impact of Very High S/N Spectroscopy on Stellar Physics*, edited by G. Cayrel de Strobel and M. Spite, Kluwer, Dordrecht, 1988, p. 83.
5. Campbell, B., and Walker, G. A. H., *PASP*, **91**, 540 (1979).
6. Campbell, B., Walker, G. A. H., Johnson, R., Lester, T., Yang, S., and Auman, J., *Proc. Soc. Photo-opt. Inst. Eng.*, **290**, 215 (1981).
7. Campbell, B., and Walker, G. A. H., "Stellar Radial Velocities of High Precision: Techniques and Results," in *I. A. U. Colloquium No. 88, Stellar Radial Velocities*, edited by A. G. D. Phillip and D. W. Latham, L. Davis Press, Schenectady, 1985, p. 5.
8. Marcy, G. W., and Butler, R. P., *PASP*, **104**, 270 (1992).
9. Valenti, J. A., Butler, R. P., and Marcy, G. W., *PASP*, **107**, 966 (1995).
10. Butler, R. P., Marcy, G. W., Williams, E., McCarthy, C., Dosanjh, P., and Vogt, S. S., *PASP*, **108**, 500 (1996).
11. Bouchy, F., Pepe, F., and Queloz, D., *A&A*, **374**, 733 (2001).

Pushing Down the Limits of the Radial Velocity Technique

C. Lovis[1], M. Mayor[1], F. Pepe[1], D. Queloz[1], and S. Udry[1]

Geneva Observatory, University of Geneva, 51 ch. des Maillettes, 1290 Sauverny, Switzerland, christophe.lovis@obs.unige.ch

Summary. We present results from the first three years of operations of the HARPS spectrograph installed on the ESO-3.6m telescope at La Silla Observatory, Chile. This instrument, primarily built to detect extrasolar planetary systems, has demonstrated a long-term accuracy below the 1 m s^{-1} level, exploring a new regime in RV precision and discovering several very low-mass planets. We present recent improvements in the wavelength calibration process, including the creation of a new ThAr reference atlas and the use of a much larger number of lines to fit the wavelength solution. Other instrumental error sources such as guiding accuracy and photon noise are discussed and a global error budget is presented. Finally, ongoing studies to obtain an "ideal" wavelength calibration are briefly reviewed.

1 On the RV Precision of HARPS

The present capabilities of HARPS [1] are best illustrated by the recent discoveries of planetary systems made with this instrument. Table 1 gives the list of all planets orbiting stars from the HARPS high-precision sample. Particularly noteworthy are the very low residuals around the orbital solutions, which are in several cases below 1 m s^{-1}.

Table 1. HARPS planets from the high-precision sample of stars.

Name	$m \sin i$ [$M_\oplus$]	P [days]	Year of Discovery	O-C [m s^{-1}]
μ Ara c	14.0	9.55	2004	0.9
HD 93083 b	118	144	2005	2.0
HD 101930 b	95.4	70.5	2005	1.8
HD 102117 b	44.5	20.7	2005	0.9
HD 4308 b	14.9	15.56	2005	1.3
HD 69830 b	10.2	8.67	2006	
HD 69830 c	11.8	31.6	2006	0.8
HD 69830 d	18.1	197	2006	
μ Ara d	166	311	2006	1.4

As a specific example we consider the discovery of three Neptune-mass planets around the star HD 69830 ([2], see Fig.1). These planets induce RV signals of only ~2 m s^{-1} on their parent star, which make them extremely difficult to detect. The dispersion of the residuals around the best-fit orbital solution amounts to 0.6 m s^{-1}, and goes down to 0.2 m s^{-1} when binning the data (one point per 10-20 days) to study long-term RV variations.

Such a level of precision, if reproducible on a sufficiently large sample of stars, would permit the detection of a 3 $M_\oplus$ planet orbiting at 1 AU. We are therefore very close to the discovery of terrestrial planets in the habitable zone of Sun-like stars.

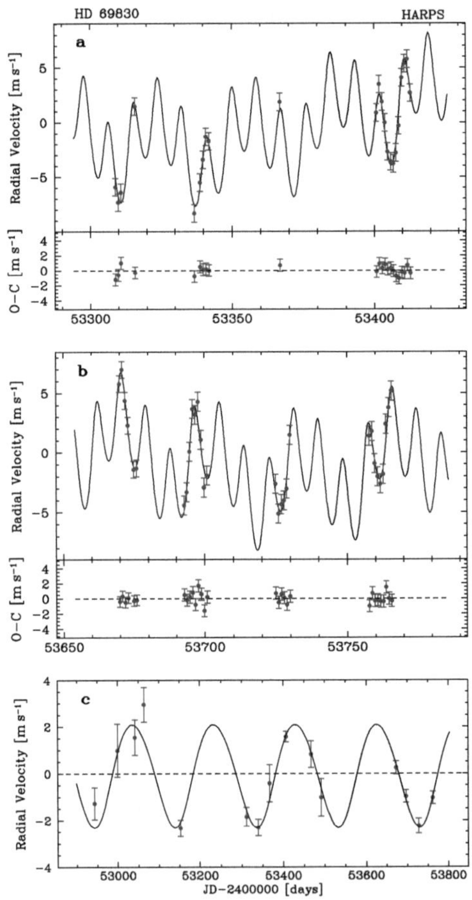

Fig. 1. Top and middle panels: radial velocity curve of the star HD 69830. The data reveal the presence of three low-mass planets. Lower panel: RV signal due to the third, longer-period planet only. Data points were binned to one per observing run (~10 days).

2 Error Budget

The main error sources in HARPS radial velocity measurements are briefly presented below, along with some strategies to minimize them.

- Wavelength calibration: several improvements are possible, such as the use of more precise laboratory wavelengths and more lines (see next section).
- Guiding accuracy: an accurate guiding software has been developed on the ESO-3.6m telescope, permitting to keep the guiding RV noise below $\sim$0.3 m s^{-1}.
- Stellar oscillation "noise": an exposure time between 10-15 minutes is necessary to keep the oscillation signal below $\sim$0.5 m s^{-1} on solar-type stars.
- Stellar activity: a careful selection of the target stars is necessary to minimize the influence of RV jitter. Line shape diagnostics are used to distinguish between planetary and activity-related RV signals.

3 A New ThAr Reference Atlas for Wavelength Calibration

The widely used reference for thorium laboratory wavelengths is the Atlas of the Thorium Spectrum published by [3]. This atlas contains about 3,000 usable lines within the HARPS spectral range and gives their wavelengths at a precision between 15-150 m s^{-1}, depending on line strength. This relatively low precision leads to a high dispersion (50-70 m s^{-1}) of the residuals around the wavelength solutions on HARPS spectra. This makes wavelength solutions somewhat unstable.

On the other hand, HARPS ThAr spectra reveal lots of unidentified lines and allow us to measure the position of a given line at a precision of 5-10 m s^{-1}, limited by CCD pixel inhomogeneities.

We therefore used a series of HARPS ThAr spectra to build a new ThAr atlas containing more lines and more precise wavelengths (Lovis et al. 2007, in prep.). The intrinsic stability of the instrument allowed us to combine many ThAr spectra to minimize photon noise on line position measurements. We then performed a systematic search for lines in the spectra. As most lines are blended to some level, we fit neighboring lines simultaneously with multiple Gaussians. This is necessary because unrecognized blends easily lead to systematic offsets exceeding 10 m s^{-1} on the line positions.

In the last step of the new atlas construction, we fit a global wavelength solution through all spectral orders and find the systematic offset of each line due to the uncertainties in the input wavelengths. We are then able to correct the wavelengths of all known lines and to assign a precise wavelength to all unidentified lines. At the end of the process we obtain a list of 8,600 lines between 3,800 and 6,900 Å, with an internal precision on individual wavelengths of 5-10 m s^{-1} for most lines. The global wavelength scale of the new atlas is the same as in [3].

The use of the new atlas in the HARPS data reduction software allowed us to reduce the wavelength calibration error to ~20-30 cm s^{-1}.

4 Future Developments

The next generation of high-precision velocimeters (see for example the CODEX project [4]) that will be installed on Extremely Large Telescopes (ELTs) will require even better calibrators to achieve accuracies on the wavelength zero point of ~1 cm s^{-1}. To meet this challenging requirement, new technical developments are needed to obtain an "ideal" calibrator, which should have the following properties:

- High density of lines (up to one per 2-3 resolution elements)
- All wavelengths precisely known, accurate and stable
- All line intensities about equal

Laser frequency combs linked to an atomic clock might represent a viable solution that fulfills all these requirements. Studies are ongoing to develop such a new, ultra-accurate calibration system.

References

1. F. Pepe, M. Mayor, D. Queloz et al: The ESO Messenger **120**, 22 (2005)
2. C. Lovis, M. Mayor, F. Pepe et al: Nature **441**, 305 (2006)
3. B.A. Palmer, R. Engleman: *Atlas of the Thorium Spectrum* (Los Alamos National Laboratory 1983)
4. L. Pasquini, S. Cristiani, H. Dekker et al: CODEX: measuring the acceleration of the universe and beyond. In: *Proceedings of the 232nd Symposium of the International Astronomical Union*, ed by P. Ann Whitelock, M. Dennefeld, B. Leibundgut (Cambridge University Press 2006) pp 193–197

Transiting Planets: Follow the FLAMES...

C. Melo[1], N.C. Santos[2,3,4], F. Pont[4], M. Mayor[4], S. Udry[4], D. Queloz[4], and F. Bouchy[5]

[1] European Southern Observatory, Chile
[2] Centro de Astronomia e Astrofísica da Universidade de Lisboa, Portugal
[3] Centro de Geofísica de Évora, Portugal
[4] Observatoire de Genève, Switzerland
[5] Laboratoire d'Astrophysique de Marseille, France

Summary. The identification of transiting planets candidates and their a posteriori confirmation and characterization by means of spectroscopic follow-up is allowing us to study the internal physics of these planets. Here we briefly discuss the spectroscopic follow-up of the OGLE transiting candidates carried out at the ESO/VLT.

1 Introduction

Although powerful as detection method, radial velocity alone can only give the planetary minimum mass $M_p \sin i$ telling nothing about the internal structure (size, density, temperature) of the planet itself. Therefore, the knowledge of the internal physics of these new planets remain mostly unknown.

In the rare configuration where the inclination of the planetary orbital plane is aligned with the line of sight, i.e., $i \sim 90°$, the planet transits across the stellar disk blocking a tiny fraction of the stellar light causing the star to dim for a given duration of time. As the planet moves out of the stellar disk, the stellar light increases again producing a U-shape light curve. Analysis of a planetary transiting light curve is a valuable piece of information. Combined with the radial velocity orbit, the true planetary mass and radius, and therefore, the mean density of the planet can be found. The planet radius and its mean density are directly related to the planet internal physics.

Using the 1.3-m Warsaw University Telescope at Las Campanas Observatory (Chile), the OGLE-III survey (Optical Gravitational Lensing Experiment) has realized an extensive photometric search for planetary and low-luminosity object transits. In two seasons, about 200,000 stars were monitored with photometric accuracy better than 1.5% and analyzed for periodic eclipse signals with depth from a few per cent down to slightly below one per cent. Altogether 177 stars with multi-transiting low-luminosity objects were detected and announced [1, 2, 3, 4].

The estimated radii of these objects range from 0.5 Jupiter radius to 0.5 solar radius and their orbital periods from 0.8 to 8 days. The smallest objects could be suspected to be extrasolar giant planets, but the radius estimated from the photometric signal is not sufficient to conclude on the planetary nature of the objects. They could as well be brown dwarfs or low-mass stars, since in the low mass regime (M < 0.1 $M_\odot$) the radius becomes practically independent of the mass. Some configurations of grazing binary eclipses and of eclipsing binaries in multiple systems can also mimic a planetary transit signal (see next section).

Here we briefly recall why a spectroscopic follow-up of the transiting candidates is needed in order to access the true nature of the transiting body. The main results concerning the properties of the transiting bodies obtained by our team are summarized and the main references are given. We also refer to some recent results on the physical of the transiting planets. Finally, we point the reader for the recent review by Charbonneau et al. on the potential of transiting planets in the study the internal structure of extrasolar planets [5].

2 Why a Spectroscopic Follow-up is Needed?

The spectroscopic follow-up of OGLE transit candidates presented a complete panorama of the configurations that can mimic the photometric signal of a planetary transit. These fall into four categories:

- *Grazing eclipsing binaries.* Two large stars, when eclipsing at a high angle, can produce shallow transit-like dips in the light curve. These cases are the easiest to discriminate. Several hints are usually present in the light curve itself, such as a V-shaped transit curve, or ellipsoidal modulations in the light curve due to tidal effects and reflected light. Nevertheless, at low signal-to-noise such systems can also be mistaken for good planetary transit candidates. They are easy to resolve with spectroscopic observations, thanks to the presence of two sets of lines in the spectra with large velocity variations.
- *M-dwarf transiting larger companions.* A small M-dwarf transiting a larger star can produce a photometric signal closely similar to a planetary transit. If the companion is not larger than a Hot Jupiter, and the orbital distance is too large for tidal and reflection effects to be detectable in the light curve, then the photometric signal is strictly identical to that of a planetary transit. In both cases, an opaque, Jupiter-size object transits the target star. These cases can only be resolved by Doppler observations, the amplitude of the reflex motion of the star revealing the mass of the transiting companion. Two nice example of planet-size transiting stellar companions were found among the OGLE candidates, OGLE-TR-122 and OGLE-TR-123 [6, 7].
- *Multiple systems.* An eclipsing binary can produce shallow transit-like signals if the deep eclipse is diluted by the light of a third star. There are many possible configurations for such systems, and as a result they can be very difficult to disentangle, even with Doppler informations. In

some cases, the conjuration of the parameters can be so good as to mimic not only the light curve of a planetary transit, but also induce planet-like variations of the inferred radial velocity, produced by the blending of several sets of lines in the spectra. OGLE-TR-33 [9] is such a case.

- *False detections.* Stellar variability, and systematic trends in the photometry, can produce fluctuations in the light curve interpreted as a possible transit signal, especially as one tries to detect shallower signals near the detection threshold. Several cases of low signal-to-noise detections showed no significant radial velocity variations, making them suspected false positives of the photometric transit detection procedure. Further photometric observations at the epoch of the detected transit are needed in these cases to distinguish bona fide transits from false positives.

3 Main Results

Since 2002, our team has been carrying out the spectroscopic follow-up of the transiting candidates detected by OGLE in the Galactic Bulge [10] and in the Carina arm [11]. The observations were carried out using the multi-fiber facility FLAMES [12] attached to Kueyen (VLT/UT2) telescope at Paranal Observatory. Eight dedicated fibers of FLAMES feed the high resolution spectrograph UVES installed at the Nasmyth-B platform allowing the observations of 8 simultaneous objects or 7 objects along a simultaneous calibration fiber fed with the spectrum of ThAr lamp. The resolution of the FLAMES/UVES achieved mode is about $R \sim 47,000$. The typical precision reached is around 30-50 m/s for a 45 min exposure of a $I = 16$ non-rotation G-K dwarf star.

The follow-up of the 60 most promising OGLE candidates led to the characterization of: 24 low-mass-star transiting companions; 5 grazing eclipsing binaries; 12 low-mass-star transiting companions in triple or quadruple systems; 7 false positives of the transit detection; 5 exoplanets.

Seven cases were left unsolved, because of the early type or faintness of the primary (none of them promising planet transit candidates).

Concerning the planets, two planets were found in the region of the Galactic bulge, OGLE-TR-56b [13] and OGLE-TR-10b (see the latest news on its somewhat controversial radius by [14]), whereas in the Carina field three other planets were confirmed, namely, OGLE-TR-111 [18], OGLE-TR-113 and OGLE-TR-132 [19]. Most of the candidates turned out to the low-mass spectroscopic binaries. Of particular interest tough was the discovery of OGLE-122 [6] and OGLE-123 [22]. The secondaries of both systems are stars close the hydrogen burning limit of $0.08 M_{\odot}$ with radii similar to the Jupiter. These systems are interesting case studies exemplifying the need of spectroscopic follow-up in order to confirm the planetary nature of the transiting body.

4 Beyond the Follow-up

The derived radii and masses of the transiting planets depend not only on the quality of the light-curve and radial velocity data but also on the stellar parameters of the primary star. Therefore much effort is put in an important second step which consists in acquiring new photometric and high-resolution, high S/N spectroscopic observations of the transiting planets in order to refine the derived masses and radii [21, 20]. As a matter of fact, in a recent paper on OGLE-TR-10b by Pont et al., these authors put in evidence the relevance of the a posteriori improvement of the derived quantities on the risk of misinterpreting the results and the internal physics of the giant planets [14]. These additional observations are also carried out at the VLT mainly with FORS1 (spectro-imager) and UVES/slit-mode both attached to Kueyen (VLT/UT2) telescope.

Thanks to the growing number of transiting planets, a number of complementary studies can now be carried out focusing in the physical aspects of these systems and their host stars rather than on their pure detection. For instance, Mazeh and colleagues [15] draw attention to a possible relation between the orbital period and the mass of the transiting planets. They suggest that such relation is set by the fact the lightest planets in very close-in orbits disappear in time due to strong evaporation. On the other hand, Melo et al. [16] show that the host stars of transiting planets are not younger than 1 Gyr indicating that the transiting planets are not experiencing runaway evaporation.

Guillot et al. [17], using stellar parameters derived by Santos et al. [20], show that the amount of heavy elements necessary to be added to the planet composition in order to explain the observed planetary radii is correlated to the Iron abundance of the parent star. This remarkable result suggests that the wide range of radii observed among the transiting planets can be explained in a single framework. Moreover, this result is in-line with the higher frequency of planets found around metal-rich stars.

References

1. Udalski, A., Paczynski, B., Zebrun, K. et al.: Acta Astronomica **52**, 1 (2002a)
2. Udalski, A., Zebrun, K. et al.: Acta Astronomica **52**, 115 (2002b)
3. Udalski, A., Zebrun, K. et al.: Acta Astronomica **52**, 317 (2002c)
4. Udalski, A., Pietrzynski, G. et al.: Acta Astronomica **53**, 133 (2003) 313 (2003)
5. Charbonneau, D., Brown, T., Burrows, A., Laughlin, G.,: When Extrasolar Planets Transit Their Parent Stars. To appear in: *Protostars and Planets V*, Edited by B. Reipurth, D. Jewitt, and K. Keil, University of Arizona Press, Tucson, 2006
6. Pont, F., Melo, C., Bouchy et al. : A&A **433**, L21 (2005a)
7. Pont, F., Moutou, C., Bouchy, F., et al. : A&A **447**, 1035 (2006a)
8. Pont, F., Zucker, S., Queloz, D. : MNRAS **373**, 231 (2006b)

9. Konacki, M., Torres, G., Sasselov, D. et al.: ApJ **609**, L37 (2004)
10. Bouchy, F., Pont, F., Melo, C. et al.: A&A **431**, 1105 (2005)
11. Pont, F., Bouchy, F., Melo, C. et al.: A&A **438**, 1123 (2005b)
12. Pasquini, L., Avila, G., Blecha, A., et al. : The Messenger **110**, 1 (2002)
13. Konacki, M., Torres, G., Jha, S., Sasselov, D. D. : Nature **421**, 507 (2003)
14. Pont, F, Moutou, C, Gillon, M et al. : A&A, accepted, astro-ph/0610827
15. Mazeh, T., Zucker, S., Pont, F. : MNRAS **356**, 955 (2005)
16. Melo, C., Santos, N.C., Pont, F. et al. : A&A **460**, 251 (2006)
17. Guillot, T., Santos, N. C., Pont, F., et al. : A&A **453**, L21 (2006)
18. Pont, F., Bouchy, F., Queloz, D., et al. : A&A **426**, L15 (2004)
19. Bouchy, F., Pont, F., Santos, N. C., et al. : A&A **421**, L13 (2004)
20. Santos, N. C., Pont, F., Melo, C. et al. : A&A **450**, 825 (2006)
21. Moutou, C., Pont, F., Bouchy, F., Mayor, M. : A&A **424**, L31 (2004)
22. Pont, F., Moutou, C., Bouchy, F. et al. : A&A **447**, 1035 (2006)

Planet Detection Around M Dwarfs: New Constraints on Planet Formation Models

T. Forveille[1], X. Bonfils[2], X. Delfosse[1], J.-L. Beuzit[1], C. Perrier[1], D. Ségransan[3], S. Udry[3], M. Mayor[3], F. Pepe[3], D. Queloz[3], F. Bouchy[4], and J.-L. Bertaux[5]

[1] Laboratoire d'Astrophysique, Observatoire de Grenoble
Thierry.Forveille@obs.ujf-grenoble.fr
[2] Centro de Astronomia e Astrofísica da Universidade de Lisboa
[3] Observatoire de Genève
[4] Institut d'Astrophyisque de Paris
[5] Service d'Aéronomie du CNRS

Summary. The dependence of planetary statistics on the physical conditions during the initial stages of the system (i.e. in the proto-planetary disk) represent an essential constraint on planetary formation mechanisms. The planets orbiting M dwarfs were formed in a lower-mass disk, with an ice boundary much closer to the star, and a longer orbital period (for the same separation) than the planets formed around solar-type stars. Planetary formation is extremely sensitive to all three parameters, and the relative statistical properties of planets around G dwarfs and very low mass stars therefore represent a very sensitive diagnostic. In this talk we will present the first results of a very sensitive search for planets around southern M dwarfs with the HARPS spectrograph (as part of the HARPS GTO program). The northern counterpart of the survey will be conducted with SOPHIE at OHP, within the SOPHIE Exoplanets Consortium. These first results suggest that that planets around M dwarfs have lower mass than around solar type stars.

1 Introduction

The compared statistical properties of exoplanets around host stars with different masses provide quantitative constraints on models of planet formation. The physical conditions in the proto-planetary disk during the initial stages of the planetary system depend strongly on the central star. Changing the central star directly affects:

- the central gravity, and thus the orbital period at a given separation, which influences the accretion rate of the proto-planet,
- the temperature in the disk, and therefore the sound speed, which affects planetary migration, and the position of the ice boundary,
- most likely the mass of the disk, and thus the quantity of material available for planet formation,
- ...

Each step of the planetary formation process (core growth, accretion rate, migration, etc...) therefore depends directly or indirectly on the central star, and the relative statistical properties of planets around G dwarfs and very low mass stars represent a very powerful diagnostic.

The two leading candidate mechanisms for planet formation, core accretion and disk instabilities, have different sensitivities to this control parameter, and observations could thus discriminate them. Core accretion models (initial formation of solid core, followed by gas accretion) predict than Jupiter mass planets are rare around M dwarfs (Laughlin et al. 2004; Ida & Lin 2005) but that Neptune mass planets at small orbital separations could be numerous (Ida & Lin 2005; Kornet et al. 2006). By contrast, disk instability models seem unable to form Neptune mass planets (except in very particular environments; Boss 2006a) but readily form Jupiter mass (and above) planets (Boss 2006b)

2 Planets Around M Dwarfs

Radial velocity surveys have to date unveiled ~170 planetary systems. They were initially searched for, and thus found, around solar-type stars (e.g. Mayor & Queloz 1995). As consequence of this bias, the vast majority were detected around late-F, G and K dwarfs stars.

To date 6 planets around M dwarfs have been detected by radial velocity surveys: 1.9 M_{Jup}, 0.6 $M_{Jup}$and 0.42 M_{Nep}around Gl 876 (Delfosse et al., 1998; Marcy et al., 1998, 2001; Rivera et al., 2005)); 0.8 M_{Jup}around Gl 849 (Butler et al, 2006); 1.2 M_{Nep}around Gl 436 (Butler et al. (2004)) and 0.97 M_{Nep}around Gl 581 (Bonfils et al. (2005).

Figures 1 and 2 respectively plot the planet mass versus separation and planet mass vs mass of the host stars for all extrasolar planets detected by radial velocity. A strong observational bias must be taken into account, since the parent samples only contain 200 or 300 M dwarfs, compared to a few thousands of solar type stars. With due account for this, these first detections suggest that planets around very low mass stars have rather distinct characteristics:

- A large fraction (~30%) of all currently known Neptune mass planets orbit M dwarfs (Figure 1),
- None of the large population of hot-Jupiter planets (with separation around 0.04 a.u.) orbits an M-dwarf,
- planets with mass from 7 earth mass (Gl876d; Rivera et al. (2005)) to 2 Jupiter mass (Gl876b; Delfosse et al. (1998)) are formed around M-dwarfs.

Monte Carlo simulations of the observational biases confirm that the hot Jupiter deficit is intrinsic (Bonfils et al. (2006))

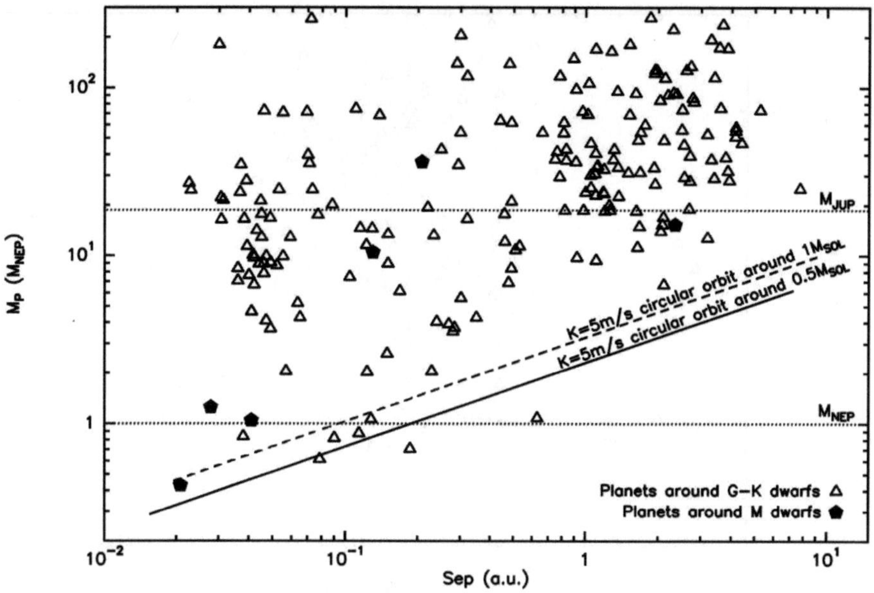

Fig. 1. Planet mass versus orbital separation. Dark pentagons represent planets around M dwarfs while open triangles represent planets around solar type stars. The two lines represent a a radial velocity semi-amplitude $K_1 = 5m.s^{-1}$ for respectively 1 and 0.5 M⊙central stars.

3 Our HARPS and SOPHIE Programs

To pinpoint the statistical characteristics of planets around M-dwarfs we monitor over 200 very low mass stars with an accuracy of a few meters per seconds. Since 2003 (as part of the HARPS GTO program) we measure 100 M-dwarfs with HARPS, the new ESO high-resolution (R = 115 000) fiber-fed echelle spectrograph. HARPS has proved by far the most precise spectro-velocimeter to date, reaching an instrumental RV accuracy significantly better than 1 m s^{-1} (Mayor et al. 2003). The M dwarfs are typically too faint to reach the stability limit of HARPS, but the median rms dispersion (left part of figure 3) which we obtain for M dwarfs remains excellent: ∼2 m/s. Because K (semi-amplitude of radial velocity variation) varies with stellar mass as $M_*^{2/3}$, a 3 m/s accuracy for a 0.3$M_\odot$ M dwarf can detect the same planet than a 1.5 m/s accuracy for a solar type star. Our accuracy is this sufficient to detect planets of a few earth masses on short period orbits around M dwarfs.

The 0.97 M_{Nep} planet around Gl 581 (Bonfils et al. (2005), cited above) is the first planet discovered during this program and illustrates this capability. The right part of Fig. 3 shows our phased radial velocities, and shows

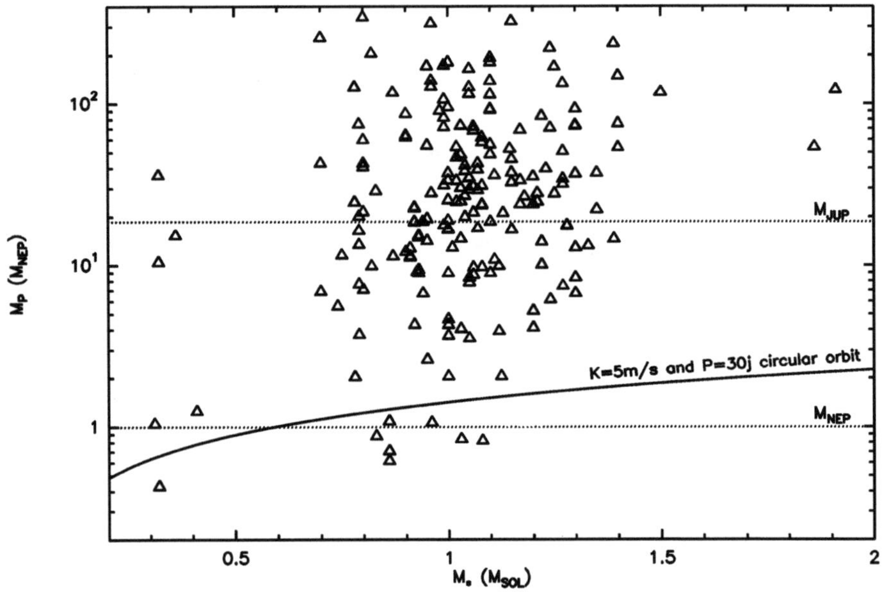

Fig. 2. Planet mass vs mass of the host star for all currently known planets. A large fraction of all M<0.01M_{Jup}planets know to date orbit M dwarfs, while a vast majority of the Jupiter-mass planets orbits solar type stars.

that radial velocity orbits with a semi-amplitude half of that of Gl 581 would be fairly easily detected. Such detections however need good temporal coverage, and the HARPS observations are just now starting to deliver their first candidates.

At the end of 2006 we will start a new program with SOPHIE at 1.93-m OHP telescope (Loeillet et al. 2006). 150 additional M-dwarfs will be monitored with an accuracy better than 3-4 m/s. With ~250 M-dwarfs monitored with such radial velocity accuracy we will obtain good statistics on the frequency of Jupiter-mass planets around very low mass stars for all periods under a few years, and of super-earth planets with periods under ~ 50 days.

4 Habitability and Transits

Discovering habitable planets is another important motivation in searching for planets around M-dwarfs: the habitable zone is much closer to the star around an M-dwarf, and a planet in that zone therefore produces a larger velocity variation than around a solar type star. An 8 earth-mass planet in the habitable zone of a 1 M⊙ stars (1 a.u.), for example, produces a radial velocity semi-amplitude of 0.7 m/s, while the same planet habitable around

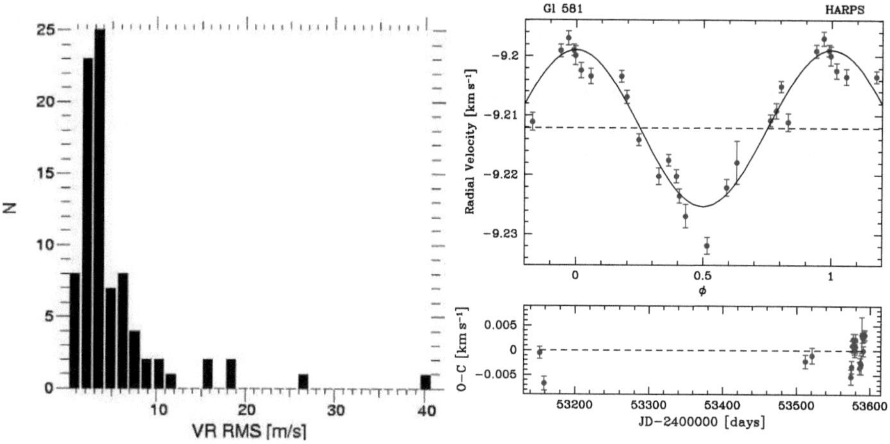

Fig. 3. Left: rms dispersion of our HARPS radial velocity measurements. Right: *Upper panel:* Phased radial velocities for Gl 581; *Lower panel:* Residuals of the fitted solution versus time.

an M2 dwarf ($0.4 M_\odot$) will be at 0.1 a.u. and cause a 3.5 m/s semi-amplitude radial velocity variation. This amplitude is detectable today with HARPS.

Finally, the radius ratio between the planet and its star makes transits much deeper across M dwarfs. A Neptune mass planet eclipsing an M3 dwarf produce a deeper transit than a Jupiter mass planet eclipsing a solar-type star. Ground telescope can detect a Neptune radius planets eclipsing an M-dwarf fairly easily, when space-based observation are needed to detect the same event around solar type stars. The hot Neptune discovered by radial velocity around very low mass stars are clearly strong candidates for photometric monitoring.

References

1. Bonfils X., Forveille T., Delfosse X., Udry S., Mayor M., Perrier C., Bouchy F., Pepe F., Queloz D., Bertaux J.-L., 2005, A&A 443, L15
2. Bonfils X., Delfosse X., Udry S., Forveille T., Naef D., 2006, in "Tenth Anniversary of 51 Peg-b: Status of and prospects for hot Jupiter studies", L. Arnold, F. Bouchy and C. Moutou. (eds) p111.
3. Boss A.P., 2006a ApJ 644, L79
4. Boss A.P., 2006b ApJ 643, 501
5. Butler R.P., Vogt S.S., Marcy G.W., Fischer D.A., Wright J.T., Henry G.W., Laughlin G., Lissauer J.J., 2004, ApJ 617, 580
6. Butler R.P., Johnson J.A., Marcy G.W., Wright J.T., Vogt S.S., Fischer D.A., 2006, PASP 118, 1685
7. Delfosse X., Forveille T., Mayor M., Perrier C., Naef D., Queloz D., 1998a, A&A 338, L67.

8. Ida S., Lin D.N.C., 2005, ApJ 626, 1045
9. Kornet K., Wolf S., Rozyczka M., 2006, A&A 458, 661
10. Laughlin G., Bodenheimer P., Adams F.C., 2004, ApJ 612, L73
11. Loeillet B., et al. 2006, SF2A proceedings.
12. Marcy G.W., Butler R.P., Vogt S.S., Fischer D., Lissauer J.J., 1998, ApJ 505, L147
13. Marcy G.W., Butler R.P., Fischer D., Vogt S.S., Lissauer J.J., Rivera E.J., 2001 ApJ 556, 296.
14. Mayor M., Queloz D., 1995, Nature 378, 355
15. Mayor, M., et al. 2003, The Messenger, 114, 20
16. Rivera E.J., Lissauer J.J., Butler R.P., Marcy G.W., Vogt S.S., Fischer D.A., Brown T.M., Laughlin G., Henry G.W., 2005, ApJ 634, 625

Planets Around Giant Stars

A.P. Hatzes[1], M. Döllinger[2], L. Pasquini[2], J. Setiawan[3], L. Girardi[4], and L. da Silva[5]

[1] Thüringer Landessternwarte Tautenburg artie@tls-tautenburg.de
[2] European Southern Observatory, Karl-Schwarzschild-Strasse 2, 85748, Garching bei München, Germany
[3] Max-Planck-Institut für Astronomie, Konigstuhl 17, 69117, Heidelberg, Germany
[4] INAF-Trieste, Italy
[5] Observatório Nacional, Rio de Janeiro, Brasil

1 Introduction

Long-period radial velocity (RV) variations were discovered over 13 years ago in the K giant stars β Gem, α Tau, and α Boo by [1]. One hypothesis for these variations was that they were caused by substellar companions with masses of 2–11 $M_{Jupiter}$ and orbital semi-major axes of $\approx$ 2 AU. At the time, however, it could not be excluded with any certainty that these RV variations were due to rotational modulation since the expected rotation periods for these stars were comparable to the RV periods.

Since then it has been established that K giant stars can indeed host extrasolar planets. Frink et al. [2] discovered an extrasolar planet around the K giant star HD 137759 (ι Dra). The planet hypothesis for this star was largely accepted due to the high eccentricity of the orbit. Stellar companions to giant stars have also been discovered around the giant stars HD 47536 [3], HD 104985 [4], HD 11977 [5], and HD 13189 [6].

More recently [7] confirmed that the initial RV variations found by HC93 in β Gem were in fact due to a planetary companion. Using RV data from 4 different telescopes they demonstrated that the 590-day RV period in this star was constant in phase and amplitude over 26 years. Furthermore, there were no variations in the Ca II K emission, spectral line shapes, or Hipparcos photometry with the RV period. A planetary companion with a mass of 2.3 $M_{Jupiter}$ was the only logical explanation. The mass of β Gem is estimated to be 1.7 $M_\odot$

Although over 200 extrasolar planets have been discovered orbiting other stars very few have been found around stars more massive than our sun. RV surveys have avoided early-type, more massive stars because of the poor measurement precision. This is illustrated in the top panel in Figure 1 which shows a spectral region of a main sequence A7 star. There is only one spectral line in this region that is broadened by a high rotational velocity. Consequently the RV measurement precision for such a star would be poor.

Giant stars, on the other hand, are ideally suited for precise stellar radial velocity measurements. The lower panel of Figure 1 shows the same spectral

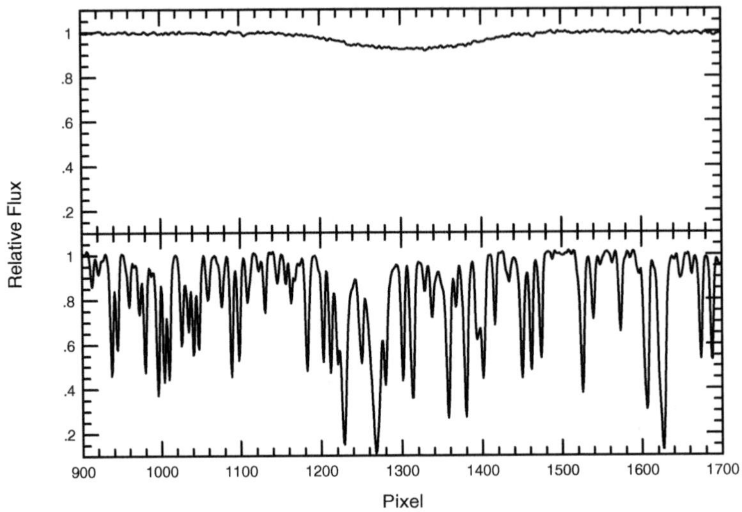

Fig. 1. (top) Spectral region for an A7 main sequence star. (bottom) Same spectral region for a K giant star whose progenitor was most likely an A7 main sequence star.

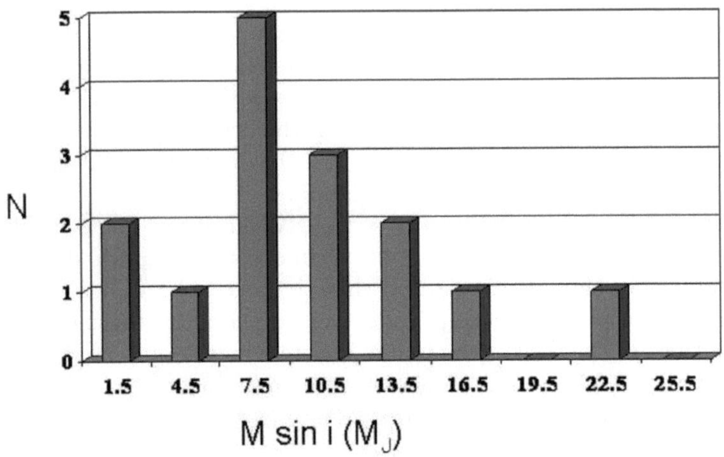

Fig. 2. The mass distribution of sub-stellar companions to giant stars.

region of a K giant star whose progenitor was most likely an A7 main sequence star. There is a plethora of stellar lines which are narrow due to the slower rotational velocity of this star. K giant stars thus offer astronomers a way to study extrasolar planets around more massive stars.

2 Properties of Planets Around Giant Stars

To date over a dozen giant planets have been found around G-K giant stars. This number includes all published planets around giant stars as well as unpublished candidates from the Tautenburg program (see [8] for a description of the program). Although this number is small, we can begin to look at the orbital properties, mass distribution and metallicities. The stellar masses for the host giant stars range from 1.05–3.5 $M_\odot$ and are thus for the most part intermediate mass stars. This stellar mass distribution should be compared to main sequence exoplanet host stars where over 90% of the stars have stellar masses less than about 1.2 $M_\odot$.

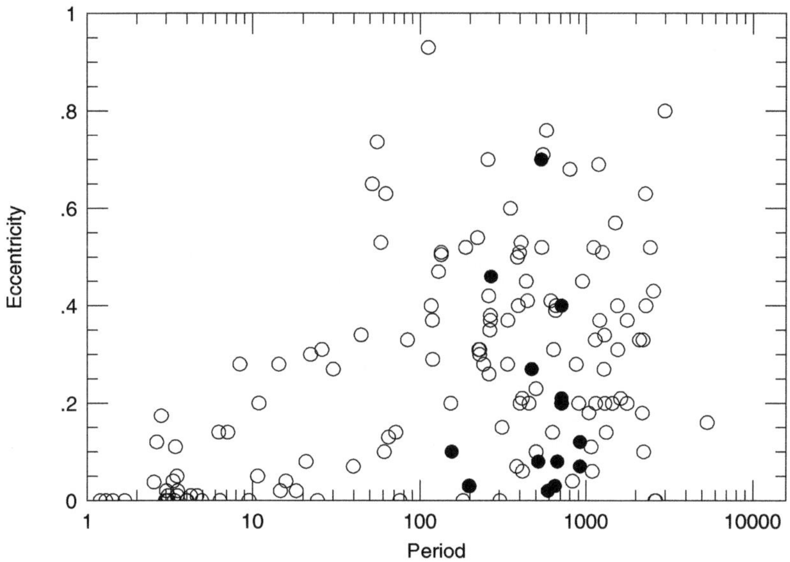

Fig. 3. The eccentricity versus period relationship for planets around main sequence stars (open circles), and planets around giant stars (filled circles).

Figure 2 shows the histogram of the mass distribution for planets around giant stars. Over 80% of the extrasolar planets around giant stars have masses greater than about 5 $M_{Jupiter}$. This is in contrast to main sequence stars where almost 90% of the planets have masses less than 5 $M_{Jupiter}$. Preliminary indications are that intermediate mass stars tend to have more massive planets. We caution the reader that this may be an observational bias. Short period planets around giants most likely do not exist due to the large radius of the star, and it is easier to detect low mass companions in short period orbits. Also, giant stars exhibit stellar oscillations with periods of days and amplitudes of up to 100 m s^{-1} (e.g. [9] [10]). These can hinder one's ability to detect low mass companions around giant stars.

Figure 3 shows the orbital eccentricity versus period for planets around main sequence stars (open circles) and for giant stars (solid circles). Over the period range for which planets around giant stars have been discovered the distribution of eccentricities seems to be similar to those found among main sequence stars. Preliminary indications are that planets around intermediate mass stars show a similar distribution of eccentricities.

Abundance studies of planet hosting main sequence stars indicate that these tend to have higher metallicities than stars that do not possess giant planets [11]. Figure 4 shows the distribution of metallicities for the planet hosting giant stars. In contrast to main sequence solar mass stars, relatively metal poor (and more massive) K giant stars can host giant extrasolar planets. The search for planets around giant stars may hold important clues to understanding the dependence of both stellar mass and metallicity on the process of planet formation.

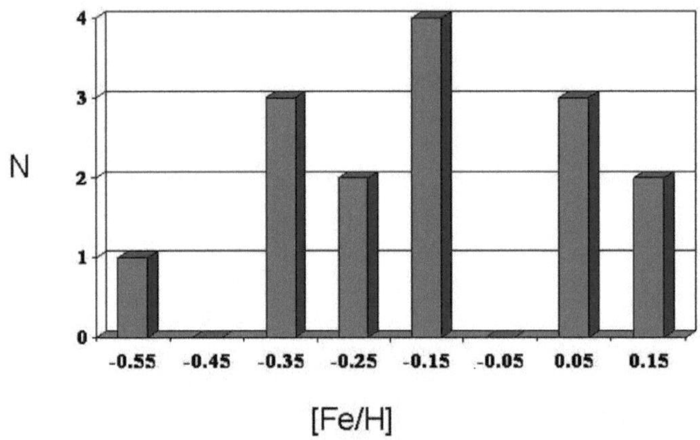

Fig. 4. The distribution of metallicities for host giant stars of extrasolar planets.

References

1. Hatzes, A.P. & Cochran, W.D. 1993, ApJ, 413, 339 (HC93)
2. Frink, S., and 5 co-authors 2002, ApJ 576, 478
3. Setiawan, J., and 8 co-authors, 2003, A&A 398, L19
4. Sato, B., and 18 co-authors, 2003, ApJ 597, L157
5. Setiawan, J., and 8 co-authors, 2005, A&A, 437, L31.
6. Hatzes, A. P., and 5 co-authors 2005, A&A, 437, 743.
7. Hatzes, A. P., and 10 co-authors 2006, A&A, 457, 335.
8. Döllinger, M. and 6 co-authors, A&A, submitted
9. Hatzes, A.P. & Cochran, W.D. 1994, ApJ, 422, 366
10. Setiawan, J., and 7 co-authors, 2004, A&A 421, 241
11. Santos, N.C., Israelian, G., Mayor, M. 2004, A&A, 415, 1153

Planets Around Active Stars

J. Setiawan[1], P. Weise[1], Th. Henning[1], A.P. Hatzes[2], L. Pasquini[3], L. da Silva[4], L. Girardi[5], O. von der Lühe[6], M.P. Döllinger[3], A. Weiss[7], and K. Biazzo[3,8]

[1] Max-Planck-Institut für Astronomie, Heidelberg, Germany setiawan@mpia.de
[2] Thüringer Landessternwarte, Tautenburg, Germany
[3] European Southern Observatory, Garching bei München, Germany
[4] Observatório Nacional, Rio da Janeiro, Brazil
[5] Osservatorio Astronomico di Padova-INAF, Padova, Italy
[6] Kiepenheuer-Institut für Sonnenphysik, Freiburg (Brsg), Germany
[7] Max-Planck-Institut für Astrophysik, Garching bei München, Germany
[8] Osservatorio Astrofisico di Catania-INAF, Catania, Italy

Summary. We present the results of radial velocity measurements of two samples of active stars. The first sample contains field G and K giants across the Red Giant Branch, whereas the second sample consists of nearby young stars ($d < 150$ pc) with ages between 10 and 300 Myrs. The radial velocity monitoring program has been carried out with FEROS at 1.52 m ESO telescope (1999 – 2002) and continued since 2003 at 2.2 m MPG/ESO telescope. We observed stellar radial velocity variations which originate either from the stellar activity or the presence of stellar/substellar companions. By means of a bisector technique we are able to distinguish the sources of the radial velocity variation. Among them we found few candidates of planetary companions, both of young stars and G-K giants sample.

1 Introduction

Precise radial velocity (RV) technique by using high-resolution spectrographs has been very successful in detecting extrasolar planets around inactive solar-like stars. Recently, this method has also some success in planet detections around several active stars, e.g., in G and K giants [2],[4],[5],[8],[10]. However, the number of planets discovered around such active stars is still very low, compared to those of solar-like stars. This is due to the fact that either the stellar activity, like fast stellar rotation, prohibits accurate RV determination or the measured RV variation is indeed not caused by the presence of low-mass companions, instead of by stellar pulsations or starspots.

Therefore, detailed investigations of spectral line profiles by the bisector and other activity indicators are mandatory in order to avoid false detections. Beside G-K giants, young stars also belong to samples which are still not well explored with the precise RV technique because of their high activity level. However, for less active (slow rotating F, G, K) young stars RV method is still applicable and will be able to find close orbiting giant planets which cannot be detected by the current direct imaging methods. The detection of young planetary systems will certainly improve our understanding of the planet formation.

In this paper we present results of our RV survey, which include the stability performance of FEROS, detection of stellar activity (non-radial pulsations and rotational modulation) and detections of a planet around a young star HD 70573 ($t \approx$100 Myrs) and a second planet candidate of an old metal poor K giant star HD 47536 ($t \approx$9 Gyrs, [Fe/H]=-0.68).

2 Observations

Since 1999 we have been carrying out RV monitoring program of G-K giants with FEROS at 1.52 m ESO in La Silla observatory. The results were presented in [7] and [8]. After FEROS was moved to 2.2 m MPG/ESO telescope follow-up observations of several G-K giant targets with long-period RV variation have been done. In 2003 we added nearby young stars (t = 10–300 Myrs) to our FEROS sample to be observed within the MPG time. To our surprise, about 30% of young stars in the sample show low-level stellar activity and slow stellar rotation, making them suitable for accurate RV measurements. The accuracy of FEROS at 2.2 m MPG/ESO from Dec 2003 until Oct 2006 is 10 m/s (Fig. 1), which is a significant improvement than at 1.52 m ESO.

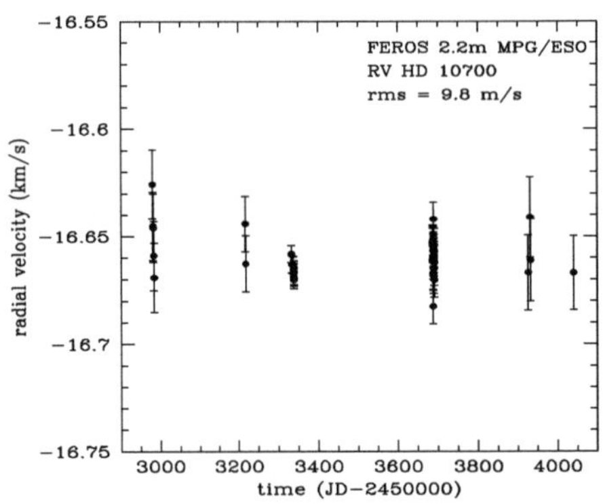

Fig. 1. RV measurements of a standard star (HD 10700) from Dec 2003 - Oct 2006 with FEROS at 2.2 m MPG/ESO telescope, La Silla.

We used a cross correlation technique to measure the stellar RVs. Basic stellar parameters (M, $T_{\rm eff}$, [Fe/H], $\log g$ and R) for our G-K giant targets have been determined from the spectroscopic analysis [1]. For the young stars sample we are currently working on the method for the stellar parameters measurement.

3 Stellar Activity

From our RV survey we found short-period and long-period RV variations in both samples. To distinguish intrinsic stellar activity, e.g. non-radial pulsations or stellar rotational modulation due to starspots, from the signature of substellar companion we used the bisector technique [3],[12]. In Fig. 2 we show an example of a star from our young star target list, where the bisector velocity spans correlate with the RVs. Thus, the RV variation is more likely due to the stellar activity rather than the presence of a substellar companion.

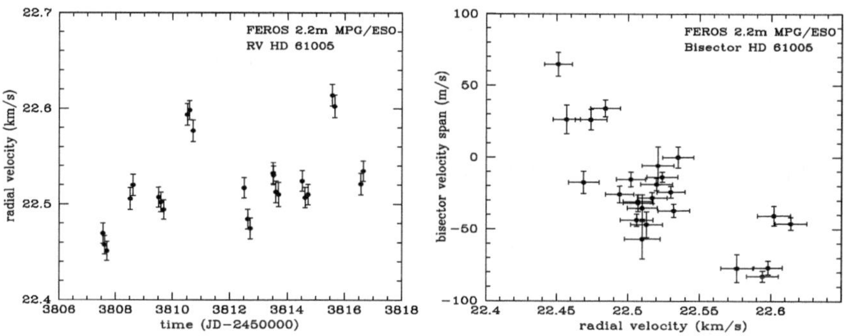

Fig. 2. RV measurements of a young star HD 61005 (left). The right panel shows a correlation between the bisector and RV. Thus, the source of the RV variation is possibly the stellar activity.

In the same way, non-planet detections (starspots and pulsations) in K giants have been found in HD 78647 [9] and HD 81797 [11].

4 Planetary Companions and Future Works

Fig. 3 (left panel) shows RV measurements of a nearby young star HD 70573 ($d = 46$ pc, $t \approx 100$ Myrs). The RV variation of this star shows a periodicity of 843 days which is significantly longer than the stellar rotational period of 3.296 days [6]. Thus, rotational modulation as the source of RV variation is unlikely. Moreover, we found no correlation between bisectors and RVs. We also did not find any correlation between the RVs and variation in Ca II K emission lines. We concluded, that the RV variation is caused by the presence of a substellar companion. Assuming a one solar-mass for HD 70573, we calculated a companion's minimum mass of $\approx$6 $M_{\rm jup}$ and a preliminary orbit with a semi-major axis of $\approx$ 1.7 AU and an eccentricity of $e = 0.4$ (Setiawan et al., in preparation).

One particular star of our G-K giants sample is HD 47536, which has been detected to harbor a planetary companion [8]. Further spectroscopic analysis

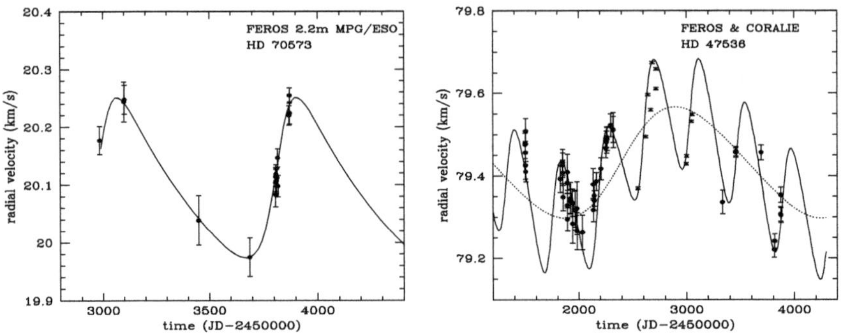

Fig. 3. Evidence for a planetary companion around a young star HD 70573 (left panel). Right panel: RV measurements of HD 47536 (see text), dots: FEROS measurements, triangle: CORALIE measurements.

[1] yields a primary mass of 0.94 M_{Sun} and a sub-solar metallicity [Fe/H]= -0.68. Together with HD 13189 [5], HD 47536 belongs to the most metal poor stars, which have planetary companions.

To our surprise, further RV follow-up observations with CORALIE and FEROS at 2.2 m MPG/ESO (Fig. 3, right panel) show an evidence for another planet candidate in the system. The second planet has an orbit with a longer period than the first inner planet. According to this detection, a revision of the orbital solution in [8] has been made. We obtained an orbital period of 431 days for the inner planet and ~2400 days for the outer planet. From the revised orbital solution we calculated companion's minimum masses of 3.8 M_{Jup} and 8.1 M_{Jup}. More detailed analysis will be presented in upcoming papers (Setiawan et al., in preparation).

References

1. L. da Silva, L. Girardi, L. Pasquini et al: A&A **458**, 609 (2006)
2. S. Frink, D.S. Mitchell, A. Quirrenbach et al: ApJ **576**, 478 (2002)
3. A.P. Hatzes: PASP **108**, 839 (1996)
4. A.P. Hatzes, W.D. Cochran, M. Endl et al: ApJ **599**, 1383 (2003)
5. A.P. Hatzes, E.W. Guenther, M. Endl et al: A&A **437**, 743 (2005)
6. G.W. Henry, F.C. Fekel, D.S. Hall: AJ **110**, 2926 (1995)
7. J. Setiawan, L. Pasquini, L. da Silva et al: A&A **397**, 1151 (2003a)
8. J. Setiawan, A.P. Hatzes, O. von der Lühe et al: A&A **398**, L19 (2003b)
9. J. Setiawan, L. Pasquini, L. da Silva et al: A&A **421**, 241 (2004)
10. J. Setiawan, J. Rodmann, L. da Silva et al: A&A **437**, L31 (2005)
11. J. Setiawan, M. Roth, P. Weise, M.P. Döllinger: Multi-periodic oscillations of HD 32887 and HD 81797 in: *Memorie della Societa Astronomica Italiana*, vol 77, 2006, p. 510
12. D. Queloz, G.W. Henry, J.P. Sivan et al: A&A **379**, 279 (2001)

A Catalogue of Nearby Exoplanets

Hugh R.A. Jones[1], R. Paul Butler[2], Jason T. Wright[3], Geoff W. Marcy[3], Deborah A. Fischer[4], Steve S. Vogt[5], Chris G. Tinney[6], Brad D. Carter[7], Jon A. Johnson[3], Chris McCarthy[4], and Alan J. Penny[8]

[1] Centre for Astrophysics Research, University of Hertfordshire, Hatfield AL10 9AB, UK h.r.a.jones@herts.ac.uk
[2] Department of Terrestrial Magnetism, Carnegie Institute of Washington, 5241 Broad Branch Road NW, Washington, DC 20015-1305, USA
[3] Department of Astronomy, University of California, Berkeley, CA 94720-3411, USA
[4] Department of Physics and Astronomy, San Francisco State University, San Francisco, CA 94132, USA
[5] Lick Observatory, University of California, Santa Cruz, CA 95064, USA
[6] Anglo-Australian Observatory, PO Box 296, Epping 1710, Australia
[7] Faculty of Sciences, University of Southern Queensland, Toowoomba, QLD 4350, Australia
[8] SETI Institute, 515N. Whisman Road, Mountain View, CA 94043, USA

Summary. We report on our online catalogue (http://exoplanets.org/planet.shtml) which contains all the known exoplanets with masses and orbits established by the Doppler measurements of stars with 200pc. It contains updated information including many whose orbits have not been revised since their announcement. Both the new and previously published velocities are more precise due to improvements in our data reduction pipeline, which we applied to archival spectra.

1 Introduction

It has now been more than 10 years since the discovery of the first objects that were identified as planets orbiting normal stars. The epochal announcement in 1995 October of 51 Peg b (Mayor & Queloz 1995) was confirmed within a week (Marcy et al. 1997) and followed shortly by two other planets 47 UMa b and 70 Vir b (Butler & Marcy 1996; Marcy & Butler 1996). The unexpected diversity and mass distribution of exoplanets was represented well by those first three planets, as the first one orbits close-in, the second orbits beyond 2 AU, and the last resides in a very eccentric orbit.

This distance threshold of 200 pc serves several purposes. Nearby planets and their host stars are amenable to confirmation and follow-up by a variety of techniques, including high resolution imaging and stellar spectroscopy with high signal-to-noise ratios, and astrometric follow up. In addition, nearby planet-host stars are bright enough to permit precise photometric and chromospheric monitoring by telescopes of modest size, permitting careful assessment of velocity jitter, starspots, and possible transits, e.g., Henry et al. (2000).

In addition, to recording orbital properties, we use the latest estimates of stellar mass to improve the precision of the minimum planet mass, M sin i (Valenti & Fischer 2005) as well as to provide detailed stellar parameters. The catalogue contains Doppler measurements for the planet host stars in our database, allowing both further analyses of these velocities and novel combinations with other measurements. Full details of the catalogue are described in Butler et al. (2006).

2 Data Reduction

We have revised our entire reduction pipeline, including an overhauled raw reduction package which includes corrections for cosmic rays and an improved flat-fielding algorithm, a more accurate barycentric velocity correction which includes proper-motion corrections, and a refined precision velocity reduction package which includes a telluric filter and a more sophisticated deconvolution algorithm. The very slight non-linearity in the new HIRES CCD and the characterization of the charge transfer inefficiency in the old CCD are corrected.

Uncertainties in orbital parameters were derived following Marcy et al. (1995) by subtracting the best-fit orbital solution from the data and interpreting the residuals as a population of random deviates with a distribution characteristic of the noise in the data. We randomly selected deviates from this set, with replacement, and added this noise to the velocities calculated from the best-fit solution at the actual times of observation. We then found the best-fit orbital solutions to this mock data set. Repeating this procedure 100 times, we produced 100 sets of orbital parameters. We report the standard deviation of each individual parameter over the 100 trials as the 1σ errors.

References

1. Butler R.P. et al. 2006, ApJ, 646, 505
2. Butler R.P., Marcy G.W., 1996, ApJ, 464, L153
3. Henry G.W. et al. 2000, ApJ, 531, 415
4. Marcy G.W., Butler R.P., 1996, ApJ, 464, L147
5. Marcy G.W. et al. 2005, ApJ, 619, 570
6. Mayor M., Queloz D.,1995, Nature, 378, 355
7. Valenti J., Fischer D., 2005, ApJS, 159, 141

Determination of the Orbital Parameters of a System with $N + 1$ Bodies using a Simple Fourier Analysis of the Data

Alexandre C.M. Correia

Departamento de Física da Universidade de Aveiro, Campus Universitário de Santiago, 3810-193 Aveiro, Portugal. acorreia@fis.ua.pt

Summary. Here we show how to determine the orbital parameters of a system composed of a star and N companions (that can be planets, brown-dwarfs or other stars), using a simple Fourier analysis of the radial velocity data of the star. This method supposes that all objects in the system follow keplerian orbits around the star and gives better results for a large number of observational points. The orbital parameters may present some errors, but they are an excellent starting point for the traditional minimization methods such as the Levenberg-Marquardt algorithms.

1 Radial Velocity

The radial velocity of a star with N companions is given by $v(t) = \gamma + v_0(t)$, where γ is a drift due to the global shift of the system center of mass, and[1]

$$v_0(t) = \sum_{j=1}^{N} K_j \left(e_j \cos \omega_j + \cos(\omega_j + \nu_j) \right) \ . \tag{1}$$

For each companion j, e_j is the eccentricity, ω_j the longitude of the perihelium, $\nu_j = \nu_j(t)$ the true longitude of the date and

$$K_j = n_j a_j \frac{m_j}{\mathcal{M}} \sin I_j \, (1 - e_j^2)^{-1/2} \tag{2}$$

the amplitude of radial velocity variations. n_j is the mean motion, a_j the semi-major axis, I_j the inclination of the orbital plane with respect to the line of sight, m_j the companion mass and $\mathcal{M}$ the total mass of the system. The orbital period of each companion is given by $P_j = 2\pi/n_j$.

1.1 Elliptic Expansions

There is no explicit expression for the true anomaly $\nu_j(t)$, but making use of the Kepler equation we can expand it in power series of e_j such that[2]:

$$\mathrm{e}^{i\nu_j} = \sum_{k=-\infty}^{+\infty} C_k(e_j) \, \mathrm{e}^{ikM_j} \ , \tag{3}$$

where $M_j = n_j(t - T_{0j})$ is the mean anomaly, T_{0j} the date for the passage at the perihelium and

$$C_k(e_j) = \frac{1}{2\pi} \int_0^{2\pi} \left(\cos E - e_j + i\sqrt{1 - e_j^2} \sin E \right) \mathrm{e}^{-ik(E - e_j \sin E)} \, dE \ . \quad (4)$$

To the fifth order in eccentricity, $C_k(e_j)$ for $k = 1$ and $k = 2$ becomes:

$$C_1(e_j) = \left(1 - \frac{9}{8} e_j^2 + \frac{25}{192} e_j^4\right) + i \left(1 - \frac{7}{8} e_j^2 + \frac{17}{192} e_j^4\right) , \quad (5)$$

$$C_2(e_j) = \left(1 - \frac{4}{3} e_j^2 + \frac{3}{8} e_j^4\right) e_j + i \left(1 - \frac{7}{6} e_j^2 + \frac{1}{3} e_j^4\right) e_j \ . \quad (6)$$

Replacing expression (3) in (1) we can finally rewrite the radial velocity of the star as the real part of:

$$v_0(t) = \sum_{j=1}^{N} K_j \, \mathrm{e}^{i\omega_j} \sum_{k=1}^{+\infty} C_k(e_j) \, \mathrm{e}^{-ikn_j T_{0j}} \, \mathrm{e}^{ikn_j t} \ . \quad (7)$$

2 Fourier Analysis

In this paper we will use an ordinary FFT transform of the radial velocity,

$$F(\phi) = \frac{1}{2\pi} \int_{-\infty}^{+\infty} v(t) \mathrm{e}^{-i\phi t} \, dt \ , \quad (8)$$

but other frequency analysis that make use of weight functions to ensure a better convergence with the data are possible. However, the calculus become more complicated, and harder to follow. Notice also that in the real case we cannot compute the FFT using the previous expression, because we are restricted to a discrete number of observations N_{obs} in a time span $[0, T]$:

$$F(\phi) \approx \frac{1}{T} \int_0^T v(t) \, \mathrm{e}^{-i\phi t} \, dt \approx \frac{1}{T} \sum_{k=2}^{N_{\mathrm{obs}}} v_k \, \mathrm{e}^{-i\phi t_k} (t_k - t_{k-1}) \ , \quad (9)$$

where v_k is the star radial velocity measured for the date t_k.

2.1 Determination of γ

Replacing $v(t) = \gamma + v_0(t)$ in expression (8) with $\phi = 0$ we get:

$$\gamma = F(0) \ . \quad (10)$$

It is then possible to have an estimation of γ using $\phi = 0$ in expression (9). Once we have γ it is preferable to subtract its value from the data v_k and then continue the Fourier analysis (Eqs.8,9) with the expression of $v_0(t)$ (Eq.7).

2.2 Determination of n_j

The orbital frequency n_1 corresponding to the companion with largest amplitude K_1 is given by the frequency ϕ corresponding to the highest peak in the power spectrum, that is,

$$n_1 : \quad \forall\phi \ , \ |F(n_1)| \geq |F(\phi)| \ . \tag{11}$$

After finding n_1 it is easy to determine the remaining orbital parameters (see next section). Once the orbit of the first companion is completely established, it is recommended to subtract its contribution from the data v_k and then continue the Fourier analysis (Eqs.8,9) with the expression of

$$v_0 - K_1 \left(e_1 \cos\omega_1 + \cos(\omega_1 + \nu_1)\right) \ . \tag{12}$$

We then have to repeat this procedure for all the other $N-2$ remaining companions of the star. Thus, the n_j orbital frequencies are always given by the highest peak in the spectrum (Eq.11) after subtracting the signal from the already detected companions (Eq.12).

2.3 Determination of the Remaining Orbital Parameters

Replacing expression (7) in (8) with $\phi = n_j$ and $\phi = 2n_j$ we have

$$F(n_j) = K_j \mathrm{e}^{i\omega_j} C_1(e_j)\, \mathrm{e}^{-i n_j T_{0j}} \ , \tag{13}$$

$$F(2n_j) = K_j \mathrm{e}^{i\omega_j} C_2(e_j)\, \mathrm{e}^{-i 2 n_j T_{0j}} \ , \tag{14}$$

where the quantities $F(n_j)$ and $F(2n_j)$ can be computed from the data using expression (9). Multiplying the above expressions by their conjugates, we get

$$|F(n_j)| = K_j |C_1(e_j)| \quad \text{and} \quad |F(2n_j)| = K_j |C_2(e_j)| \ , \tag{15}$$

which gives an implicit condition for the eccentricity,

$$f(e_j) = \frac{|C_2(e_j)|}{|C_1(e_j)|} = \frac{|F(2n_j)|}{|F(n_j)|} \ , \tag{16}$$

where e_j can be determined using the bisection or the Newton's method[3]. After determining e_j it is now straightforward to compute K_j from Eqs.(15):

$$K_j = \frac{|F(n_j)|}{|C_1(e_j)|} = \frac{|F(2n_j)|}{|C_2(e_j)|} \ . \tag{17}$$

From expressions (13) and (14) we finally have

$$\mathrm{e}^{i n_j T_{0j}} = \frac{F(n_j)\, C_2(e_j)}{F(2n_j)\, C_1(e_j)} \tag{18}$$

and

$$\mathrm{e}^{i\omega_j} = \frac{F(n_j)}{K_j C_1(e_j)}\, \mathrm{e}^{i n_j T_{0j}} = \frac{F^2(n_j)\, C_2(e_j)}{K_j F(2n_j)\, C_1^2(e_j)} \ . \tag{19}$$

3 Conclusion

For a single companion of a star, we are able to determine its orbital parameters directly from the observational data by computing the FFTs for three different frequencies, namely $F(0)$, $F(n)$ and $F(2n)$. We chose n and $2n$, but according to expression (7) we could have chosen any frequency multiple of n. However, unless the eccentricity is extremely high, the two chosen frequencies correspond to the highest peaks produced by the companion in the spectrum and are therefore easier to identify. Moreover, if the eccentricity is close to zero (which is often the case for "hot Jupiters" and close binaries), $F(kn) \approx 0$, except for $k = 0$ and $k = 1$. In this case ω_j and T_{0j} cannot be determined, but it is still possible to establish the position of the planet in the orbit, $\lambda_j = \omega_j - n_j T_{0j}$, as

$$\mathrm{e}^{i\lambda_j} = \frac{F(n_j)}{K_j C_1(e_j)} \,. \tag{20}$$

The orbital parameters determined with our method present errors that are proportional to the precision of the instrument and inversely proportional to the number of data points, since a large number of points increases the convergence between expressions (8) and (9). The agreement between the Fourier parameters and the true parameters can be increased if we perform a χ^2 minimization after determining the orbit of each companion. This procedure should be fast using a standard method such as a Levenberg-Marquardt algorithm[3], since the Fourier parameters are already close to the minimum value of χ^2. Even though the FFT method is established for keplerian orbits, it also works on realistic systems for which planet-planet interactions are weak. Indeed, this method has already been tested with success in the determination of the orbital parameters of three different planetary systems[4, 5, 6], where we obtained the same results as other classical alternative methods.

References

1. R.W. Hilditch: *An Introduction to Close Binary Stars*, (Cambridge University Press 2001)
2. C.D. Murray, S.F. Dermott: *Solar System Dynamics*, (Cambridge University Press 1999)
3. W.H. Press, S.A. Teukolsky, W.T. Vetterling, B.P. Flannery: *Numerical recipes in FORTRAN*, (Cambridge University Press 1992, 2nd Ed.)
4. A.C.M. Correia, S. Udry, M. Mayor et al: Astron. Astrophys. **440**, 751 (2005)
5. C. Lovis, M. Mayor, F. Pepe et al: Nature **441**, 305 (2006)
6. F. Pepe, A.C.M. Correia, M. Mayor et al: Astron. Astrophys. **462**, 769 (2007)

Extrasolar Comets

Roger Ferlet, Jérémie Boissier, Alain Lecavelier des Etangs, and Alfred Vidal-Madjar

Institut d'astrophysique de Paris ferlet@iap.fr

Summary. Gaseous and dusty circumstellar disks are commonly thought to be the birth place of planetary systems. Once planets are formed, the so-called "debris disks" are still active complex evolving systems in which gas and dust are interacting and replenishing. This was notably demonstrated through visible and UV absorption spectroscopy of the most famous debris disk around the main sequence star β Pictoris. Hourly variable absorbing components are explained by a multitude of sublimating comets when grazing the star. This intense cometary activity is the proposed source for the gas and one possible origin for the replenishment of the dust. Hundreds of cometary-like events are observed, from different comet families.

1 The Existence of Comets Around β Pictoris

Seen nearly edge-on, the gaseous counterpart of the famous β Pictoris circumstellar disk has been extensively observed through high resolution absorption spectroscopy in both the visible and the UV. One of the absorption features appears almost identically in all the observations and is thus named the "stable" absorption gas component. All the others are variable in time and strength, from over hours or even less, to weeks, and most of them are more or less strongly redshifted with respect to the stellar velocity, from about 10 to at most 400 km s^{-1} for the highly ionized species Al III and C IV (see e.g. [8], [11], [16]).

There is only one simple way to produce such features, namely the evaporation of grains from kilometer size, "cometary-like" bodies moving toward the star [8]. Extensive numerical simulations of this scenario have succeeded in reproducing the observations by considering two types of infalling comets: those with large periastron producing narrow lines at smaller velocities, and those with small periastron producing broad lines at larger velocities (see e.g. [1] and references therein). This so-called Falling Evaporating Bodies (FEB) scenario explains many characteristics of the gas such as: redshifted radial velocity, variability, clumpiness, temperature of hot gas, highly ionized species (formed through collisions in the very coma surrounding the FEBs [2]).

Clumpiness was a very nice test of the model. Although the 2800 Mg II doublet has an intrinsic oscillator strength ratio of two, the measured ratio is exactly one, even for unsaturated lines. This proves that the absorbing gas cloud is optically thick but does not cover the total stellar disk [15].

Thanks to HST, the molecule CO has been detected in absorption using electronic bands at 1400-1500 Å, together with one of its dissociation products, C I [15]. Their mere presence in the β Pic environment is challenging since they have similar very short life-times (< 200 years), being destroyed by the ambient interstellar UV photons. The CO level population implies gas at about 25 K, and the $^{12}CO/^{13}CO$ is found to be < 20 ([10], [13]). Moreover, more recent searches for H_2 in the β Pic disk with the FUSE satellite [12] have resulted in a column density upper limit of 10^{18} cm^{-2}, inducing thus a CO/H_2 ratio $> 6.10^{-4}$, instead of about 10^{-6} in interstellar translucent clouds. We conclude that CO is not protected from UV radiations by H_2 and should be continuously replenished; it must originate from a frozen source, the evaporation of comets.

2 Solving the Stability of the Gaseous Disk

Lines from the stable gaseous disk of β Pic have also been observed in emission, indicating that the gas is in keplerian rotation and extends to at least 300 AU (e.g. [7]). This is in apparent contradiction with the strong radiation pressure from β Pic which should rapidly blown away the gas from the disk. FUSE observations allowed to measure the volatile species C^+, C^{++} and O I, resulting in $C/Fe_{gas} = 16$ solar and $C/O_{gas} = 18$ solar [14]. The strongest absorption lines of oxygen and carbon are in the far-UV where the star is faint and therefore these species do not feel strong radiation pressure. By contrast, iron feels extremely strong radiation pressure through lines in the near-UV and could be blown out of the system. While this effect can produce an apparent C overabundance relative to Fe, it cannot explain the measured C/O ratio, which thus implies a real large overabundance of carbon relative to other species.

The carbon overabundance may solve the problem of the stable gaseous disk of β Pictoris ([14], [9]). Because C is abundant and moderately ionized, it is an important constituent of the ionic fluid which is kept as a single fluid through Coulomb interactions between the ions. Because C feels negligible radiation pressure, the ionic fluid will be self-braking if carbon is enhanced by at least a factor of 10 over the solar abundance. This condition being fulfilled, the gas disk is stable and kept in keplerian rotation.

3 Two Families of Comets?

Taking advantage of the new HARPS instrument at the 3.6 m ESO telescope, we were able to follow hundreds of "comets" around β Pic through the 3934-3969 Å Ca II doublet at high resolution and high S/N ratio, and thus to perform a statistical analysis of an homogeneous sample of comets [6]. Using a profile fitting method (Fig.1), we put forward the existence of two distinct

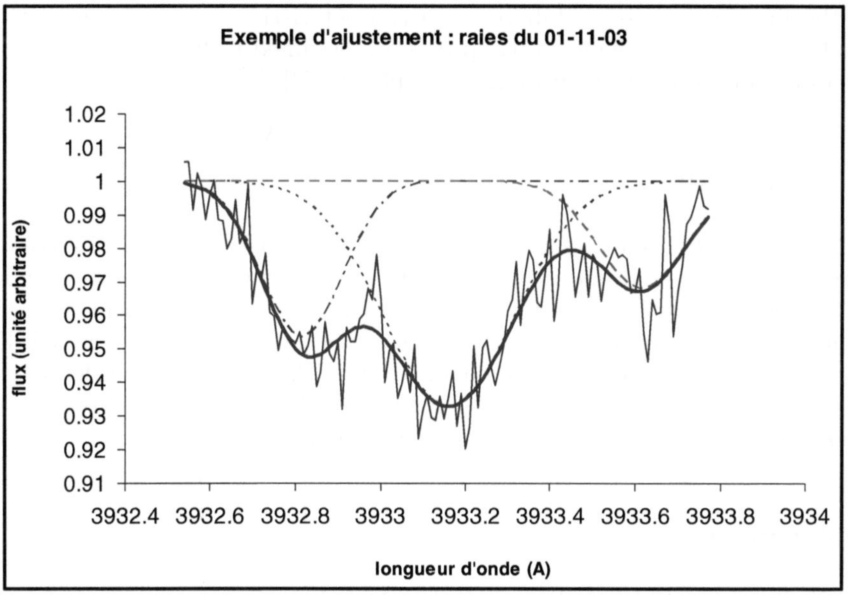

Fig. 1. Example of Ca II absorption features observed with HARPS toward β Pictoris. Data are normalized to unity. The solid line is the convolved theoretical spectrum of three individual absorption components, each represented by a gaussian velocity distribution. Profile fitting has been done simultaneously in both lines of the calcium doublet. Parameters for more than 400 such variable absorption features led to put forward the existence of two distinct families of comets responsible for the observed FEBs events.

families of comets: one producing deep absorption lines (FWHM $= 21 \pm 2$ km s^{-1}) at velocities of 25 ± 5 km s^{-1}, and the other producing shallow lines (FWHM $= 70 \pm 7$ km s^{-1}) at larger velocities of 48 ± 5 km s^{-1}.

Assuming that the absorptions are due to FEBs, the distance r of an FEB to the star can be derived from the observed line width Δv (r $\propto \Delta \mathrm{v}^{-6/5}$). For a body on a keplerian orbit, there is a relation between r, the radial velocity and the angle ω of the periastron with respect to the line of sight [3]. For FEBs producing deep absorptions, we preliminary find that ω remains constant whatever the periastron distance is. It is thus possible to express the observed radial velocity as a function of r. This model seems to closely account for the deep lines data points. We conclude that these deep features could be due to a family of bodies resulting from the recent break-up of one or two parent comets.

On the contrary, the shallow absorptions data points show a much larger dispersion in the plot radial velocity versus distance r, and do not follow the above $\omega =$ constant relationship. Instead, a relation with a varying ω as a function of the periastron seems more appropriate. Furthermore, about 20%

of the lines appear blue-shifted. Both characteristics are consistent with a model in which the shallow absorptions are assumed to be due to comets in mean motion resonance with a massive planet ([4], [5]).

4 Conclusion

While the dusty components of young planetary disks are well observed in the optical, near- and mid-infrared, their gaseous components can be studied in details with optical and UV spectroscopy. Very tight constraints on the composition and physical characteristics of the gas can thus be derived. Especially for the famous system surrounding β Pictoris, amazing details are now gathered on gas, dust, comets, planets, and even more importantly, on their mutual interaction within a complete planetary system.

References

1. H. Beust, A. Vidal-Madjar, R. Ferlet, A.M. Lagrange-Henri: A&A **241**, 488 (1991)
2. H. Beust, M. Tagger: Icarus **106**, 42 (1993)
3. H. Beust, A.M. Lagrange, F. Plazy, D. Mouillet: A&A **310**, 181 (1996)
4. H. Beust, A. Morbidelli: Icarus **120**, 358 (1996)
5. H. Beust, A. Morbidelli: Icarus **143**, 170 (2000)
6. J. Boissier, A. Lecavelier des Etangs, A. Vidal-Madjar, R. Ferlet: A&A in preparation (2006)
7. A. Brandeker, R. Liseau, G. Olofsson, M. Fridlund: A&A **413**, 681 (2004)
8. R. Ferlet, L.M. Hobbs, A. Vidal-Madjar: A&A **185**, 267 (1987)
9. R. Fernandez, A. Brandeker, Y. Wu: ApJ **643**, 509 (2006)
10. A. Jolly et al.: A&A **329**, 1028 (1998)
11. A.M. Lagrange-Henri, A. Vidal-Madjar, R. Ferlet: A&A **190**, 275 (1988)
12. A. Lecavelier des Etangs et al.: Nature **412**, 706 (2001)
13. A. Roberge et al.: ApJ **538**, 904 (2000)
14. A. Roberge et al.: Nature **441**, 724 (2006)
15. A. Vidal-Madjar et al.: A&A **290**, 245 (1994)
16. A. Vidal-Madjar, A. Lecavelier des Etangs, R. Ferlet: P&SS **46**, 629 (1998)

Measuring Winds in Titan's Atmosphere with High-precision Doppler Velocimetry

David Luz[1,2] and Régis Courtin[2]

[1] CAAUL, Observatório Astronómico de Lisboa, Tapada da Ajuda, 1349-018 Lisboa, Portugal dluz@oal.ul.pt

[2] LESIA, Observatoire de Paris–Section de Meudon, 5, Place Jules Janssen, 92195 Meudon CEDEX, France Regis.Courtin@obspm.fr

Summary. The technique of absolute velocimetry has been applied to the backscattered solar spectrum of Titan in order to measure the Doppler shift associated with the zonal circulation in the equatorial region. Eastward winds were measured, with lower limits of 62 and 50 ms^{-1} at altitudes near 200 and 170 km, and significant statistical evidence was observed for stronger winds at the higher altitudes. The same technique was applied in January 2005 as part of an international effort in support of the European Space Agency's Huygens Probe mission and led to wind speed values quite consistent with our previous results.

1 Background

Saturn's moon Titan is one of the four terrestrial-type planetary bodies in our Solar System possessing a dense atmosphere. Its extended atmosphere is composed of $\sim 98\%$ nitrogen and $\sim 2\%$ methane. Its obliquity ($\sim 26°$) makes seasonal effects an important aspect of its atmospheric dynamics. Despite the long Titan year (29.5 y), modelling suggests a strong coupling between atmospheric dynamics, photochemistry and microphysics of the haze which occults the surface from direct observation in the visible [18, 12, 7]. At the onset of our observing program (2002) the winds on Titan were constrained from two sets of data: the Voyager latitudinal temperature gradients, providing an inference of the zonal winds [6], and an estimate of the atmospheric oblateness from stellar occultation measurements [8]. These methods, however, only provided the wind magnitude, leaving the wind direction undetermined.

2 Observations

We observed Titan with UVES in 2002-03, with the aim of determining the direction and magnitude of its stratospheric winds. Similar observations were carried out in January 2005 in coordination with the Huygens Probe entry on Titan's atmosphere. A detailed description of the campaigns and their results is given in two previous papers [14, 15]. In this technique, the Titan haze, which is the main source of scattering at visible wavelengths, plays in our favor. Since the haze is transported by the wind, observing Titan's backscattered solar spectrum allows us to determine the Doppler shift (DS) associated with the zonal circulation. Measurements were made in the wavelength range 4200–6200 Å, with a 0.3" slit width (Titan's diameter $\simeq$ 0.88"). The field was de-rotated, to align the aperture perpendicularly to Titan's spin axis, so that information in the East-West direction was preserved in the direction perpendicular to dispersion. Spectra being acquired simultaneously across the disk, this technique is not sensitive to Titan's orbital motion, and the retrieved DSs between different spectra correspond directly to relative motions between air masses at different locations on the disk. Respectively 39 and 32 exposures were obtained in the two sets of observations of 2002–03 and 2005. Incidentally, we have extended this program to Saturn in 2004, with the same objective of monitoring its atmospheric circulation as a function of latitude [13].

3 Analysis and Interpretation

The analysis consisted of three main stages: extraction of the disk spectra, computation of the DS with a radial velocity algorithm, and statistical analysis of the results. We used an extraction window of 31 pixels, which is sufficient to cover the disk and obtain the background level at the order extremities. Since UVES red-arm spectra are recorded on a mosaic of two CCDs (one EEV and one MIT CCD, for $\lambda \simeq$ 4200–5200Å and $\lambda \simeq$ 5200–6200Å), these two sets are extracted and analyzed separately. A radial velocity algorithm [3] is used to compute the DSs of the extracted spectra.

For each exposure, the velocity curve plotted as a function of spatial position on the disk provides a signature of the rotation. The interpretation of this signature is complicated by two major factors: first, turbulence in the terrestrial atmosphere, which mixes photons with different shifts, coming from different locations on the disk. Secondly, in the Titan atmosphere photons with different wavelengths are backscattered at different altitudes, with a contribution function varying with altitude and from center-to-limb. We carried out an extensive validation of the method [2], based on observations of Jupiter's moon Io and on modelling of the velocity signal.

The main conclusions drawn from the two campaigns are: first, the large majority of results indicate prograde rotation of Titan's atmosphere; second, the results are characterized by a high dispersion, indicating that these observations are sensitive to the variability of observing conditions, namely the seeing. We interpret the dispersion as consequence of poorer quality spectra (in terms of velocity signature) rather than physical variability of Titan's winds. In order to deal quantitatively with the dispersion, we introduced a quality parameter, defined as $\varepsilon = R_s/R_{eff}$, where R_s is the symmetry radius of the velocity curve and R_{eff} is the effective radius of the spectral orders. Poorer quality data either stem from widening of the orders due to a large seeing or from a lack of symmetry of the velocity curves originating from perturbations, such as target drifts. When the velocities are plotted against the quality parameter (Fig. 1), the "retrograde" results tend to cluster at low ε. The "best estimate" method consists of a ε-weighted running average of ΔV for $\varepsilon > \varepsilon_0$. Since all sources of error we could identify only tend to diminish the measured DS, we take as ε_0 the value which maximizes the running average, and interpret the resulting average value as a lower limit to the equatorial zonal wind velocity. Applying this method to the 2002–03 data set, we obtain lower limits of $u = 63 \pm 13$ ms^{-1} and $u = 50 \pm 11$ ms^{-1} from the lower and upper part of the spectrum, respectively. From the peak of the contribution function for the two wavelength domains $\lambda \simeq$ 4200–5200Å and $\lambda \simeq$ 5200–6200Å, obtained from radiative transfer modelling of Titan's atmosphere, we determine a predominance of backscattered photons from the levels $z = 197 \pm 72$ km and $z = 172 \pm 65$ km. A statistical test based on the Wilcoxon signed ranks method [4], applied to the two sets of paired measurements in the EEV and MIT CCDs, confirms that winds are stronger at the shorter wavelength range. The observed vertical variation of the wind, with stronger winds at higher altitudes, is in agreement with general circulation modelling (GCM) [19]. The 2005 data set yields similar results, with $u = 50 \pm 10$ ms^{-1}, but there is no significant difference between the results from the two wavelength ranges. The ensemble of Titan wind measurements from various techniques forms a complementary data set, since no technique is able to provide a vertical wind profile from the mesosphere to the surface (Fig. 2). Our results are consistent with other determinations in this set [16, 9, 10, 11, 5, 17, 1].

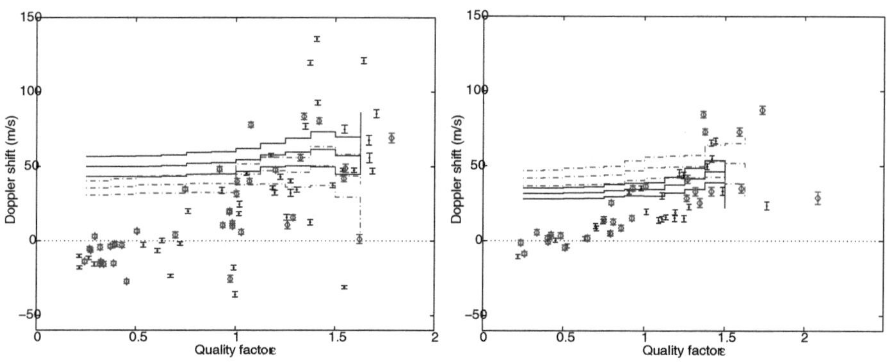

Fig. 1. Equatorial zonal wind velocity from the EEV (dots) and MIT (diamonds) CCD data sets. Left and right panels show 2002–3 and 2005 results. Measurements are plotted as a function of the quality factor. Lines are running averages for $\varepsilon > \varepsilon_0$ (step of 0.125) with the upper and lower lines at $\pm\sigma$.

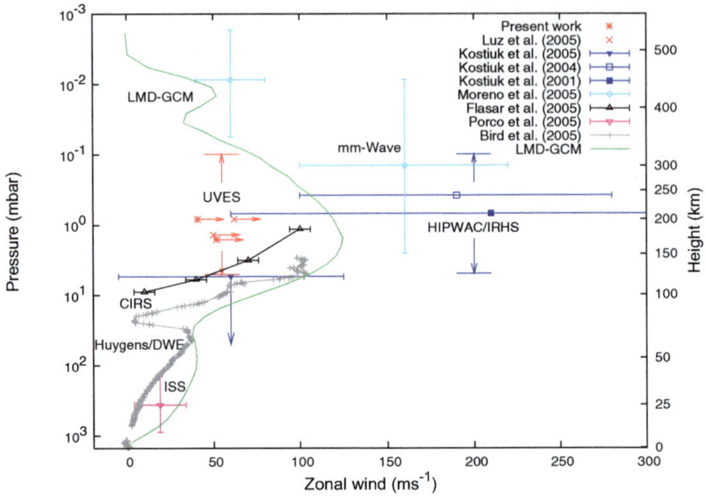

Fig. 2. Comparison of Titan wind measurements obtained with different ground-based techniques [14, 15, 16, 9, 10, 11], Cassini orbiter [5, 17], entry probe tracking [1] and with the results of a GCM [19].

Acknowledgements: DL acknowledges financial support from *Fundação para a Ciência e a Tecnologia*, Portugal (fellowship PRAXIS XXI/BPD/-3630/2000, and from FCT project POCI/CTE-AST/57655/2004.

References

1. M.K. Bird, M. Allison, S. Asmar et al: *Nature*, 438:800–802, 2005.
2. T. Civeit, T. Appourchaux, J.-P. Lebreton et al: *A& A*, 431:1157–1166, 2005.
3. P. Connes. *Astrophys. and Space Sci.*, 110:211–255, 1985.
4. W. J. Conover. *Practical non-parametric statistics.* Wiley, 3rd edition, 1999.
5. F. M. Flasar, R. Achterberg, B. Conrath et al: *Science*, 308:975–978, 2005.
6. F. M. Flasar, R. Samuelson, and B. Conrath. *Nature*, 292:693–698, 1981.
7. F. Hourdin et al: *J. Geoph. Res.*, 109:E12005, 2004.
8. W. B. Hubbard, B. Sicardy, R. Miles et al: *A& A*, 269:541–563, 1993.
9. T. Kostiuk, K. Fast, T. Livengood et al: *Geoph. Res. Lett.*, 28:2361–2364, 2001.
10. T. Kostiuk, T. Livengood, T. Hewagama et al: *ArXiv Ap. e-prints*, Dec. 2004.
11. T. Kostiuk et al: *J. Geoph. Res.*, 111:E07S03, 2006.
12. D. Luz, F. Hourdin, P. Rannou, and S. Lebonnois. *Icarus*, 166:343–358, 2003.
13. D. Luz, T. Civeit, R. Courtin et al: *Geoph. Res. Abstracts*, 7:08888, 2005.
14. D. Luz, T. Civeit, R. Courtin et al: *Icarus*, 179:497–510, 2005.
15. D. Luz, T. Civeit, R. Courtin et al: *J. Geoph. Res.*, 111:E08S90, 2006.
16. R. Moreno, A. Marten, and T. Hidayat. *A & A*, 437:319–328, 2005.
17. C. C. Porco, E. Baker, J. Barbara et al: *Nature*, 434:159–168, 2005.
18. P. Rannou, F. Hourdin, and C. McKay. *Nature*, 418:853–856, 2002.
19. P. Rannou, F. Hourdin, C. McKay, and D. Luz. *Icarus*, 170:443–462, 2004.

Part VII

Future Developments

The European Large Telescope and its Spectroscopic Instrumentation

Sandro D' Odorico[1]

European Southern Observatory, Garching bei München, Germany
sdodoric@eso.org

A European ELT in the 30-60m diameter class is currently being studied by ESO in collaboration with the astronomical community. Two designs are under consideration in this phase, to be concluded by the end of 2006. The baseline is a 42m, 5 mirror telescope, fall back is a Gregorian Design like in the American 30m TMT proposal. Both designs include an adaptive mirror in the optical train. The gain expected from an ELT with respect to the current generation of 8-10 m telescopes are most significant for the programs which require high angular resolution or are photon-starved. Concept studies have been carried out for 1st generation instruments. They include high resolution spectrographs for the visual and the infrared and a medium resolution, multi-object spectrograph for the NIR.

1 The E-ELT Programme

The 10m Keck telescope started operation in 1994. The ESO VLT project, an array of 4 8m telescope, saw first light in 1998 and it has been fully operational since 2001. Many other 6-8m telescopes have been or are about to be completed (Fig.1). In the last decade, interest has grown in the USA and more recently in Europe about developing an optical-infrared telescope of a new diameter class. In the USA, efforts are now focused on two projects. The TMT, Thirty Meter Telescope, project (http://www.tmt.org/) is a 30m segmented mirror telescope being studied by AURA, CIT and UC in the USA together with ACURA in Canada. The GMT, Giant Magellan Telescope, project (http://www.gmto.org/) is an ensemble of seven 8m mirrors mounted on a single structure, and with a common secondary, providing an angular resolution corresponding to a diameter of 24.5m. It is being studied by the Carnegie Institution of Washington in partnership with other American universities and the Australian National University.

In Europe, the Observatory of Lund, in partnership with a few other European institutes within the EC-sponsored network OPTICON, has studied a 50 m telescope (2003), see: www.astro.lu.se/~torben/euro50/index.html.

The European Southern Observatory (ESO) has explored an 100m telescope concept, known as OWL
(2005, www.eso.org/projects/owl/Phase_A_Review.html). A review of the

OWL study was held in November 2005. In the final recommendation, later on taken up by the ESO Council, the Review Board recommended to bring the diameter of the telescope to 60m or below to bypass some specific technical difficulties (e.g. a segmented secondary coupled with the segmented primary) and to reduce the risks associated to the project budget. ESO has thus rebaselined the project in 2006, in strict cooperation with the European astronomical community. The new E-ELT programme has explored the scientific case for a telescope of 42m and derived requirements for the telescope, its adaptive optics systems and the instruments (www.eso.org/projects/e-elt/). By the end of 2006, a reference design and a cost estimate for a 42m telescope will be presented to the ESO Council.

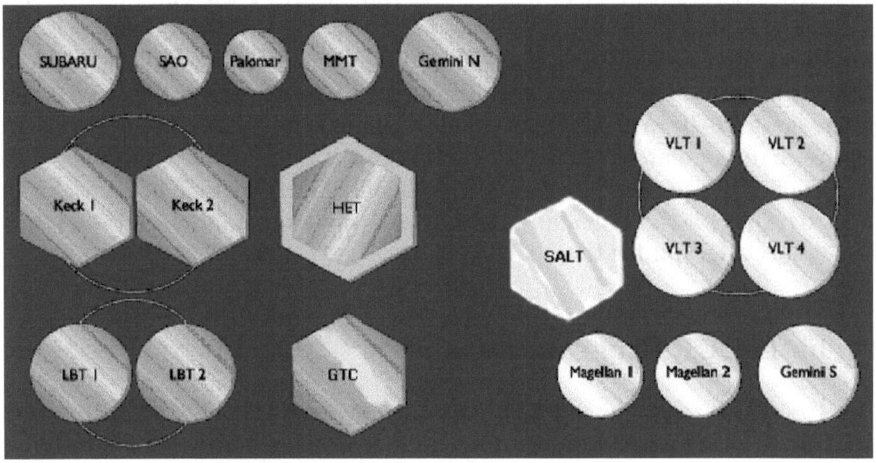

Fig. 1. 5-10m telescopes in operation or about to start in the northern hemisphere (left) and southern hemisphere (right)

2 Advantages of an ELT

Fig 2 (Table 1) presents the gain to be obtained at visual and infrared wavelengths as a function of telescope diameter for different types of observations. The gain is expressed as better accuracy, limiting flux of a target detected at a given S/N in a given time and speed , that is the time to complete an observation at a given S/N and magnitude limit. The values are derived from the standard formula which relates S/N to observed flux and time. The largest gain (that is proportional to the fourth power of the diameter) is achieved for observations of faint stellar targets observed at the diffraction limit (or anyway with a large Strehl).The study of stellar populations in crowded regions of other galaxies fits well in this category. There the extremely large telescope provides both the larger flux and its exquisite angular concentration,decreasing the effect of sky noise. A very significant advantage (that is proportional to the square of the diameter) is also present in the other observing regimes. A spectroscopic survey of QSOs at high S/N and high resolution can be completed 25 times faster at a 40m than at an 8m. A program like the CODEX measurement of the universe expansion becomes feasible , while at an 8m it would require 2500 nights.The same type of gain is expected for searches for planet-induced wobbling in the radial velocity curves of relatively bright stars.

TABLE 1. Gain dependency on telescope diameter in various observing regimes

	Stellar Object, sky noise limited, best seeing (a)	Stellar Object, sky noise limited, diffraction limit (b)	Photon Noise limited (c)	Detector noise limited (d)
Accuracy (= S/N, at given magnitude m_λ and integration time t)	D	D^2	D	D^2
Limiting flux (at given t and S/N)	D	D^2	D^2	D^2
Speed (=t to reach given S/N for a target of magnitude m	D^2	D^4	D^2	D^4

Notes: (a) This is the case of a deep imager or, e.g. a MOS at visual and NIR wavelengths. In principle, a very large field instrument at an 8m could be more effective that a 5' x 5' instrument at an ELT for survey work.

(b) This is the case of an instrument for imaging and spectroscopy over a limited field with an AO system which provides high Strehl. This is possible at IR wavelength only in the foreseeable future. In both photometry and spectroscopy the reduction of the image size (less underlying sky) coupled with the increase in flux results in the largest gains. In case of extended sources (e.g. high z galaxies with 0.1 – 0.2" sizes) the gain can be intermediate between (a) and (b) and depends on how well the AO system can improve the Encircled Energy within those image sizes.

(c) For sources much brighter than the sky, in all observing modes.

(d) This is the case of very high resolution spectroscopy of faint targets at visual wavelengths and in J and H IR bands between the sky lines.

Fig. 2. Table 1

3 Status of E-ELT Programme

3.1 The Telescope and the Adaptive Optics Systems

The telescope designs which have been explored in 2006 are both based on the use of a large adaptive optics mirror in the telescope optical train. In the case of a gregorian design the AO mirror is M2, in the case of the 5M design (see Fig.3) it is M4 with M5 used for fast tip-tilt correction. Many of the E-ELT prominent science cases call for angular resolution at the diffraction limit or close to it. This can only be achieved by adaptive optics corrections.

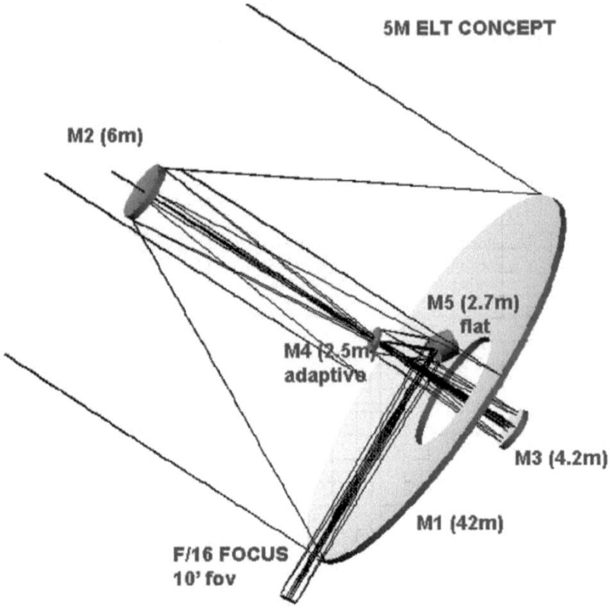

Fig. 3. Main optical components of the 5M E-ELT Concept

Even for seeing limited observations corrections of telescope induced errors need low-order adaptive optics over the whole field and field stabilization. Adaptive optics in the telescope reduces the complexity of all post-focal AO system, is sufficient for Single Coniugate AO, and mandatory for Ground Layer AO, Multi Object AO. Multi Conjugate AO over fields of views as large as 2 arcmin requires a large DM in the instrument if it is not available in the telescope, in addition to smaller AO in the instrument. The necessity of using large deformable mirrors for large stroke corrections behind a non-adaptive telescope leads to large optical setups for the AO associated with the instrument which, in addition, might have to be duplicated for all focal stations. This would reduce the available space at the focal stations and thus would reduce the number of instruments which could rapidly be brought online.

The overall conclusion is that a 1st stage adaptive system incorporated into the telescope is needed to meet the basic performance requirements imposed by the science case for first light. While it increases the risk to the project, it improves the overall performance, reduces the overall complexity and reduces the total cost of the observatory over the lifetime of the project.

3.2 Instrumentation Studies

The first instrumentation studies for an ELT have been carried out in Europe within the OWL framework (2004-2005). More recently 8 instrument concepts have been developed within the FP6 ELT Design Study supported by the European Community. The spectroscopic instruments which are more interesting to this workshop are an high resolution, ultra-stable visual spectrograph, CODEX ([1] and the contribution by L.Pasquini in this conference), IR HISPEC and MOMSI. IR HiSPEC [2] is an high resolution (R=150000) echelle spectrograph. The spectral range 1-5 micron is split over two separate arms, one operating at 200 K in the range 0.8 - 1.8 micron, the other fully cryogenic for the 1.8 - 5 micron range. Preliminary performance estimate indicates that a S/N ratio of 150 for a star of J mag 15 can be reached in one hour. This mode could be used to search for planets through the wobbling effect on the radial velocity curve of the parent star. This method has proven very successful with observations at visual-blue wavelengths but it is still to be proven in the infrared range. With the same instrument in an exposure time again of 1 hr but at a resolution of R=50000 a predicted S/N of 1000 can be reached for a star of H=11.5, as it would be required to detect the atmosphere of a transiting planet. MOMSI [3] is a multi-object spectrometer (multiplexing 20) which could work at the E-ELT in a compact field (¡ 2' squared) together with a Multi Conjugate AO system for spectroscopy of faint stars in crowded fields. A resolution mode of 20000 would permit to extend stellar abundance work in Dwarf galaxies in the Local Group and to nearby galaxy groups. At this resolution a limiting magnitude of K 23 at a S/N=20 should be obtained according to the MOMSI team in 10 hrs integration time. The instrumentation activities combined with the prominent science cases in the report by the E-ELT by the Science Working Group set-up by ESO in 2006 (see http://www.eso.org/projects/e-elt/publications.html) have led a preliminary list of high priority E-ELT 1st generation instruments which is shown in Fig. 4 (Table 2). The instrument concepts are listed together with their potential location in the telescope, the flavour of AO they need to operate, their properties and the prominent science cases they are associated to. In the next months , this list will be refined in interaction with the community with the goal to arrive to a first version of the E-ELT instrumentation plan by the middle of 2007. In the M5 telescope concept up to 8 instruments could be permanently mounted at the different focal stations. The Nasmyth platforms are designed to host up to 5 instruments including one mounted in a gravity- invariant location. Two large spectrographs can be located in the stable, thermally insulated coude' laboratories. One small Mid IR instrument could be placed in the intermediate focus to minimize the thermal emissivity.

TABLE 2 Preliminary list of 1st generation E-ELT Instruments

INSTRUMENT	OBS. MODES	FOCUS / AO	WAV. RANGE (μm)	FIELD	PIXEL SIZE (mas)	$\Delta\lambda / \lambda$	PROMINENT SCIENCE CASE*	REF. STUDY
DL, NIR Imager	imaging	Nasm. /LTAO or MCAO	0.9-2.5	Max 1'	4	Wide, narrow bands	~ most	ONIRICA @ OWL
Single Field Spectrograph	spectroscopy	Nasm./SCAO or LTAO	0.6- 2.5	1"; 10":	20 -50	3000, 20000	~ all	Not studied
High Resolution Vis Spectrograph	spectroscopy	coude/ GLAO	0.4 -0.8	Point source	=	150000	C2, C7	CODEX
Planetary Imager Spectrograph	imaging, spectroscopy	Nasm/ EXAO	0.5-1.5	~2" V ~4" H	>= Nyquist	>15	S3, S9	EPICS
NIR MOS	spectroscopy multiplex.20	Grav. Inv./ MOAO	0.8-2.5	>=5'	30 -50	3000, 10000	C4, C10	WFSPEC, MOMSI
NIR MOS ~DL	Spectroscopy, Multiplex.>10	Grav.Inv./MOAO	0.8- 2.5	0.5' -1'	10- 30	3000, 20000	G4, G9	MOMSI
MIR Imager	imaging (+limited spectroscopy)	Nasm. or IF/ SCAO or LTAO	3- 20	30"	6 -20	Wide/narrow bands	S3, S9, S5, G9, C10	MIDIR

* : PROMINENT SCIENCE CASES: S3- Detection,Characterization and Evolution of Exoplanets; S9 -Circumstellar Disks; S5- Young Stellar Clusters (incl. Galactic Centre); G4- Resolved Stellar Populations in Galaxies; G9- AGN/BlackHoles ; C2 – A Dynamical Measurement of the Expansion History of the Universe; C4- First Light, The Highest Redshift Galaxies; C7- High z IGM Properties; C10- Physics of High z Galaxies. They are taken from the ELT Science WG Report dated 30.4.2006, available at: http://www.eso.org/projects/e-elt/publications.html

Fig. 4. Table 2

4 Acknowledgements

Studies on ELT instruments have been supported by the European Community (Framework Program 6, ELT Design Study, Contract Number 011863).

References

1. L. Pasquini et al. **ELT-TRE-ESO-11200-0001**, (2006)
2. W.Dent, D. Gostick, E. Atad, M. Strachan **ELT-TRE-UKA-11200-0001**, (2006)
3. C. Evansm et al. **ELT-TRE-UKA-11200-0002**(2006)

CRIRES: A High Resolution Infrared Spectrograph for ESO's VLT

Hans Ulrich Käufl[1]

European Southern Observatory, Karl-Schwarzschild-Str. 2
D-85748 Garching, Germany
hukaufl@eso.org

Summary. CRIRES, a pre-dispersed CRyogenic Infrared Echelle Spectrograph, provides a resolving power $\lambda/\Delta\lambda \approx 10^5$ (or $\Delta v \approx 1.5\frac{km}{s}$ per pixel) between 1000 and 5000 nm at the 8m ESO VLT-UT 1. A curvature sensing adaptive optics system feed is used to minimize slit losses and to provide 0.2" spatial resolution along the slit. A mosaic of 4 Aladdin InSb-arrays packaged on custom-fabricated ceramics boards provides for an effective $5k$ x $0.5k$ pixel focal plane array. Remote insertion of gas cells to measure high precision radial velocities is possible. A linear and circular polarization mode for magnetic Doppler imaging of stellar surfaces is foreseen with motorized retarders in combination with a Wollaston prism. The major design features of CRIRES and a glimpse tutorial preview of astronomical data are given.

1 Introduction

CRIRES is the last of the first generation VLT instruments [1]. It started in 1999 when feasibility was established. In April 2006 the adaptive optics part was commissioned stand-alone at the VLT [2] while the cryostat underwent last modifications and tests in Garching. In May the complete spectrograph was integrated at its final destination (c.f. [3]) for 1st light in June 2006.

2 Main Characteristics of CRIRES

Figure 1 shows the general lay-out of CRIRES. For commissioning reports and pictures see [2] and [3]. The main characteristics are summarized in table 1. CRIRES and its operations concept are described in some detail in [4]. For the latest state see the ESO user documentation. The calibration unit comprises of a Ne arc-lamp, $ThAr$ hollow cathode lamp[1], a halogen lamp and an infrared glower in combination with an integration sphere for flat fielding and spectral calibration. Gas-cells can be moved into the beam for calibration purposes and for searches for very small radial velocity changes similar to the Iodine-cell technique[2]. Retarders in motorized mounts can be

[1]A precision infrared $ThAr$-lamp spectral atlas was created, c.f. [5]

[2]CRIRES a classical long slit spectrograph without image slicer or scrambler is fully sensitive to tracking problems (e.g. $5mas$ yield a spectral shift of $\approx 75\ m/s$).

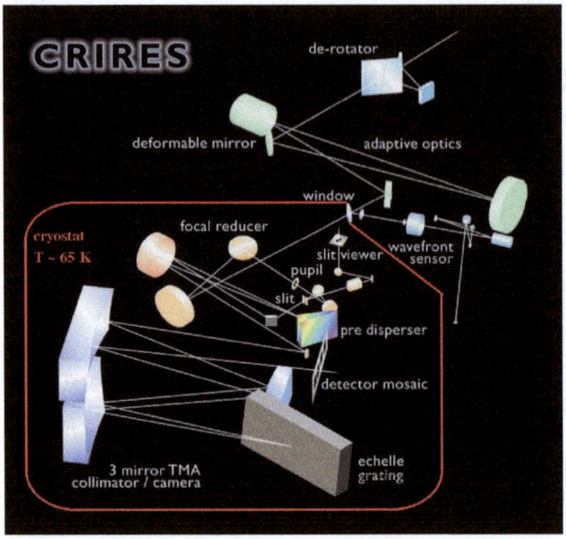

Fig. 1. CRIRES Optical Design The VLT Nasmyth focus (*f15*) is close to the first mirror of the de-rotator assembly. There is a calibration unit beneath the de-rotator (for more details, see text). The shaded area is the cryogenic part.

inserted to use CRIRES for spectro-polarimetry. The de-rotator is followed by a curvature sensing adaptive optics system [6], [7] with the deformable mirror on a kinematic gimbal mount. The entrance window to the cryostat is a dichroic, separating the visible light with high efficiency for the AO wavefront control. The cryogenic optical bench is cooled by three Closed Cycle Coolers to $\approx 65K$. Temperatures are stabilized, in critical areas to the mK level. An all-reflective re-imager with a cold-pupil stop reduces the f-ratio to *f7.5*. Close to the cold pupil a Wollaston prism (MgF_2) can be inserted. The slit-viewer has a pixel scale of 0.05 $arcsec/pixel$ and an unvignetted field-of-view of 25x50 $arcsec^2$. The main slit is continuously adjustable up to several $arcsec$. The pre-disperser has a collimated beam diameter of $100mm$ and uses a $ZnSe$ prism in retro-reflection. The collimator mirror can be slightly tilted with a Piezo actuator for vernier adjustment to compensate for stick-slip effects in the grating and the prism drives. For order selection after the pre-disperser there is a second motorized intermediate slit (close to the small folding mirror next to the prism). The main collimator, a three-mirror anastigmat, produces a $200mm$ collimated beam which illuminates a R2 Echelle grating ($31.6gr/mm$). Whenever necessary, preference in design was given to stability.

Table 1. CRIRES Main Characteristics

Spectral Coverage	$\lambda \approx 950 - 5300nm(\nu \approx 56 - 315\ THz)$
Spectral Resolution	$\frac{\lambda}{\Delta\lambda} \approx 10^5$ or $\Delta v \approx 3\frac{km}{s}$; 2 pixel Nyquist sampling
Array Detector Mosaic	4 x 1024 x 512 Aladdin III *InSb* mosaic; λ-coverage 2%
	pixel-scale 0.1 "/pixel
Dark Current	$0.05 - 0.1e^-/s$ per pixel
Infrared Slit Viewer	Aladdin III *InSb* with *J*, *H* and *K* filters
Precision	calibration and stability (goal) $\approx 75\frac{m}{s}$
	i.e. $1/20^{th}$ of a pixel or $5mas$S tracking error
Adaptive Optics	curvature sensing ESO-MACAO system
	60 sub-apertures, *R*-band wave front sensor

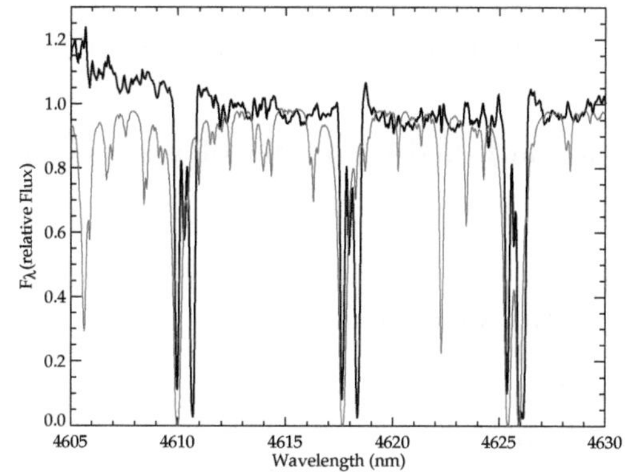

Fig. 2. CO-fundamental Band Spectrum from W33A: In this sample - a fraction of one of the four detectors - the solid line is the spectrum from the compact HII-region at full spectral resolution, corrected as well as possible for telluric absorption; the gray line is a reference illustrating telluric absorption. The parabolic shape of the continuum from W33A is due to *CO*-ice. (data courtesy R. Siebenmorgen, K. Menten and the CRIRES Science Verification Team). *CO* is the most abundant molecule in the universe readily observable, i.e. emitting dipole radiation.

3 Example Spectra from Commissioning

Two examples, both of scientific and tutorial value, are picked from the cornucopia of data taken during commissioning and early science verification. Fig. 2 shows a *CO*-gas envelope around W33. Telluric interferences are minimized by scheduling such that Earth's orbital motion yields the "right" Doppler-shift. The *CO*-spectra provide for new and extremely detailed constraints on the conditions of the molecular cloud and the co-existence of gas, dust and ices. Fig. 3 gives an extragalactic example. Observing a broad line with a resolution of $3km/s$ enables rigorous rejection of telluric absorption. The Si IX line falls into a band of narrow telluric molecular absorption lines which need to be resolved for correction. Only then is it possible to correct and to get

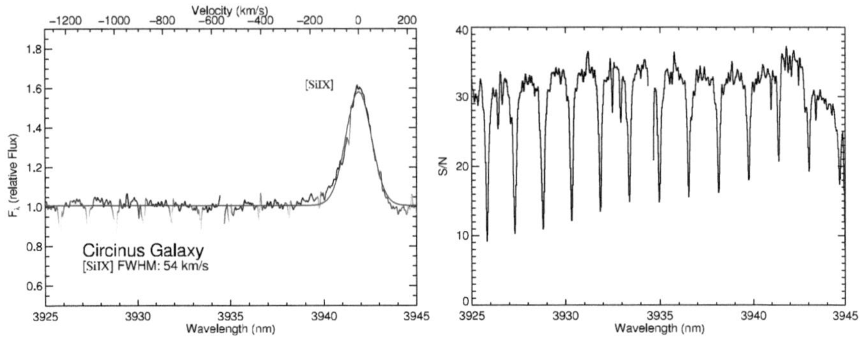

Fig. 3. Sample Spectrum of an Active Galaxy: The left spectrum shows a coronal line (Si IX) in Circinus (data and Gaussian fit, FHWM 54km/s). The right spectrum shows the raw-data, i.e. without telluric correction (see text).

both the equivalent width and the center of gravity and thus the redshift right.

4 Conclusions and Outlook

CRIRES, an unique instrument allowing to access a largely unexplored parameter space, is now fully integrated into the VLT observatory. Various small improvements are envisaged to improve operations. In the next step polarimetry will be implemented. The extremely low dark current in the instrument makes a detector upgrade programme desirable. The implemementation of a Bowen-Walraven image slicer is being studied.

Acknowledgments: Big thanks to the CRIRES team and the Paranal observatory staff for their hard work. Thanks to the dedication of all, CRIRES could be made available to all ESO users by now.

References

1. S. D'Odorico, A.F.M. Moorwood. and J. Beckers: Journal of Optics **22**, pp. 85-98 (1991)
2. H. U. Käufl et al.: The Messenger, **124**, pp. 2-4 (2006)
3. H. U. Käufl et al.: The Messenger, **126**, pp. 32-26 (2006)
4. H. U. Käufl et al.: Proc. SPIE **5492**, p. 1218 (2004)
5. F. Kerber et al.: these proceedings (2007)
6. R. Arsenault et al.: Proc. SPIE **4839**, p. 174 (2003)
7. J. Paufique et al.: Proc. SPIE **5490**, p. 216 (2004)

Stellar Oscillations Network Group: Asteroseismology and Planet Hunting

Frank Grundahl[1]

Department of Physics and Astronomy University of Aarhus Ny Munkegade, 8000 Århus C Denmark. fgj@phys.au.dk

1 Introduction

During the past 10–15 years the search for extrasolar planets has been a driving force in the development of techniques for measuring high precision radial velocities for stars. The application of these methods has led to the discovery of many extrasolar planets and openend the avenue for developing asteroseismology ie. the study of solar-like oscillations in other stars.

SONG (Stellar Oscollations Network Group) is a new initiative which aims at designing and building a global network of small telescopes dedicated to the study of solar-like oscillations in other stars and searching for low-mass planets in short-period orbits around other stars using high precision spectroscpic velocities.

2 Asteroseismology and Planets

During the past 25 years the study of global Solar p-mode oscillations (helioseismology) has dramatically improved our knowledge of the Sun and its interior structure (Christensen-Dalsgaard 2004). If similar methods could be applied to other stars this could lead to a major advance in our understanding of stellar structure and produce strong tests of stellar evolution theory.

Solar-like oscillations in other stars have very low amplitudes and are thus very difficult to observe – in the Sun the intensity fluctuations due to oscillations are at a level of roughly 5 parts per million. The corresponding surface velocity is of the order 20-23cm/s for the highest amplitude modes. The superposition of many oscillation modes causes the total velocity amplitude to be of the order 3-5m/s. In comparison to the background "noise" signal due to mainly granulation and activity the amplitude (ie. contrast) of the oscillation signal is significantly better for velocity measurements than for intensity measurements – this is illustrated by the fact that the direct detection of oscillations in other stars is primarily done with the radial-velocity technique (Bedding and Kjeldsen, 2006)

In the context of this meeting the search for extrasolar planets has already been addressed extensively by several authors and we will not repeat the discussion here. We note here that there is a possible synergy between asteroseismology and the search for exoplanets. This arises from the need for high duty-cycle observations for asteroseismology (due to the typical oscillation periods of 5-50min) which at the same time will reveal the presence of short period planets as clearly demonstrated in the study of μArae by Santos et al. (2004). In addition to this asteroseismology offers the possibility to determine the basic stellar parameters such as mass, radius and age with a precision significantly higher than conventional metohds.

However, there are several limiting factors in the application of high-precision radial velocities to the asteroseismic exploration of stars and the search for low-mass planets. Firstly, for the extraction of oscillation spectra the window function of the observations is severely affected by the fact that stars can only be observed for 8-10 hours at a time from a single observatory causing aliasing in the observed oscillation power spectra. Secondly, the solar-like oscillations are stocastically excited (by convection) and damped. This causes a given oscillation mode have a slithtly different frequency for each excitation – if the observing periods are short compared to the lifetime of the mode the frequency precision that can be obtained will be limited – one way to remedy this is to observe for longer times. Thirdly, it is exceedingly hard to ensure instrumental stability (accuracy) to better than 1m/s over long (months/years) timescales. Any drift of the velocity zeropoint is very hard to monitor and evaluate with observations from a single site. Fourth, in the search for low mass planets, which have small radial-velocity variations, the planetary signal is comparable to the stellar oscillation signal which will act as a non-white noise source. In Fig. 1 we show the solar amplitude spectrum as obtained with the GOLF instrument on SOHO – and how the background "noise" due to granulation and activity may limit the detection of low-mass plantes.

The best way to overcome these difficulties is to establish a network of telescopes dedicated to obtaining long term high-precision velocity time-series observations of bright stars in order to study them asteroseismically and to search for low-mass planets in short period orbits and study their host stars – this is exactly the purpose of SONG.

3 Outline of SONG

After the discovery of the global solar oscillations in the 1970'ties it was quickly realized that long, continuous observations were needed to obtain the best possible oscillation spectrum for the Sun thereby leading the way for several networks of solar observatories, eg. GONG and BiSON.

SONG aims as constructing a similar network dedicated to asteroseismology and planet hunting. The network is intended to have 8 identical nodes distributed globally at existing sites (to avoid the costs of constructing new infrastructure). In order to obtain full sky coverage there will be four sites in

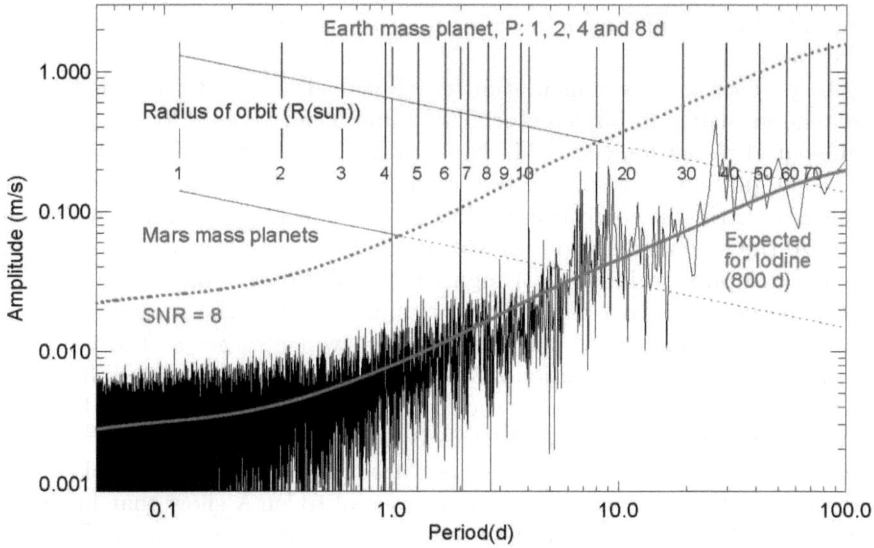

Fig. 1. Amplitude spectrum for an assumed 800 d solar observation using Iodine. The figure show the sensitivity of planet detections (observing for a period of 800 days). A signal from 4 Earth mass planets was added to the simulated Iodine time series. The periods of those planets were 1, 2, 4 and 8 days. Shown is also the SNR = 8 curve. Note that SNR = 8 corresponds to SNR = 4 for a 200 d observing period (more realistic for SONG). Shown are the amplitudes for Earth and Mars mass planets as well as lines indicating the period for planets orbiting the Sun in distances of 1, 2, 3, 4, ... 10, 20 .. times the solar radius. Mars mass planets can be found at periods below 1.1 days (distance: 4.5 R(sun)) and Earth mass planets at periods below 8 days (distance: 16.8 R(sun) = 0.078 AU). The intrinsic stellar noise will not allow detection of a true Earth (P: 1 year, d: 1 AU). In order to get the stellar noise background down to about 2 cm/s (needed for detection of the velocity signal from a true Earth) one would need to collect data for 200-300 years.

each hemisphere. The telescopes and their instrumentation should be operated remotely from a single institution.

At each site the instrumentation will consist of an 80cm telescope with a spectrograph located at a Coudé focus. The spectrograph will have a working spectral resolution of 100000 and employ an iodine cell for velocity reference similar to the method developed by Butler et al. (1996). A preliminary study of the spectrograph design has been carried out at the Anglo Australian Observatory which indicates that by optimizing the instrument for the spectral region covering the wavelengths of the iodine absorption (5000-6000Å) it will be possible to obtain a very high efficiency. The detector will be a 2K×2K detector, possibly with frame-transfer to obtain a high dutycycle.

The motivation behind using an iodine cell for velocity reference is that this is a proven technique (eg. Butler et al. 2004) and the demands on spectrograph stability are much lower than for eg. the ThAr method – this helps to reduce costs.

As part of the instrumentation there will also be a CCD camera with fast readout using an L3 detector. This camera can be used for imaging with high spatial resolution (Law, Mackay and Baldwin, 2006) in the VRI and for telescope guiding. Due to the fast readout capability it is planned to also implement a tip-tilt correction facility for the spectrograph input which will stabilize the stellar image on the slit and help to maintain a high observing efficiency in poor seeing and/or high wind conditions. A dichroic mirror will direct the light at wavelengths shorter than 670nm to the spectrograph and the longer wavelengths to the imager which will also act as the wavefront sensor for the tip-tilt correction ensuring that this can be used for all targets.

It is important to note that the high velocity precision achieved at VLT (UVES), HARPS or KECK (HiRES) is due to the exellent performance of their instruments and not the large aperture of the telescope. A simple scaling between the Butler et al. 2004 observations of αCen A show that for bright stars ($V < 3$) SONG will produce velocities of comparable precision as the VLT – mainly due to a much higher dutycycle from a shorter CCD readout and lower slit losses.

4 Implementation

Currently (autumn 2006) SONG is in the Conceptual Design Phase. It is anticipated that detailed design of a prototype will commence during 2007. An extended prototype phase is foreseen in order to solve all major problems for upscaling to a full network. The price range expected per site will be of the order 1 million Euros.

References

1. T.R. Bedding, H. Kjeldsen: SOHO 18 / GONG 2006 / HELAS conference, "Beyond the Spherical Sun: a new era of helio- and asteroseismology", 7-11 August 2006, Sheffield, England
2. R.P. Butler, G.W. Marcy, E. Williams et al: PASP **108**, 500 (1996)
3. R.P. Butler, T.R. Bedding, H.Kjeldsen et al: ApJ **600**, L75-L78 (2004)
4. J. Christensen-Dalsgaard: SOHO 14 / GONG 2004 Workshop (ESA SP-559). "Helio- and Asteroseismology: Towards a Golden Future". 12-16 July, 2004. New Haven, Connecticut, USA. Editor: D. Danesy., p.1
5. N.M Law, C.D. Mackay, and J.E. Baldwin: A&A **446**, 739 (2006)
6. N.C. Santos, F. Bouchy, M. Mayor: A&A, **426**, L19 (2004)

Interferometric Spectroscopy

Andreas Quirrenbach[1,2] and Simon Albrecht[2]

[1] ZAH, Landessternwarte, Königstuhl 12, D-69117 Heidelberg, Germany
[2] Sterrewacht Leiden, P.O. Box 9513, NL-2300RA Leiden, The Netherlands

Summary. Interferometers operating with high spectral resolution at visible and infrared wavelengths can provide information in spectral lines and velocity-resolved data on the milliarcsecond and sub-milliarcsecond scale. This enables completely new observational approaches to many open problems regarding the physics of stars and circumstellar matter. Using fibers to couple the output from a relatively simple beam combiner to existing spectrographs offers a fast and cost-effective way to implement interferometric high-resolution spectroscopy.

1 Introduction

For more than one century, spectroscopy has been the most important tool of stellar astrophysics. However, despite of its enormous diagnostic power, traditional stellar spectroscopy suffers from one major drawback: it averages the light from all parts of the stellar disk, entangling or even destroying useful information about rotation, surface structure, and the dependence of atmospheric parameters on optical depth. These shortcomings can be overcome by interferometers equipped with high-resolution spectrographs as back-end instruments, which can provide spatially resolved data at high spectral resolution (Quirrenbach 2004).

2 Astrophysics with Interferometric Spectroscopy

In this chapter, we give a few examples of astrophysical questions that can be addressed with interferometric spectroscopy.

Late-Type Giant Stars: Stellar diameter variations with wavelength, or even better wavelength-dependent limb darkening profiles, provide a sensitive probe for the structure of strongly extended atmospheres of cool giant stars. Such data can be directly compared with predictions of theoretical models, and provide qualitatively new tests of three-dimensional stellar model atmospheres (Quirrenbach et al. 1993, Quirrenbach & Aufdenberg 2004).

Cepheids: Uncertainties in the projection factors, which relate the true velocity of the pulsation to the observed radial velocity curve, are a serious limiting factor in current estimates of Cepheid distances with the Baade-Wesselink method (e.g. Sabbey et al. 1995, Marengo et al. 2002). At present, these "p factors" have to be computed from theoretical models. They could be measured directly from limb darkening curves in spectral lines obtained with interferometric spectroscopy.

Generalized Doppler Imaging: The chemical and magnetic properties of stellar photospheres can be mapped with classical Doppler imaging (e.g. Rice 2002, Kochukhov et al. 2004). However, the reconstruction of stellar surface features from line profile variations alone is plagued with ambiguities. These can largely be resolved by the additional phase information contained in interferometric data (Jankov et al. 2001). For stars with inhomogeneous surface properties, interferometry enables studies of individual surface regions.

Rotational Axes: Stellar rotation induces a phase difference between the red wings and the blue wings of stellar absorption lines. Measuring the position angle of the phase gradient allows determining the orientation of the stellar axis on the sky (Petrov 1989, Chelli & Petrov 1995). More detailed modeling of the interferometric signal can also provide the inclination of the stellar rotation axis (Domiciano de Souza et al. 2004). Determining the orientation of stellar rotational axes in space is of interest in double or multiple star systems, where one can check whether the rotation axes are aligned with each other, and with the orbital rotation axes of the systems. For stars which harbor planets, the mutual inclination between the orbital plane of the companion and the rotation axis of the star contains information about the early history of the planetary system.

Differential Rotation: Along with the oscillation spectrum, differential rotation is a powerful diagnostic of the interior structure of a star. Unfortunately, it is difficult to observe differential rotation with classical spectroscopy because of degeneracies between inclination, limb darkening, and differential rotation (e.g. Gray 1977). These degeneracies can be resolved by the additional information from interferometric spectroscopy (Domiciano de Souza et al. 2004). High spectral resolution is essential for this application.

3 Implementation of Interferometric Spectroscopy

The most important observable in an interferometer is the *complex visibility*, i.e., the amplitude and phase of the coherence function of the radiation received by the two telescopes (e.g. Quirrenbach 2001). Full information on the complex visibility is contained in the four fringe quadratures; therefore one has to count the photons at phases 0, $\pi/2$, π, and $3\pi/2$. A convenient way of generating these four bin counts A, B, C, and D, is shifting one polarization by $\pi/2$ with respect to the other, as shown schematically in Fig. 1. For any wavelength λ, the full interferometric information is thus contained in the intensities $A(\lambda)$, $B(\lambda)$, $C(\lambda)$, and $D(\lambda)$ carried in the four output beams of the beam combiner. These can be routed to a spectrograph with optical fibers.

In the case of the VLTI, one could take advantage of the existing UVES instrument, which provides a resolution of $R \approx 60,000$ when a fiber with

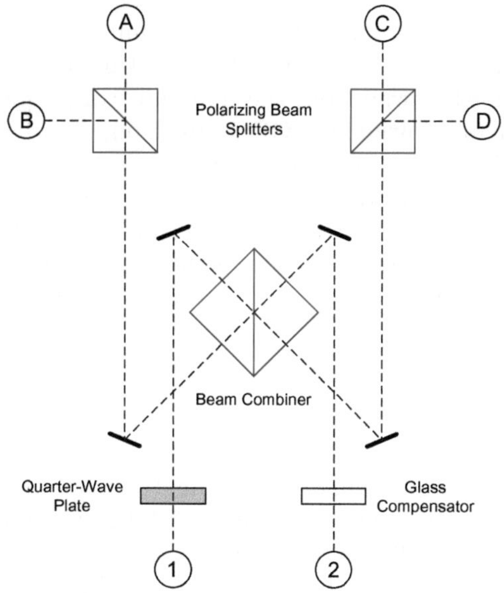

Fig. 1. Sketch of a four-output beam combiner. An external fringe tracker stabilizes the pathlength difference between the beams from two telescopes, which are relayed to the inputs at the bottom of the figure. The quarter-wave plate produces an achromatic $\pi/2$ phase shift between the two polarization states in one beam. The beams from both telescopes are combined at the main beam combiner; light emanating from the right-hand side of the beam combiner is shifted by π with respect to the left side. The two polarizations in each arm are separated be polarizing beam splitters, and the four outputs coupled into fibers for relay to detectors or spectrograph inputs. The phases in the four outputs A, B, C, and D are shifted by 0, $\pi/2$, π, and $3\pi/2$, respectively.

core diameter of 70 μm is chosen. The required fiber length is ~ 150 m; fibers of this length and of the type used for the multi-object FLAMES link have a transmission ≥ 80 % over the wavelength range from 0.6 μm to 1 μm. The interface between the fibers and the spectrograph could be very similar to that of FLAMES.

Interferometry at visible wavelengths requires a good wavefront quality. At present, this could be achieved at the VLTI only by stopping down the apertures to diameters of ~ 75 cm, which corresponds to $\sim 3\, r_0$ at 800 nm. However, the full apertures of the 1.8 m telescopes could be used in the visible together with modest adaptive optics systems.

Since the VLTI delay lines are filled with air, there is an imbalance between the pathlength in air between the two interferometer arms, which gives rise to a variation of the optical pathlength difference with wavelength due to the dispersive nature of air. For a short interferometric observation this leads to a slow variation of the phase with wavelength, which can easily be calibrated out. During the course of a several-minute spectroscopic exposure, however, the amount of air imbalance changes; this leads to a loss of fringe contrast at wavelengths away from the nominal fringe tracking wavelength. During a 15-minute integration, the OPD can change by several meters, which leads to a differential fringe rotation of several μm between the blue and red edges of a 0.6 μm to 1.0 μm wavelength band. This dispersion problem could be circumvented by restricting the exposure time and/or the baseline geometry, but it is better to foresee an atmospheric dispersion compensator. A combination of SF10 and BK7 glasses can provide compensation to ≤ 70 nm even for extreme delay rates.

4 Conclusions

The combination of long-baseline interferometry with high-resolution spectroscopy in one instrument will give access to hitherto unobservable properties of stellar surfaces, stellar rotation, and circumstellar matter. The infrastructure available at ESO's VLT facility on Cerro Paranal — the VLTI with fringe-tracking capabilities, the 1.8 m Auxiliary Telescopes, and the high-resolution spectrograph UVES — provides an outstanding opportunity to implement interferometric high-resolution spectroscopy with only a few additional components: a simple beam combiner and a fiber link. Since none of the items that make most instrumentation projects time-consuming and expensive (detector, dewar, read-out electronics, spectrograph optics, ...) are required, the implementation could be done quickly and at low cost. Similar opportunities of coupling interferometers to existing visible-light or near-infrared spectrographs exist at other observatories.

References

1. Chelli, A., Petrov, R.G. (1995). A&AS 109, 401-415
2. Domiciano de Souza, A., et al. (2004). A&A 418, 781-794
3. Gray, D.F. (1977). ApJ 211, 198-206
4. Jankov, S., et al. (2001). A&A 377, 721-734
5. Kochukhov, O., et al. (2004). A&A 424, 935-950
6. Marengo, M., et al. (2002). ApJ 567, 1131-1139
7. Petrov, R.G. (1989). In *Diffraction-limited imaging with very large telescopes.* Eds. Alloin, D.M., Mariotti, J.M., NATO ASI Vol. 274, p. 249-271
8. Quirrenbach, A. (2004). In *New frontiers in stellar interferometry.* Ed. Traub, W.A., SPIE Vol. 5491, p. 146-153
9. Quirrenbach, A. (2001). ARAA 39, 353-401
10. Quirrenbach, A., Aufdenberg, J. (2004). In *Modelling of stellar atmospheres.* Ed. Piskunov, N., Weiss, W.W., Gray, D.F., IAU Symp. 210, p. E68.
11. Quirrenbach, A., et al. (1993). ApJ 406, 215-219
12. Rice, J.B. (2002). AN 323, 220-235
13. Sabbey, C.N., et al. (1995). ApJ 446, 250-260

A Global Network of 2 m-class spectroscopic telescopes

Mkrtichian D. E.,[1], Hatzes A. P.[2], Lehmann H.[2], Han I.[3], Lee B. C.[3], Kim K.-M.[3], Sergeev A.[4], Kameswara Rao N.[5], and Plachinda S.[6]

[1] Astrophysical Research Center for the Structure and Evolution of the Cosmos, Sejong University, Seoul 143-747, Korea, davidm@sejong.ac.kr
[2] Thüringer Landessternwarte Tautenburg, Sternwarte 5, D-07778, Tautenburg, Germany, artie@tls-tautenburg.de
[3] Korea Astronomy and Space Science Institute, Daejeon 305-348, Korea, iwhan@kasi.re.kr
[4] International Center for Astronomical and Medico-Ecological Researches, Terskol, Russia, sergeev@terskol.com
[5] Indian Institue of Astrophysics, Bangalor 560 034, India, nkrao@iiap.res.in
[6] Crimean Astrophysical Observatory, Nauchny, Crimea 98409, Ukraine, plach@crao.crimea.ua

1 Introduction

Nowadays, the accuracy of individual radial velocity measurements of the best spectroscopic instruments is of about 0.5-1 m/s (see these proceedings). The corresponding pulsational photometric amplitude variation is expected to be of the order of 10^{-2} mmag - a level that seems impossible to achieve from the ground, but only with specialized photometric instruments in space. This fact demonstrates the superiority of ground based Doppler shift measurements for the study of pulsations compared to photometry.

Multisite Doppler shift measurements can essentially reduce the spectral window problem - the only serious one that still hinders the ground-based spectroscopic observations. A network of 2-3 m size telescopes well distributed on geographic longitude can accomplish much of what space-born (nowadays - photometric) asteroseismic observations can do, but at significantly reduced costs. Compared to large telescopes, such a coordinated network would be more efficient in planet search surveys which require a considerable amount of telescope time and flexible scheduling.

2 Network's Sites

The project of a network of 2 m class spectroscopic telescopes was initiated in 2002 [1] in order to ensure multisite precise Doppler shift measurements of pulsating stars and collaborative exoplanet surveys from individual sites during out-off campaign time. Currently, the Network cooperation consists of six groups having access to 1.8-2.6 m spectroscopic telescopes spanning 8 hours

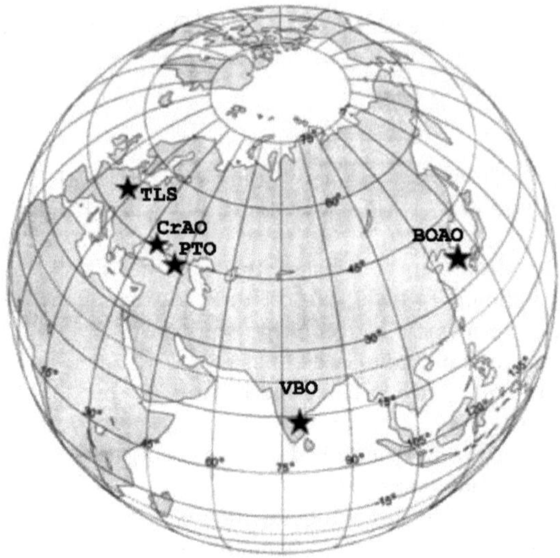

Fig. 1. The Network of 2 m-class telescopes distributed over Eurasia.

angles over Eurasia (Fig. 1). Telescopes, spectroscopic resolutions of their echelle-spectrometers, geographic locations and operating institutes are listed in Table 1. To have a complete global coverage for future campaigns, the Network's team will also apply for observing time at telescopes in North America (e.g. the 2.1 or 2.7 m telescopes of the McDonald Observatory or other suitable observatories) at which our team members have considerable experience in precise radial velocity measurements and exoplanet search. Currently, two echelle-spectrometers of Network telescopes, namely at the Crimean Astrophysical Observatory (CrAO, Ukraine) and at the Vainu Bappu Observatory (VBO, India, see [5]) are testing their echelle-spectrometers with iodine cells in a precise Doppler shift mode. The full Network will be ready for multi-site campaigns in 2007.

Table 1. The Network's sites. Operating institutes are listed according to authors affilations.

Site and Location	Longitude	Latitude	Elevation m	Tel. m	Resolution R
TLS, Tautenburg, Germany[2]	11°42.7'E	+50°58.8'	341	2.0	67,000
CrAO, Crimea, Ukraine[6]	34°0.8'E	+44°43.6'	600	2.6	70,000
Peak Terskol Obs., Russia[4]	42°30.0'E	+43°16.5'	3124	2.0	120,000
Vainu Bappu Obs., India[5]	78°49.6'E	+12°34.6'	725	2.3	70,000
BOAO, South Korea[3]	128°58.6'E	+36°09.9'	1124	1.8	90,000

3 First Results

The very recent results from Thüringer Landessternwarte Tautenburg (TLS) and Bohyunsan Optical Astronomy Observatory (BOAO) sites, operated independently, were already published [2,4].

At four sites - TLS, BOAO, PTO, and Mc Donald Observatory, networked precise radial velocity studies of pulsating stars were carried out by team members for some common candidate stars. In May-June 2005, two test campaigns on the study of K-giant pulsations and of Polaris were undertaken from McDonald Observatory and BOAO, and from BOAO and PTO, respectively. First results are in preparation now.

Here we present some additional results of the BOAO planet search and pulsation study program obtained with the high-resolution fiber-fed echelle-spectrograph BOES [3]. The short-term accuracy of the RV determination is of about 3 m/s. Figure 2 shows the BOES orbital RVs of the K-giant star β Gem that independently confirm the discovery of a planet by [2,6]. Figure 3 (left) shows the 732 day variability of the radial velocities of Arcturus obtained from a three years-long monitoring. The variation is most likely due to the rotation of the star. Figure 3 (right) presents the high-resolution segment of these observations showing low-amplitude pulsations with period of 1.9 days.

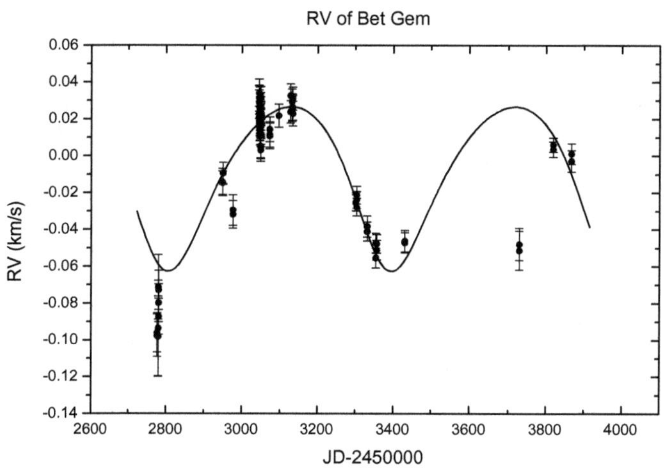

Fig. 2. The BOES radial velocities of β Gem confirming the planet orbiting the star.

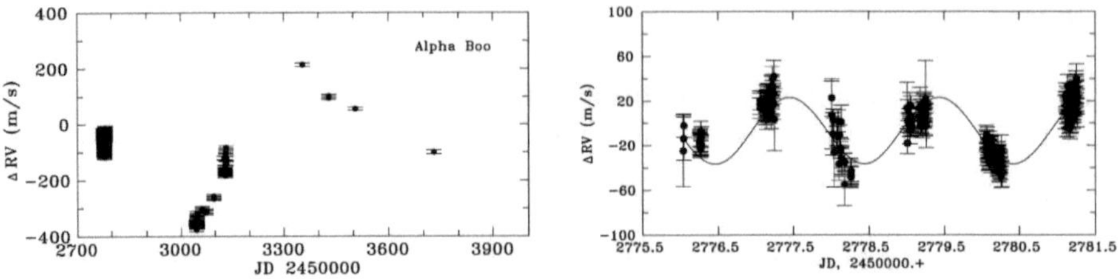

Fig. 3. Left: 413 precise relative radial velocity measurements of Arcturus obtained with BOES in 2003-2005 show the long-period 732 d variation. Right: the time series of Arcturus shows short-term oscillations with a period of 1.9 d.

4 Conclusion

We presented a brief description of the cooperation of international groups that created a Network of 1.8-2.6 m size spectroscopic telescopes. First tests of new spectrometers at the included sites show the high accuracy of Doppler shift measurements that can be achieved. First results on K-giant exoplanet and pulsation search programs undertaken at different Network sites are given. The Network will start to operate in multi-site mode in 2007.

References

1. A. P. Hatzes, D. E. Mkrtichian: KFNTS, **4**, 59 (2003)
2. A. P. Hatzes, W. D. Cochran, M. Endl, E. W. Guenther, S. H. Saar, G. A. H. Walker, S. Yang, M. Hartmann, M. Esposito, D. B. Paulson, M. P. Döllinger: A&A, **457**, 335 (2006)
3. K. M. Kim, B. H. Jang, I. Han, J. G. Jang, H. C. Sung, M. Y. Chun, S. Hyung, T. S. Yoon, S. Vogt: Journ. of Korean Astron. Society, **35**, 221 (2002)
4. K. M. Kim, D. E. Mkrtichian, B.-C. Lee, Inwoo Han, A. P. Hatzes: A&A, **454**, 839 (2006)
5. N. Kameswara Rao, S. Sriram, K. Jayakimar, F. Gabriel: J. Astrophys. Astr. **26**, 331 (2005)
6. S. Reffert, A. Quirrenbach, D. S. Mitchell, S. Albrecht, S. Hekker, D. A. Fischer, J. W. Marcy, R. P. Butler:ApJ. **652**, 661 (2006)

Possibility of Heterodyne Correlation Interferometry with a Tunable Laser and Absolute Frequency Measurements

S. Johansson[1] and V. Letokhov[1,2]

[1] Atomic Astrophysics, Lund Observatory, Box 43, SE-221 00 Lund, Sweden sveneric@astro.lu.se
[2] Institute of Spectroscopy of Russian Academy of Sciences, 142190 Troitsk, Moscow region, Russia letokhov@isan.troitsk.ru

1 Introduction

We consider the possibility of measuring the true width of the narrow optical spectral lines of astrophysical lasers. The lines should have a subDoppler spectral with of 30-100 MHz or even less. To make measurements with spectral resolution better than 10^7 and angular resolution better than 0.1 arcsec we suggest to use the ground-based Brown-Twiss-Townes optical heterodyne intensity correlation interferometry with the possibility of absolute frequency measurement. The estimates made of the S/N ratio for optical heterodyne astrophysical laser experiments imply that it should be feasible.

2 Astrophysical Lasers

The spectral lines of astrophysical lasers, in Fe II [1, 2, 3] and OI [4] observable in the range 0.9-2.0 μm can have a width, $\Delta\nu$, somewhere between the Doppler width $\Delta\nu_D$ and down to at least $0.1{\cdot}\Delta\nu_D$ (Fig. 1). The magnitude of $\Delta\nu_D$ for the FeII and OI lines from the HI region of the Weigelt blobs of η Carinae depends on the temperature T and can amount to $\Delta\nu_D \sim$ 300-1000MHz, since T is in the range 100-1000 K. Accordingly, the laser line width $\Delta\nu$ can lie between 30 and 1000 MHz. To measure such lines adequately requires a spectral resolution of R$\sim 10^7$, which is very difficult to achieve with the standard spectroscopic techniques. For this reason, it is expedient to use the Brown-Twiss-Townes correlation method [6, 7, 8], modified by utilizing the up-to-date capabilities of electronics, optics and lasers [9].

Actually, the speckle interferometry technique was used to discover gas condensations (Weigelt blobs) in the vicinity of η Carinae [10]. To master new wavelength regions and achieve high spectral resolution levels, the intensity interferometry method was modified to become heterodyne interferometry [8]. This technique uses a local monochromatic laser oscillator to produce beats between the light wave of the local oscillator. The method can be

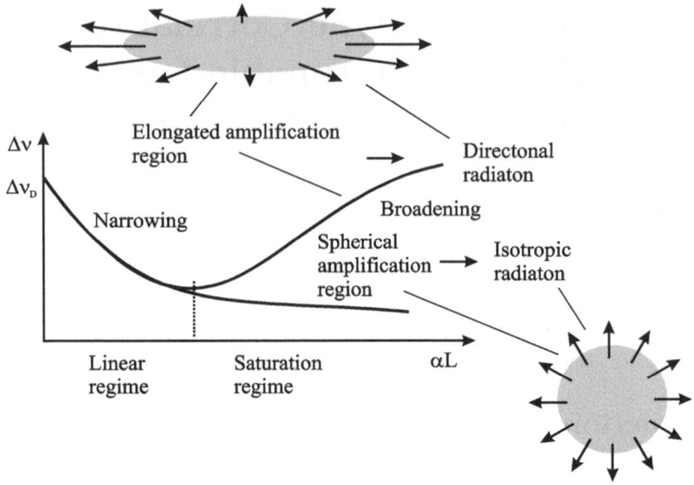

Fig. 1. The qualitative evolution of the spectral width of astrophysical laser lines in the linear and saturated regimes for elongated and spherical shapes of the lasing volumes (from [5]

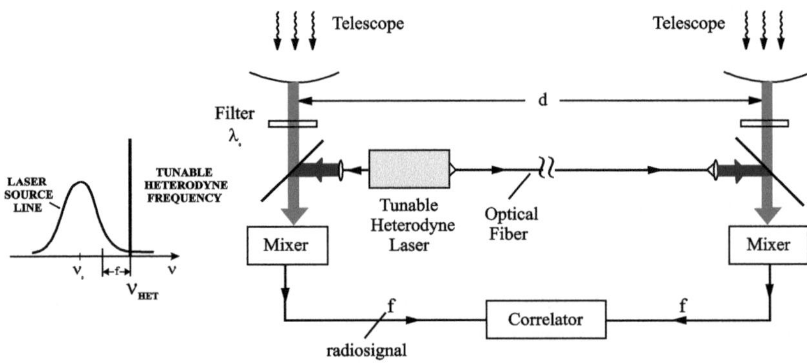

Fig. 2. Brown-Twiss-Townes optical laser heterodyne intensity correlation interferometer

considered intermediate between intensity interferometry and direct interferometry. Townes and co-workers made successful observations in the 10-μm infrared window of the atmosphere using a CO_2 laser as a local oscillator [8].

Fig. 2 presents a schematic diagram of a Brown-Twiss-Townes (BTT) optical heterodyne interferometer that can be used to measure both the angular size and the emission spectrum of the Fe II, OI laser line sources in the Weigelt blobs of η Carinae. A specific feature of such a correlation interferometer is the use of a 0.9-2 μm tunable monomode diode lasers as a local oscillator and an optical fiber to transport this monomode laser radiation. This is much easier to accomplish than to transport the space laser radiation received by the telescopes. The distance d between the telescopes should meet the requirement in angular resolution that the radiation emitted by the Weigelt blob studied is separated from the photospheric radiation of η Carinae, i.e.,

$$d \simeq \lambda_0 \frac{L}{f \cdot D} \tag{1}$$

where $L = 10^{22}$ cm is the distance to η Car, D 10^{15} cm is the diameter of the Weigelt blob, and f is the fraction of the blob region, in which the laser effect takes place to produce the radiation received by the ground- based telescopes. For estimation purposes, one can set f=1, and the necessary distance between the telescopes will then be d 10^3 cm. By choosing d 10^4 cm, one can, in principle, analyze laser radiation coming from blob regions as small as one tenth of the size of the blob.

The spectral resolution R will be determined by the reception bandwidth B (in Hz) of the photomixer:

$$R = \frac{c}{\lambda_0 \cdot B} \tag{2}$$

At B 10^6 Hz the spectral resolution R $3 \cdot 10^8$, which is sufficient to measure the emission spectrum of an astrophysical laser with a spectral width tens of times narrower than the Doppler width. The dependence of the correlation signal on the distance d between the telescopes should give information on the angular size φ of the blob region wherein the astrophysical laser of interest is active at the wavelength λ_0 under study.

The signal-to-noise ratio S/N that poses the requirement for the primary mirror diameter a_0 of the telescopes, the reception bandwidth B, and the observation time τ is quite high ($> 10^3$) for the case of an astrophysical laser in vicinity of Carinae [9].

An important feature of the considered technique is the capability of absolute frequency measurement of spectral line center of astrophysical lasers using laser frequency comb.

3 Acknowledgements

V.L. acknowledges financial support through grant (S.J.) from the Royal Swedish Academy of Sciences and the Wenner-Gren Foundation. V.L. is grateful to Lund Observatory for hospitality and to the Russian Foundation for Basic Research for support through Grant No 03-02-16377. The research project is also supported by a grant (S.J.) from the Swedish National Space Board.

References

1. S. Johansson, V. Letokhov: JETP Lett. **75**, 496 (2002)

2. S. Johansson, V. Letokhov: Phys. Rev. Lett. **90**, 01110-1 (2003)
3. S. Johansson, V. Letokhov: A&A **428**, 497 (2004)
4. S. Johansson, V. Letokhov: MNRAS **364**, 731 (2005)
5. S. Johansson, V. Letokhov: PASP **115**, 1375 (2003)
6. R. Hanbury Brown, R. Twiss: Nature **177**, 27 (1956)
7. R. Hanbury Brown: *The Intensity Interferometer* (Taylor & Francis Ltd., London 1974)
8. M.A. Johnson, A.L. Betz, C.H. Townes: Phys. Rev. Lett. **33**, 1617 (1974)
9. S. Johansson, V. Letokhov: New Astronomy **10**, 361 (2005)
10. G. Weigelt, J. Ebersberger: A&A **163**, L5 (1986)

Michel Mayor, Willi Benz and Thierry Forveille discuss planets; Roger Ferlet debates with Hans-Ulrich Käufl about comets; while Claudio Melo observes.

CODEX

Luca Pasquini[1], G. Avila[1], B. Delabre[1], H. Dekker[1], S. D'Odorico[1], J. Liske[1], A. Manescau[1], P. Bonifacio[2], S. Cristiani[2], V. D'Odorico[2], P. Molaro[2], E. Vanzella[2], P. Santin[2], M. Viel[2], M. Dessauges-Zavadsky[3], C. Lovis[3], M. Mayor[3], F. Pepe[3], D. Queloz[3], S. Udry[3], M. Haehnelt[4], M. Murphy[4], R. Garcia-Lopez[5], F. Bouchy[6], S. Levshakov[7], and S. Zucker[8]

[1] European Southern Observatory, Garching bei München, Germany
lpasquin@eso.org
[2] INAF-Trieste, Italy
[3] Observatoire de Genève, Switzerland
[4] IAC-Tenerife, Spain
[5] IoA-Cambridge, UK
[6] IAP-Paris, France
[7] Ioffe Institute, St. Petersburg, Russia
[8] Tel Aviv University, Israel

CODEX is a high resolution spectrograph for the European ELT. CODEX is conceived to reach the highest precision and stability, allowing the execution of programs spanning many years.

1 Why an Accurate H-R Spectrograph at ELT?

Many talks in this conference have shown the power of high resolution spectroscopy, coupled to high precision. A new spectrograph, fed by an Extremely Large Telescope, will have such a huge collecting power that it will allow for the first time some outstanding applications. We argue here that such a new instrument should therefore aim at the highest long term stability and precision. High resolution and high accuracy need photons; the ELTs will give us a tremendous opportunity, and we shall make use of it.

The CODEX team concentrated on a few scientific cases, considered outstanding, and most of which have been discussed in this conference. The quest for high resolution and high precision applied to *exo-planets* science has been presented by e.g. [8] which has shown as extremely accurate measurements will be required to find and to characterize terrestrial planets. The discussion as well as the controversial results on the *variability of the fine structure constant* [7] [3] [2] have emphasized that high quality data are required, and how the effects of systematics can result in misleading conclusions, enhancing the quest for high resolution, high S/N ratio, but precision as well. These studies will acquire an even stronger relevance, once they will be coupled to the study of the variability of the *electron to proton mass* [5] which also requires an optical high resolution, high accuracy spectrograph [9]. The case for a very accurate measurement of ^{7}Li *and of the* $^{6}Li/^{7}Li$ *isotopic ratio* has been well illustrated by [1]. We finally note as several speakers emphasized that, in order to reach an effective high S/N, photons must be coupled to excellent flat fielding capabilities, able to eliminate the detector imperfections (fringes) and other instrumental effects such as the blaze function in cross-dispersed echelle spectra.

All these cases call for a high resolution, super stable spectrograph, fed by a very stable input (fibre) system equipped with a superior quality calibration system.

Many other applications for such a system can be found, and several have been discussed at this conference.

1.1 Measuring the Expansion of the Universe

In addition to the cases mentioned in the previous section, the CODEX team developed a novel experiment, aiming at the direct measurement of the dynamics of the Universe. It is possible in principle to directly measure the change of the expansion rate of the Universe with time.

The CODEX experiment is conceptually very simple: by making observations of high redshift objects over a time interval of several years, we want to detect and use the wavelength shifts of spectral features of light emitted at high redshift to probe the evolution of the expansion of the Universe directly. The wavelength change is in fact directly related to the de- or acceleration of the Universe [4].

Fig. 1 shows the expected change of redshift for a range of relativistic models with no curvature as a function of redshift. The wavelength shift has a very characteristic redshift dependence. At some redshifts the wavelengths are "stretched" while in others they are "compressed". The wavelength shift corresponds to a Doppler shift of about 1-10 cm/s over a period of 10 yrs.

A priori it is not obvious which objects and which spectral features are best suited for a precise measurement of $\dot{v}$. For a given energy flux the precision of the final measurements will increase with the sharpness of the spectral features (less noise) and increasing wavelength (more photons). Another important consideration is the expected peculiar acceleration associated with peculiar motions relative to the Hubble flow, which will act as additional noise. The numerous absorption lines in the spectra of high-redshift QSOs, which make up the so-called Lyα forest, appear to be ideal targets for a measurement of $\dot{v}$.

In order to quantitatively assess the feasibility of the measurement, Monte Carlo simulations have been carried out independently by several groups. The high resolution spectra of QSOs were simulated, noise added and the process repeated for the second epoch. The pairs of spectra so produced were compared and the "measurement" performed. Fig. 2 shows the result of a full simulation, taking 36 QSOs at different redshifts.

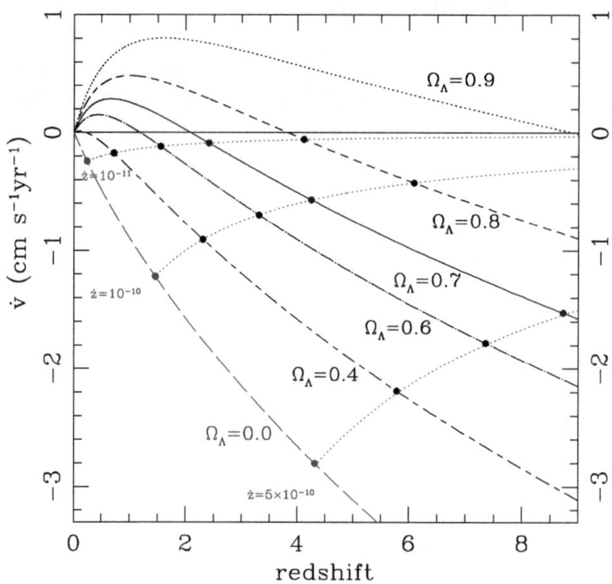

Fig. 1. Redshift drift/yr as a function of redshift. The curves refer to relativistic models with no curvature and different values of the cosmological constant.

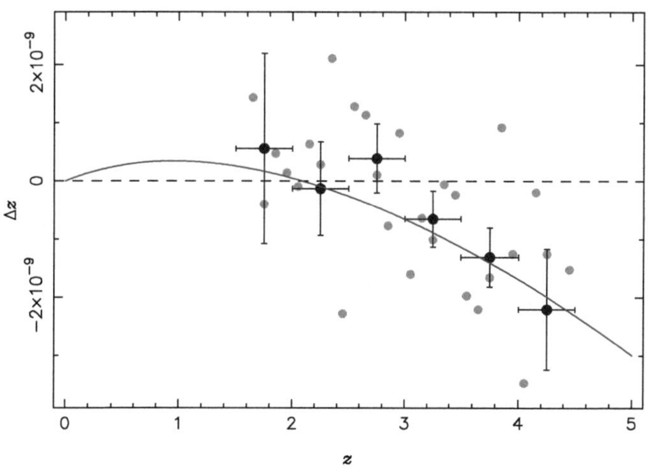

Fig. 2. Full, realistic simulation of the CODEX experiment for the measurement of the cosmic dynamics. Spectra for 36 QSOs at different redshifts have been simulated for two epochs, 10 years apart. The difference has been computed, simulating the outcome of the observations. The cosmological signal is clearly detected.

1.2 Summary of Requirements

The scientific topics summarized not only show the power of a high resolution spectrograph fed by an ELT, but also as the highest stability and precision is required. Our goal is that an accuracy of 1 cm/sec over the timescale of decades should be provided.

To define the optimal resolving power represents a difficult trade-off because on the one hand the highest R should be aimed for, while competition for photons and detector area would call for a low value. Pending a definitive decision, the best compromise has been chosen to R∼150000. Good sampling should be assured.

Similarly, we should aim at the largest spectral coverage, compatible with a high system efficiency and affordable design. The minimum coverage was set to the 380-680 nm range; an extension, in particular of the blue edge, would be highly appreciated.

2 The CODEX Design

The CODEX design is based on an extension of the positive HARPS experience [6]. The light at the coudé focus of the E-ELT is received by an image slicer and brought through (relatively short) fibres to three identical spectrographs, which are under vacuum in a mechanically and thermally controlled environment. No movable functions and no active compensations are foreseen.

The feeding system foresees to slice 1 arcsecond on 18 fibres of 120 microns diameter each.

The spectrograph design contains several novel concepts, and makes use of pupil splitting and anamorphism in order to keep the dimensions of the optics and of the echelle grating reasonable. The camera largest lens is 35 cm in diameter and the echelle grating is only twice the UVES size: 20X160 cm.

The two arm design has been chosen to optimize the efficiency and to keep each detector (and therefore cryostat) limited in size. VPHs are used as cross-dispersers. The actual design provides a minimum order separation of 1.6 mm on the detectors, which are 6Kx6K for the blue (380-548 nm) arm and an 8Kx8K for the red (550-760) arm. Further optimization will be possible in more advanced phases of the project. None of the spectrograph parameters look at present critical, so small changes can be accomodated. A picture of the CODEX optical design is given in Fig. 3.

In addition to the CODEX project, we are pursuing two important parallel developments: the first is the development of a novel calibration system based on laser frequency combs [10], which will be able to produce a super accurate, equally spaced, stable source for wavelength calibration. The second is the construction of ESPRESSO, a CODEX precursor to be fed by

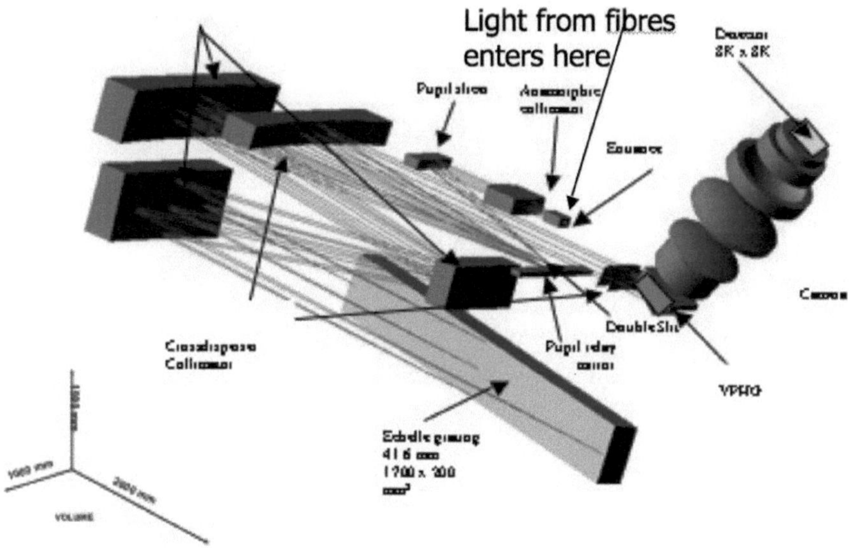

Fig. 3. Optical scheme of the CODEX spectrograph. For sake of clarity only one arm is presented. The second arm will make use of the same echelle grating.

the VLT telescopes. These developments will bring many of the requirements expressed in this conference closer to reality.

References

1. Asplund, M. Nissen, P.E. 2007, these proceedings
2. Chand, N. et al. 2007, these proceedings
3. Levshakov, S. et al. 2007, these proceedings
4. Liske, J. et al. 2007, in preparation
5. Martins, C. 2007, these proceedings
6. Mayor, M. et al. 2003, The Messenger, 114, 20
7. Murphy, M. 2007, these proceedings
8. Pepe, F. et al. 2007, these proceedings
9. Petitjean, P. et al. 2007, these proceedings
10. Udem, T., Holzwarth, R., Hänsch, T. 2002, Nature, 416, 233

Part VIII

Posters

Precision Laboratory UV and IR Wavelengths for Cosmological and Astrophysical Applications

M. Aldenius and S. Johansson

Atomic Astrophysics, Lund Observatory, Box 43, SE-221 00 Lund, Sweden
maria@astro.lu.se

1 Introduction

The quality of astronomical spectra is now so high that the accuracy of the laboratory data is getting more and more important. Both in astrophysics and in cosmology the needs for accurate laboratory wavelengths have increased with the development of new ground-based and air-borne telescopes and spectrographs. The high-resolution UV Fourier Transform (FT) spectrometer at Lund Observatory is being used for studying laboratory spectra of astrophysically important elements.

The ongoing investigations of possible variations in fundamental constants are demanding very accurate laboratory wavelengths of better than $\delta\lambda \sim 0.2\,\text{mÅ}$. One of the methods of investigating such variations in the fine-structure constant, $\alpha \equiv (1/\hbar c)(e^2/4\pi\epsilon_0)$, is the many-multiplet (MM) method, see e.g. [1]. This method requires very accurate relative laboratory wavelengths for a number of UV resonance lines from several ionic species.

With the new focus on IR spectra within astrophysics (e.g. CRIRES at VLT) the demands of accurate laboratory IR wavelengths have increased. Spectra from low density astrophysical plasmas (e.g. nebulae) contain parity-forbidden lines, which are of interest for diagnostics. These lines cannot in general be studied in laboratory spectra, but by using allowed transitions the values of the energy levels can be improved and new accurate wavelengths can be determined.

2 Resonance UV Wavelengths for Cosmology

Laboratory wavelengths and wavenumbers of in total 23 UV lines visible in high-redshift quasar absorption spectra have been measured using high-resolution FT spectrometry. To improve the accuracy of the relative wavelengths lines from Mg I, Mg II, Ti II, Cr II, Mn II, Fe II and Zn II have been measured simultaneously, using a composite light source, see Fig. 1. Emphasis has been put on wavelength calibration and possible line structure, such as isotope structure and hyperfine structure. The uncertainties of the absolute and relative wavelengths are estimated to be 0.1-0.2 mÅ and 0.03 mÅ, respectively. The results have been published in [2].

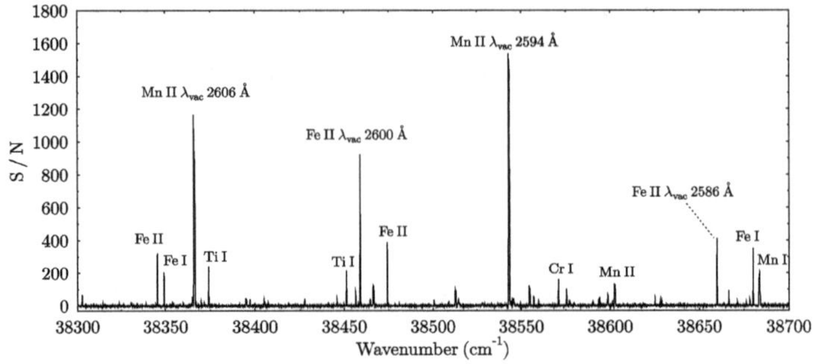

Fig. 1. A 400 cm^{-1} region of the spectrum, containing four of the UV lines

3 Parity Forbidden IR Wavelengths for Astrophysics

In low density astrophysical plasmas (e.g. nebulae) low lying metastable states can be populated. The possible radiative decay channels for these are through parity forbidden, M1 and E2, infrared transitions. Since these lines in general cannot be observed in laboratory spectra the measured UV spectra, complemented with more measurements, are used for determining accurate values of the energy levels involved. As many UV transitions as possible are used to improve the accuracy of the energy levels, see Fig. 2. The wavelengths for the forbidden lines are then determined using Ritz combination principle. Our measurements include IR lines from Ti II, Cr II, Mn II and Fe II.

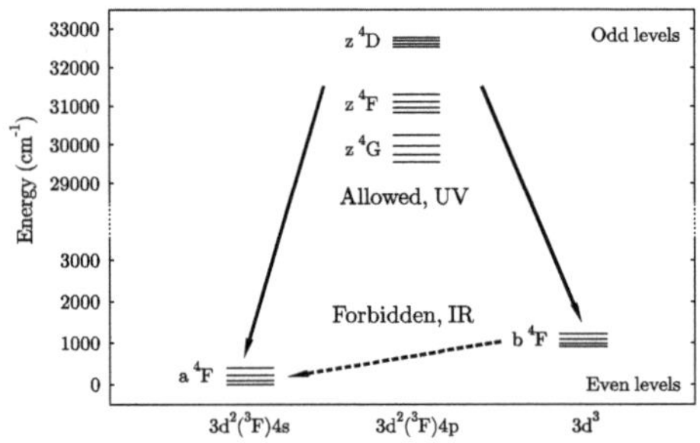

Fig. 2. Partial energy level diagram of Ti II displaying low even levels and higher odd levels, with parity forbidden IR lines and allowed UV lines

References

1. J.K. Webb, V.V. Flambaum, C.W. Churchill, M.J. Drinkwater, J.D. Barrow: Phys. Rev. Lett. **82**, 884 (1999)
2. M. Aldenius, S. Johansson, M.T. Murphy: MNRAS **370**, 44 (2006)

Abundance Analysis of α Centauri A

L. Bigot[1], F. Thévenin[1], J. Provost[1] and G. Berthomieu[1]

Observatoire de la Côte d'Azur, BP 4229, 06304 Nice, France.
lbigot@obs-nice.fr

1 Introduction

The α centauri system is a reference in stellar physics. The two components, α Cen A (G2V) and α Cen B (K1V), have been extensively studied because of their proximity (1.3pc) and similarity to the Sun. The discovery of solar-like oscillations (Bouchy & Carrier 2002) and the recent interferometric measurements (Kervella et al. 2003) make this system one of the best constrained and therefore one of the most interesting to test both stellar evolution and atmospheric models. The chemical composition of this metal-rich system is still debated ([Fe/H] $\approx +0.1$ to $+0.2$ dex) and is a source of uncertainty in asteroseismic diagnostics. In this work, we propose to improve the Iron abundance by the use of realistic 3D radiative hydrodynamical (RHD) simulations. We present preliminary results for α Cen A and found significant smaller overabundance. This result agrees well with pulsation data.

2 The RHD Models, New [Fe/H] and Asteroseismology

The numerical code used for this work has been developed for the study of solar and stellar granulation (e.g. Nordlund & Dravins 1990, Stein & Nordlund 1998) and line formation (e.g. Asplund et al. 2004). It solves in a Cartesian box the non-linear, compressible equations of hydrodynamics coupled to radiative transfer, combined with a realistic equation-of-state and opacities. We have obtained a time-dependent 3D model of the surface layer of α Cen A. We used a grid of sufficiently high resolution (x,y,z) = 125 x 125 x 82 mesh points[1] to get accurate line profiles. Our model is defined by Teff = 5804 $\pm$ 30 K and log g = 4.32. The present simulation is an improvement of the work done by Nordlund & Stein (1990) who used incompressible hydrodynamics and a much lower numerical resolution than us. The synthetic profiles were computed for numerous snapshots covering several convective turn-overs and then averaged. We assumed LTE and used the most recent quantum mechanical calculations of Van Der Waals broadening. Instrumental, rotational broadenings and gravitational redshift were taken into account.

[1] i.e. 6000 x 6000 km for the horizontal sizes and 3000 km for the vertical one.

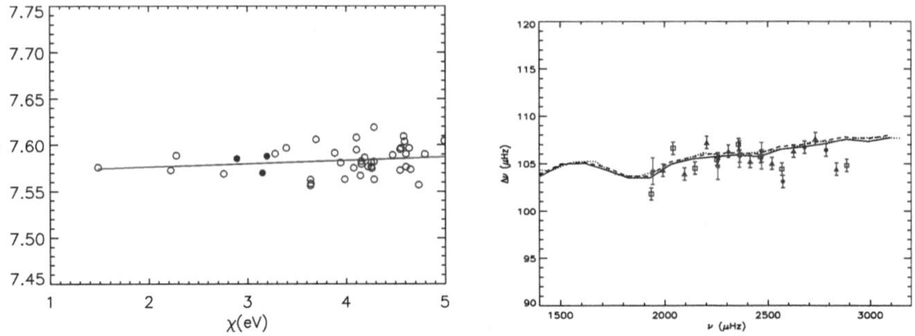

Fig. 1. (left) Individual abundances of the FeI (open circles) and FeII (full circles) lines as function of the excitation potential. (Right) Large separations (lines) as function of the frequencies compared with Bouchy & Carrier (2002) and Fletcher et al. (2005) observational data (blue and red, respectively).

The observations of α Cen A were carried out with the HARPS spectrograph installed on the ESO's 3.6m Telescope at La Silla (kindly provided by F. Bouchy). The large resolution (R $\sim 10^5$) and S/N ratio (> 500) are particularly well adapted for abundance determination. We focused on 44 Iron lines (41 FeI, 3 FeII). The linedata were taken from NIST[2]. The Iron abundances were determined through a differential analysis (with $[\mathrm{Fe/H}]_\odot = 7.46$). Thanks to the 3D RHD approach, neither micro nor macro turbulence were needed in the fit. The derived value is 7.58 ± 0.03. The new overabundance (+ 0.12 dex) is significantly lower than previous works (Neuforge-Verheecke & Magain 1997, Thévenin et al. 2002). This is a consequence of the lower temperature that comes out from the 3D RHD compared with 1D models. This new abundance is used to compute new stellar evolution models, new pulsation frequencies and re-analyze the paper of Thévenin et al. (2002). The lower metallicity led to a less evolved model and to a good agreement between predicted and observed large frequency separations (Fig 1). In a forthcoming paper we will extend this analysis to α Cen B and to more elements.

References

1. M. Asplund, N. Grevesse, A. J. Sauval, et al: A&A **423**, 1109 (2004)
2. F. Bouchy & F. Carrier: A&A **390**,205 (2002)
3. S. T. Fletcher, W. J. Chaplin, Y. Elsworth, et al: MNRAS **371**, 935 (2006)
4. P. Kervella, F. Thévenin, D. Ségransan et al: A&A **404**, 1087 (2003)
5. C. Neuforge-Verheecke & P. Magain: A&A **328**, 261 (1997)
6. A. Nordlund & D. Dravins: A&A **228**, 155 (1990)
7. R. F. Stein & A. Nordlund: ApJ **499**, 914 (1998)
8. F. Thévenin, J. Provost, P. Morel, G. Berthomieu, et al: A&A **392L**, 9 (2002)

[2]http://physics.nist.gov/PhysRefData/ASD/index.html

The SB3 Star 74 Aqr: Abundances and Magnetic Field

G. Catanzaro and F. Leone

INAF - Catania Astrophysical Observatory, Catania, Italy
gca@oact.inaf.it, fleone@oact.inaf.it

1 The System

74 Aqr (=HD 216494) is a stable hierarchical triple system composed by a spectroscopic binary, with HgMn primary, plus a single third component discovered during a speckle interferometric survey carried out by McAlister et al. [6]. The most recent set of parameters for the outer orbit has been given by Mason [4]. The solution for the SB2 pair has been improved recently by Catanzaro & Leto [2] who determined P = 3.429619 ± 0.000004 days and $M_A/M_B = 1.186 \pm 0.005$.

The presence of the third companion in the spectra of HD 216494 has been previously identified by Hubrig & Mathys [7] and then definitively confirmed by Catanzaro & Leto [2].

Detailed abundance analysis for all the three components, reported in Fig. 1, have been recently performed by Catanzaro & Leone [1]. Here we present the preliminary result of a search for magnetic field.

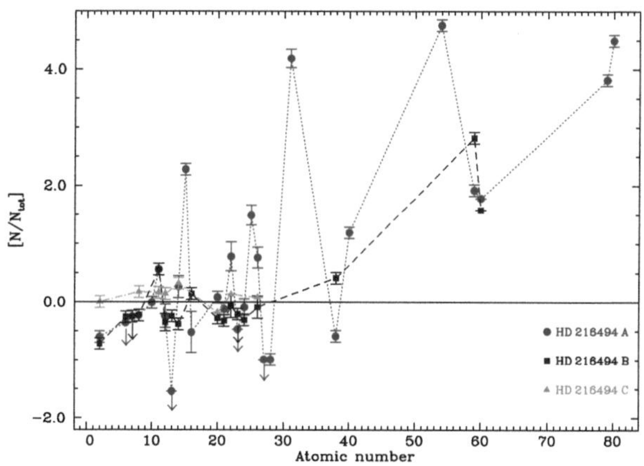

Fig. 1. Abundance patterns for the three components of HD 216494.

2 Magnetic Field

Spectro-polarimetric observations of 74 Aqr have been carried out on July 2006 at the *Telescopio Nazionale Galileo.* For a detailed description of the instruments and data reduction see Leone & Catanzaro [3].

Differently from Mathys & Hubrig, Stokes I and V/I profile reported in Fig. 2 show a null detection of circular polarization for both components, at least at the rotational phase we observed and at the noise level of our data.

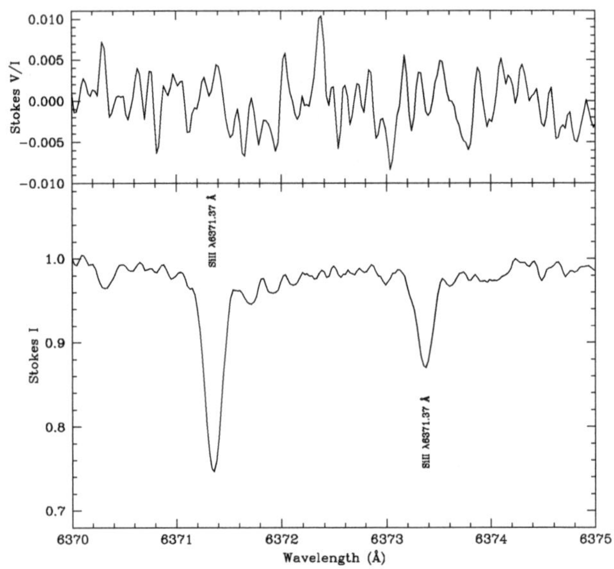

Fig. 2. Stokes I (bottom panel) and V/I (upper panel) profiles of the Si II λ6371 Å. Spectral line of the A component is the one with label above the continuum, the other belongs to the secondary.

References

1. Catanzaro G., Leone F., 2006, MNRAS, in press
2. Catanzaro G., Leto P., 2004, A&A, 416, 661
3. Leone F., Catanzaro G., 2004, A&A, 425, 271
4. Mason B. D., 1997, AJ, 114, 808
5. Mathys G., Hubrig S., 1995, A&A, 293, 810
6. McAlister H. W., Hartkopf W. I., Hutter D. J., Franz O. G., 1987, AJ, 93, 688
7. Hubrig S., Mathys G., 1995, ASP Conference Series, Vol. 81, 555, eds: A. J. Sauval, R. Blomme & N. Grevesse.

Nitrogen Isotope Ratios in Comets

Anita L. Cochran[1], Emmanuël Jehin[2], Jean Manfroid[3], Damien Hutsemékers[3], Claude Arpigny[3], Jean-Marc Zucconi[4], and Rita Schulz[5]

[1] University of Texas anita@astro.as.utexas.edu
[2] ESO
[3] Institute d'Astrophysique et de Géophysique, Liegè
[4] Observatoire de Besançon
[5] ESTEC

We used high S/N, high spectral resolution observations to determine the isotope ratios $^{12}C/^{13}C$ and $^{14}N/^{15}N$ in comets. Observations were obtained with resolving powers $R=\lambda/\delta\lambda \geq 60,000$ with telescopes at McDonald Observatory, VLT, NOT and Keck. We have now observed 13 comets on 20 different dates with heliocentric distances ranging from 0.6 au to 3.7 au. Most of these comets are from the Oort cloud; a few are members of the Jupiter family comets [2, 5, 4, 8, 6].

Cometary spectra consist of molecular emission from resonance fluorescence and a continuum resulting from sunlight scattered off dust. Before the emission spectrum may be modeled, the continuum must be removed accurately. Our isotope ratios are based on fluorescence models of the CN B-X (0,0) band [10]. The $^{12}C^{14}N$, $^{13}C^{14}N$ and $^{12}C^{15}N$ are modeled as separate components. Only unblended lines within the R-branch are used for modeling the isotopes. Typically we have 6 – 10 ^{15}N lines observed for each comet.

We found that all of the comets have the same carbon and nitrogen isotope ratios: $^{12}C/^{13}C=91\pm21$ and $^{14}N/^{15}N=141\pm29$. We find no difference in these ratios with dynamic type. Observations of comet Tempel 1 just prior to its impact with the Deep Impact spacecraft and in the following few hours show no change in these isotopic ratios, indicating that these ratios are the same in the interior and the mantle [6]. These values appear "universal" for comets.

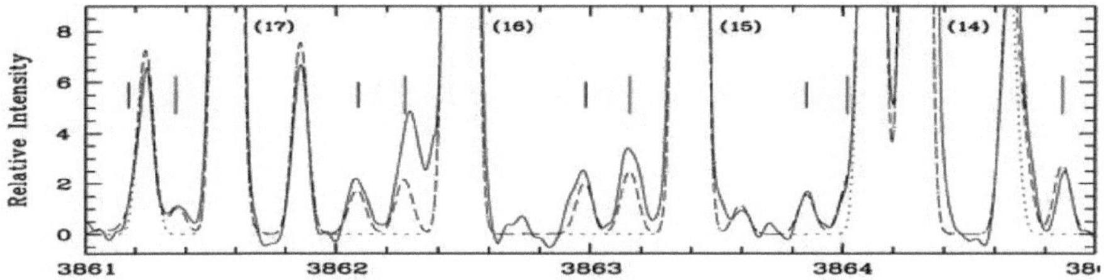

Fig. 1. This figure shows data plus fits for comet 122P/de Vico [5]. The dotted line is the fit with only $^{12}C^{14}N$ and the dashed line is the fit with the adopted isotope ratios. The long ticks mark $^{13}C^{14}N$ lines; the short ticks mark $^{12}C^{15}N$ lines. There is an unidentified line coincident with a $^{13}C^{14}N$ at 3862.3Å.

Isotope ratios in comets offer important constraints on the physical and chemical conditions in the early solar nebula. From their study in different objects and from different molecules, we can determine the relationship between comets, the protosolar nebula and the interstellar medium. The fractionation history derived from these isotopes can tell us about the formation temperature, chemical processes and whether material has been chemically evolved within the early solar nebula.

The $^{12}C/^{13}C$ ratio which we have found is the same as the Solar System value [1] and all bodies in the Solar System. The Solar System $^{14}N/^{15}N$ is not currently known since it is difficult to measure in the Sun. The telluric value of 272 has historically been adopted as the Solar System reference value (but the Earth's is not a primary atmosphere).

$^{14}N/^{15}N$ from HCN was measured in comet Hale-Bopp at sub-mm wavelengths as 323±46 [7] and 330±98 [9]. Hale-Bopp is the only comet for which this HCN ratio was measured. Our value from CN of $^{14}N/^{15}N=141\pm29$ includes observations of Hale-Bopp and is significantly different than the HCN measurements. If the Hale-Bopp HCN value applies to other comets, HCN is not the sole parent of CN in comets. A potential parent for CN is C_2N_2, but it has no easily observed transitions. If there is another parent contributing to the $^{14}N/^{15}N$ measured from CN, it must have a value which is significantly lower than the HCN value. Gas-grain effects can lead to significant ^{15}N fractionation; a major reservoir for ^{15}N in the early solar nebula might be NH_3 [3]. One explanation for the different nitrogen isotope ratios in CN and HCN in comets could be that the high relative abundance of ^{15}N originates from another source and is preferentially locked up in refractory organics.

Our observations of nitrogen isotope ratios in comets show the power of high-precision, high-spectral resolution observations for helping constrain conditions in the solar nebula.

References

1. E. Anders and N. Grevesse. *Geochim. et Cosmochim. Acta*, 53:197, 1989.
2. C. Arpigny, E. Jehin, J. Manfroid, D. Hutsemékers, R. Schulz, J. A. Stüwe, J.-M. Zucconi, and I. Ilyin. *Science*, 301:1522, 2003.
3. S. B. Charnley and S. D. Rodgers. *Ap. J.*, 569:L133, 2002.
4. D. Hutsemékers, J. Manfroid, E. Jehin, C. Arpigny, A. Cochran, R. Schulz, J. A. Stüwe, and J.-M. Zucconi. *A&A*, 440:L21, 2005.
5. E. Jehin, J. Manfroid, A. L. Cochran, C. Aprigny, J.-M. Zucconi, D. Hutsemekers, W. D. Cochran, M. Endl, and R. Schulz. *Ap. J.*, 613:L161, 2004.
6. E. Jehin, J. Manfroid, D. Hutsemékers, A. L. Cochran, C. Arpigny, W. M. Jackson, H. Rauer, R. Schulz, and J.-M. Zucconi. *Ap. J.(Letters)*, 641:L145, 2006.
7. D. Jewitt, H. E. Matthews, T. Owen, and R. Meier. *Science*, 278:90, 1997.
8. J. Manfroid, E. Jehin, D. Hutsemékers, A. Cochran, J.-M. Zucconi, C. Arpigny, R. Schulz, and J. A. Stüwe. *A&A*, 432:L5, 2005.

9. L. M. Ziurys, C. Savage, M. A. Brewster, A. J. Apponi, T. C. Pesch, and S. Wyckoff. *Ap. J.(Letters)*, 527:L67, 1999.
10. J.-M. Zucconi and M. C. Festou. *A&A*, 158:382, 1986.

Finding Stable Fits for Extrasolar Planetary Systems

J. Couetdic[1], J. Laskar[1], and A.C.M. Correia[2]

[1] Astronomie et Systèmes Dynamiques, IMCCE-CNRS UMR 8028, 77 av. Denfert Rochereau, F–75014 Paris, France couetdic@imcce.fr

[2] Departamento de Física da Universidade de Aveiro, Campus Universitário de Santiago, 3810-193 Aveiro, Portugal acorreia@fis.ua.pt

Summary. Several problems with radial velocity determination of orbital elements call for dynamical studies. One of them is that, due to uncertainties in radial velocity data and complex dynamical behavior, the best fit obtained for systems with planets experiencing strong interactions, like HD202206 and HD160691, are most often not stable. For those two systems, we show how a careful stability analysis constrained by χ^2 from radial velocity data (mostly from the Geneva Observatory research team) can lead to a stable fit.

The HD202206 system currently hosts two planets with $m \sin i = 17.4\,M_{jup}$ and $2.44\,M_{jup}$, semi-major axis of $a = 0.83$ AU and 2.55 AU, and eccentricities of $e = 0.43$ and 0.27. The best fit for coplanar orbit with minimum masses ($i = 90°$) obtained from a 3-body dynamical model is lost in forty thousand years. A study of the dynamics in the vicinity of the best fit was done by Correia et al. ([3]). Stable fits were found in the 1/5 mean motion resonance (dark region in figure 1 upper panel) : for instance, with $\omega_2 = 55.50°$, $a_2 = 2.542$ AU and $\sqrt{\chi^2} = 1.67$.

The μAra system has four known planets : μ Ara b with $P = 640$ days and μ Ara c with $P = 9.64$ days, μAra e on a wide but still poorly defined orbit, and μAra d, newly discovered (Pepe et al. [4]) using HARPS measurements. μ Ara d is on an almost circular 310 days-period orbit with a mass of $0.52\,M_{Jup}$. In this case the best fit is also unstable (the system is destroyed in 76 million years). The same kind of dynamical study as for HD202206 was done to find orbital elements for μAra d and μAra e giving a stable fit (lower panel in figure 1). This helped to constrain the period of μAra e around 4200 days and eccentricity close to 0.1. In fact, earlier fits with less data gave higher eccentricities for which we were unable to find a stable fit for μAra d.

References

1. J. Laskar: Icarus **88**, 266 (1990)
2. J. Laskar, P. Robutel: Celest. Mech. Dyn. Astron. **80**, 39 (2001)
3. A.C.M. Correia, S. Udry, M. Mayor et al: Astron. Astrophys. **440**, 751 (2005)
4. F. Pepe, A.C.M. Correia, M. Mayor et al: Astron. Astrophys. **462**, 769 (2007)

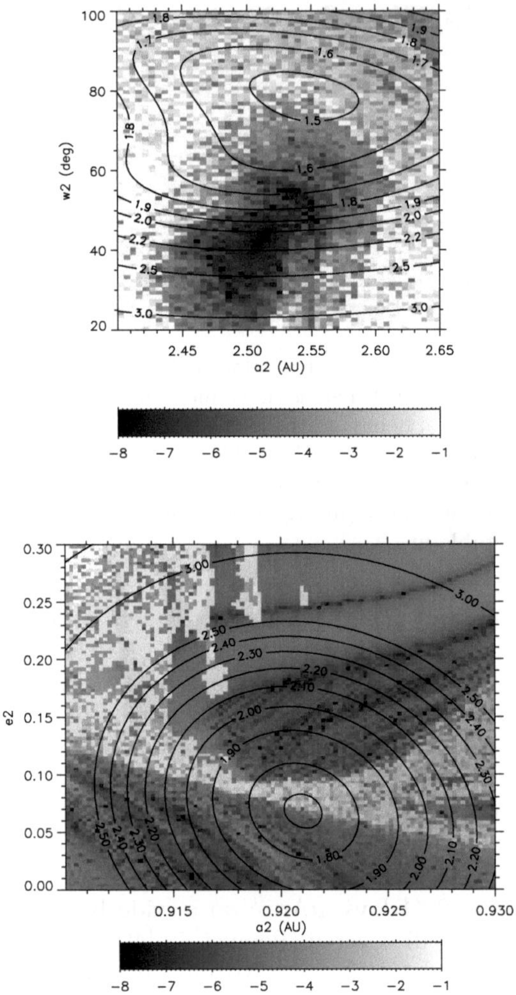

Fig. 1. The upper panel represents a global view of the dynamics of the HD202206 system for variations of the periastrum and semi-major axis of the outer planet. For each initial conditions, the orbit of the planets are integrated with the symplectic integrator SABAC4 [2]. The stability of the orbit is then measured by Laskar's frequency analysis [1]. Light grey areas correspond to high orbital diffusion (instability) and dark areas to low diffusion (stable orbits). The grey scale is the stability index (D) obtained through frequency analysis. Level curves give the value of $\sqrt{\chi^2}$ obtained for each couple of parameters. The lower panel shows the same type of study in the (a_d, e_d) plane of initial conditions for μAra. The dark grey area at eccentricities higher than the best fit (shown by the level curves) lies in the 2/1 mean motion resonance with μAra b, but is actually midly chaotic. However there exist stable orbits in the dark grey region at lower eccentricities and semi-major axis lower than 0.92 AU.

Heavy Calcium in CP Stars: A New Isotopic Anomaly

C. R. Cowley[1], S. Hubrig[2], F. Castelli[3], B. Wolff[2], and F. González[4]

[1] University of Michigan, Ann Arbor, MI 48109-1042, USA cowley@umich.edu
[2] ESO, Casilla 19001, Santigo 19, Chile shubrig@eso.org
[3] Osservatorio Astronomico di Trieste, Trieste, Italy castelli@ts.astro.it
[4] Universidad Nacional de San Juan, Argentina

Summary. We confirm that wavelengths shifts in the Ca II infrared triplet of CP stars are due to the rare, heavy isotope ^{48}Ca. We have measurements of the previously difficult component λ8542, as well as additional measurements of all three components. The shifts occur in a wide variety of CP stars, ranging from cool CP2 types to the hotter CP3 stars. The relative proportions of the Ca isotopes appear to vary from terrestrial to nearly pure ^{48}Ca.

1 Introduction

Large, isotopic wavelengths shifts in the infrared triplet of Ca II have been investigated in the laboratory (cf. [1], and references therein). Castelli and Hubrig [2] first reported these shifts in non-magnetic CP (HgMn) stars, and they were subsequently found by Cowley and Hubrig [3] in magnetic CP stars. Wavelength shifts from the dominant terrestrial isotope, ^{40}Ca to the rare ^{48}Ca isotope are nearly the same for all three lines of the triplet: +0.20, +0.20, and +0.21 Å for λλ8498, 8542, and 8662 respectively.

2 Details

All of the measurements were made with either ESO FEROS or UVES spectra. We report confirming isotopic shifts from λ8542, following a reconfiguration of the UVES spectrograph. These are shown in Fig. 1.

The Earth-sign (⨁) in the lower-left corner marks the laboratory position (an open-plus sign is superimposed). Note that the wavelengths may be meaningfully different for terrestrial materials and Sun (⊙ [4]), indicating possible isotopic abundance differences. The non-Ap star star showing the largest shift is the mild barium star, *o* Vir. For this star we find a marginally significant shift for λ8498 of +0.06 Å.

Fig. 2 shows λ8662 *vs.* λ8498, with previously unpublished measurements.

The lines of the Ca II infrared triplet are generally strong lines whose wavelengths are less subject to perturbations by blends than weaker lines. The fact that the shifts for all three lines are well correlated with one another argues against blends as a cause.

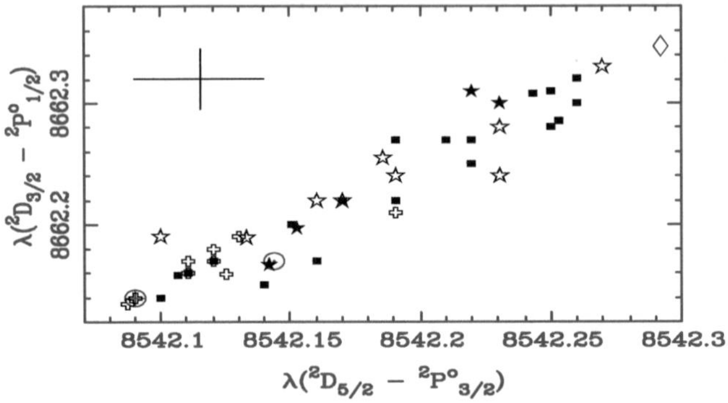

Fig. 1. Wavelength shifts relative to laboratory positions, for Ca II λ8662 correlate well with those for λ8542. The latter line was previously unavailable or poorly measured. Error bars are shown to the upper left. Filled squares are HgMn stars; filled and open stars are for roAp and noAp CP stars respectively. Open crosses are for miscellaneous types. See text for additional details.

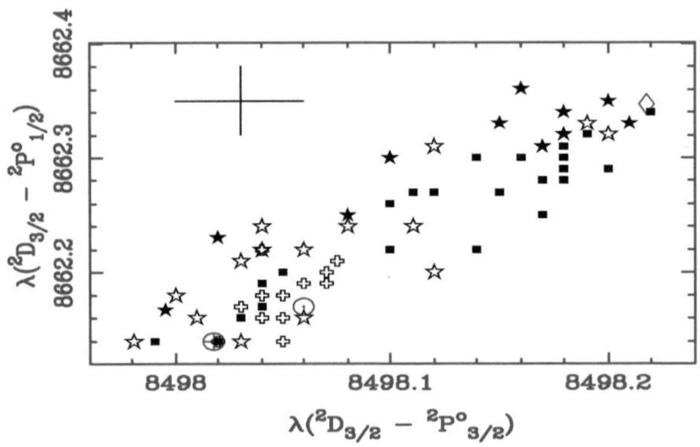

Fig. 2. Ca II λ8662 *vs.* λ8498. Symbols have the same meaning as above. This plot includes additional stars for which the λ8542-line is not available or is of poor quality.

References

1. W. Nörtershäuser, K. Blaum et al: Eur. Phys. J.D., **2**, 33 (1998).
2. F. Castelli,and S. Hubrig: A&A, **421**, L1 (2004).
3. C.R. Cowley, and S. Hubrig: A&A, **432**, L21 (2005).
4. C.H. Moore, M.G.J. Minnaert, and J. Houtgast: The Solar Spectrum 2935 Åto 8770 Å, NBS Monog., 61 (1968).

The Li Abundance and the Age of AB Dor Association

Licio da Silva[1], Carlos Alberto Torres[2], Ramiro de la Reza[1], Germano Quast[2], Claudio de Melo[3] and Michael Sterzik[3]

[1] Observatório Nacional, Rio de Janeiro, Brazil licio@on.br
[2] Laboratório Nacional de Astrofísica, Brazil
[3] European Southern Observatory, Chile

1 Introduction

In a recent paper (Torres et al. 2006) we report the results from a high-resolution optical spectroscopic survey searching for associations containing young stars (SACY), among optical counterparts of ROSAT All-Sky X-ray sources in the Southern Hemisphere. We have applied a convergence method in the (UVW) velocity space and have found several nearby young associations in the sample. As they are young and with different ages, those associations form an interesting laboratory to test the Li depletion theory, as a function of the star age. We present here our determination of Li abundance for two of them, the β Pic and AB Dor Associations.

Most of the spectroscopic observations were performed with FEROS spectrograph (resolution ~50000) at the 1.5m ESO telescope at La Silla (Chile). Another set of data (~30%) was collected at the coudé spectrograph attached to 1.60m telescope at the Observatório do Pico dos Dias, LNA, Brazil, with a RP ~ 9000. To obtain the Li abundance we used a LTE code and the MARCS models. The stars effective temperatures Teff were determined from (V-Ic) color index. The other atmosphere model parameters were fixed a priori: the metallicity as 0.1, because this is the value determined by Castilho et al. (2005) for AB Dor, and log g as 4.5 for dwarf and 4.0 for sub-giant stars. The micro-turbulence velocity also was fixed as 2 km/s.

The Age of AB DOR

The diagram of the stellar Li abundances vs Teff (Fig. 1) corresponding to program associations above shows, for Teff < 5000 K, a clear separation among the stars associations, being β Pic, with a age ~11 My, above, and TucHor, with a age ~30 My, in the besides. The AB Dor stars are in the band corresponding to the Pleiades stars (age ~100 My), given one more evidence that those associations have the same age, as it is supported by Luhman et al. (2005), contrary to Close et al. (2005), that consider AB Dor younger. This information gives also an additional support to a recent 3D Galactic dynamical analysis indicating a common origin of these two groups (Ortega et al. 2006). The result that AB Dor and the Pleiades associations have the

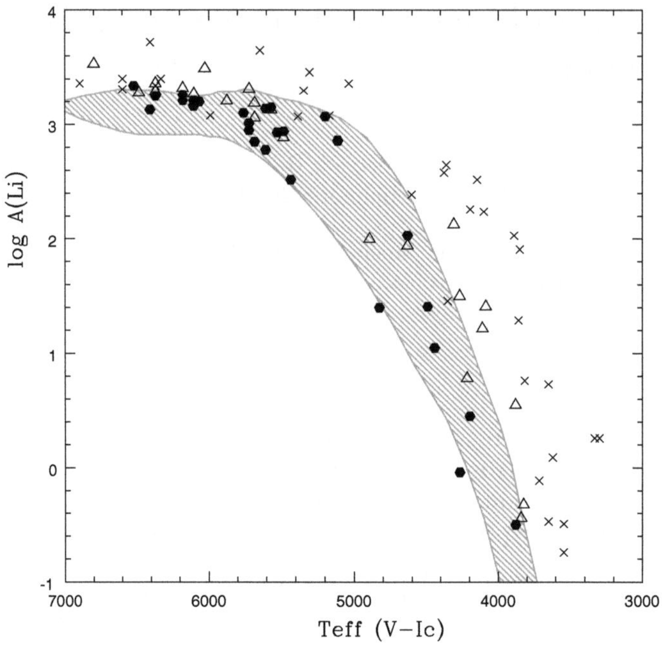

Fig. 1. Li abundances as a function of the effective temperature for the proposed members of the β Pic (crosses), Tuc Hor (triangles) and AB DOR (full circles) Associations. The hatched area shows the Li abundances with range observed for the Pleiades members

same age is very important because the models used to determine the mass of the brown dwarf AB Dor C are age dependent and this star mass has been used as zero point in the brown dwarfs mass scales. This larger age of AB Dor is used also by Luhman and Potter (2006) to confirm that the IMFs of star-forming regions are similar to the IMF of the solar neighborhood.

References

1. Castilho, B. V., Torres,C. A. O., Quast, G. R. , et al. 2005 in *IAU Symposium 228*, ed. V. Hill, P. Francois, and F. Primas, 83-84
2. Close, L. M. et al 2005, Nature, **433**, 286
3. Luhman, K. L., Stauffer, J. R. and Mamajek, E. E., 2005, ApJ, **628**, L69
4. Luhman, K. L. and Potter, D., 2006, ApJ, **638**, 887L
5. Ortega, V. G., Jilinski, E. and de la Reza, R., 2006 MNRAS submitted
6. Torres,C. A. O., Quast, G. R. , da Silva, L., de la Reza, R., Melo, C. H. F. and Sterzik, M. 2006, A&A, **460**, 695
7. Voges, W., Aschenbach, B., Boller, T. et al. 1999, A&A, **349**, 389

Si and Ca Abundances of a Selected Sample of Evolved Stars

L. da Silva[1], L. Girardi[2], L. Pasquini[3], R. De Medeiros[4], J. Setiawan[5], M. Döllinger[3], A. Hatzes[6] and A. Weiss[7]

[1] Observatório Nacional, Rio de Janeiro, Brazil licio@on.br
[2] INAF-Oss.Ast di Trieste, Italy
[3] European Southern Observatory, Chile
[4] DFTE/UFRN, Natal, Brazil
[5] Max-Planck-Institute for Astronomy, Heidelberg, Germany
[6] Thüringer Landessternwarte Tautemburg, Germany
[7] MPA Garching, Germany

1 Introduction

Recently we have published the detailed spectroscopic analysis of 72 evolved stars, previously studied for accurate radial velocity variations (da Silva et al. 2006). The spectroscopic observations were performed with the FEROS spectrograph (RP $\sim$ 50000) at the 1.5m ESO telescope at La Silla (Chile). The stellar atmosphere parameters (Teff, log g,[Fe/H]) and the micro-turbulence velocities were determined spectroscopically, using a LTE code and the MARCS models. These metallicities, together with the Teff values and the absolute V-band magnitude derived from Hipparcos parallaxes, were used to estimate basic stellar parameters (ages, masses, radii, (B-V)o and log g) using theoretical isochrones and a modified version of Jorgensen & Lindegren's (2005) method. The (B-V)o values so estimated turn out to be in excellent agreement with the observed ones, confirming the reliability of the (Teff,(B-V)o) relation used in the isochrones. The estimated stellar diameters also are in good agreement, with the ones measured by independent methods, better than 0.3 mas.

The three giants of our sample which have been proposed to host planets (HD11977, HD47536 and HD122430) are not metal rich; this result is at odds with those for main sequence stars. However, two of these stars have masses much larger than a solar mass so we may be sampling a different stellar population from most radial velocity searches for extrasolar planets. We also confirm the previous indication that the radial velocity variability tends to increase along the RGB, and in particular with the stellar radius. To better characterize the evolution stage of the sample stars, in order to compare them with the RV variability, we should determine some elements abundances, as the alpha-elements. In a first step, using the parameter values (Teff, log g, [Fe/H]) and the micro-turbulence velocities found as described above, we determined the abundances of the following neutral elements: Na, Mg, Al, Si, Ca, Ti and Ni, using the line list of Randich et al. (2006). Those

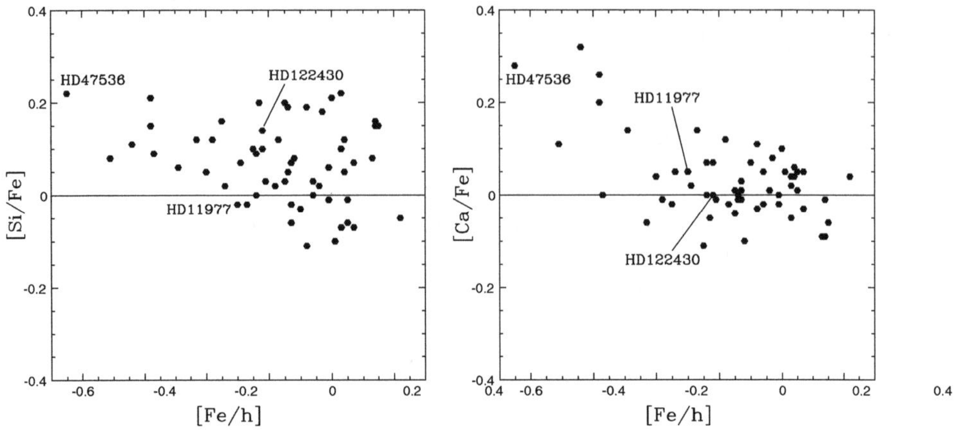

Fig. 1. Si/Fe and Ca/Fe ratio abundances as a function of the iron abundance. The indicated stars are candidates to host planets.

lines were carefully selected and used to determine the abundances of those elements in the M67 sub-giants. To keep the homogeneity, we determine also the iron abundances using the same line list and use them to compare to the other elements. Unfortunately, the spectral lines of the cold stars are large and crowed, specially of the giants, making very difficulty to have good abundance determinations and the dispersion of the values we determined is very large, especially for those elements for which only a few (2-3) lines were available. The determination of the abundance of these elements will further require a check through spectral synthesis. In Figure 1 we present the abundance of Si (9 lines) and Ca (11 lines), which are better constrained. In order to better quantify the spread in these relationship, we will further separate the stars between thin and thick disk objects. We can however see that, as far as these two elements are concerned, the planet hosting stars do not show any anomalous position in the diagrams with respect to the other sample objects.

References

1. da Silva, L., Girardi, L., Pasquini, L., De Medeiros, J. R., Setiawan, J., von der Luhe, O., Döllinger, M. P., Hatzes, A. and Weiss, A. 2006, A&A. **458**, 609
2. Jorgensen, B. R. and Lindegren, L., 2005, A& A, **436**,127.
3. Randich, S., Sestino, P., Primas, F., Pallavicini, R. and Pasquini, L. 2006, A& A **450**, 557.

Abundance Trends with Condensation Temperature in Planet-harbouring Stars: Hints of Pollution?

A. Ecuvillon[1], G. Israelian[1], N. C. Santos[2,3,4], M. Mayor[2], and G. Gilli[5]

[1] Instituto de Astrofísica de Canarias, La Laguna, Spain aecuvill@iac.es
[2] Observatoire de Genève, 1290 Sauverny, Switzerland
[3] Centro de Astronomia e Astrofisica de Universidade de Lisboa, Portugal
[4] Centro de Geofisica de Evora, 7002-554 Evora, Portugal
[5] Instituto de Astrofísica de Andalucía, 18008 Granada, Spain

Summary. We present the [X/H] trends as a function of the elemental condensation temperature T_C in 88 planet-host stars and in a volume-limited comparison sample of 33 dwarfs without detected planetary companions. We gathered homogeneous abundance results for many volatile and refractory elements spanning a wide range of T_C, from a few dozen to several hundred kelvin. We investigate possible anomalous trends of planet hosts with respect to comparison sample stars to detect evidence of possible pollution events. No significant differences are found in the behaviour of stars with and without planets. This is consistent with a "primordial" origin of the metal excess in planet-host stars. However, a subgroup of 5 planet-host and 1 comparison sample stars stands out as having particularly high [X/H] vs. T_C slopes.

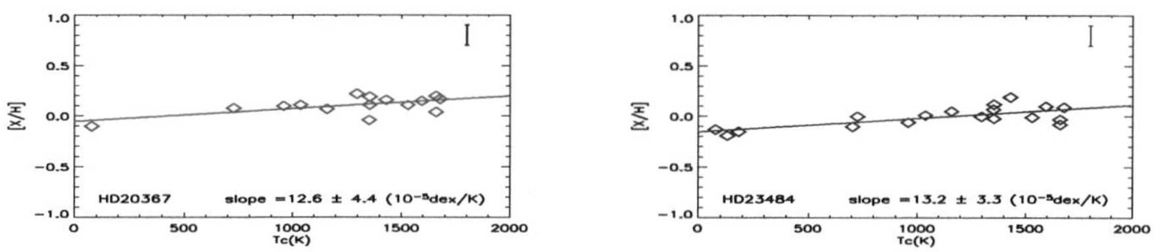

Fig. 1. [X/H] vs. T_C for the planet host HD 20367 and the "single" star HD 23484.

1 Introduction

Possible abundance patterns with the elemental condensation temperature T_C can give important clues to understand the source of the metallicity excess in planet-host stars. If the a posteriori infall of rocky planetary material was its main cause, as the "self-pollution" scenario claims (Gonzalez 1997, MNRAS,285,403), an overabundance of refractory elements with respect to

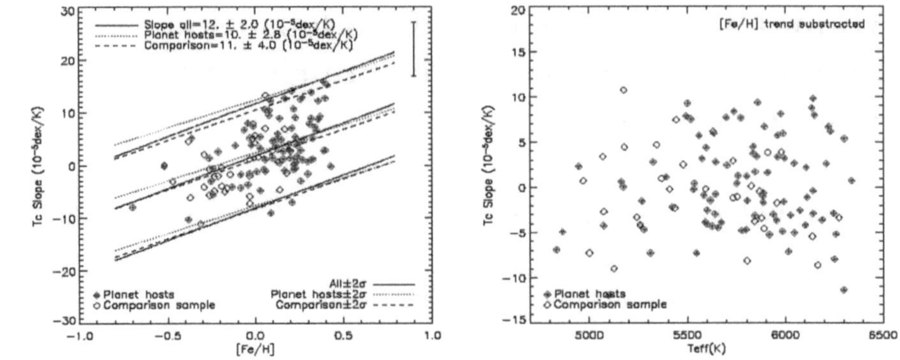

Fig. 2. *Left panel:* [X/H] vs. T_C slopes plotted vs. [Fe/H]. The solid lines represent the linear least-squares fit $\pm 2\sigma$ for all the targets, while the dotted and dashed lines indicate the linear least-squares fits $\pm 2\sigma$ for stars with and without planets, respectively. *Right panel:* [X/H] vs. T_C slopes plotted vs. $T_{\rm eff}$.

volatiles (or an increasing trend of [X/H] with T_C) should be observed, since the former might be added preferentially compared to the latter.

We refer to Ecuvillon et al. (2006, A&A, 449, 809) for a detailed description of the observational data and analysis.

2 Results and Conclusions

Our results do not reveal any striking difference in the [X/H] vs. T_C trends in planet-host stars when compared to "single" stars. There is a clear global trend of T_C slopes to increase with [Fe/H] (see Fig.2, left panel), which is an expected signature of the Galactic chemical evolution. The comparison sample target HD 23484 and 5 planet-host stars show slopes more than $2\,\sigma$ above the fit corresponding to the comparison sample. They might be interpreted as tentative candidates for having been strongly polluted and have to be submitted to further tests. Pollution effects are expected to be much more evident in stars with $T_{\rm eff}$ >6000 K. However, no peculiar trend is found in T_C slopes vs. $T_{\rm eff}$ plot, once removed the trend with [Fe/H] due to Galactic chemical evolutionary effects (see Fig.2, right panel).

Although in most cases a mainly "primordial" origin of the metallic excess in planet-host stars seems likelier, a more complex mechanism combining both scenarios may underlie the observations. More works are required to obtain more conclusive evidences on the possible cases of pollution.

Using the HeII Lyα Forest to Constrain the Temperature of the IGM

Cora Fechner[1]

Hamburger Sternwarte, Gojenbergsweg 112, 21029 Hamburg, Germany
cfechner@hs.uni-hamburg.de

The estimation of elemental abundances in the intergalactic medium (IGM) often requires ionization corrections and therefore knowledge of the radiation field. The He II/H I ratio η provides an estimate of the hardness of the ionizing radiation and can be used to study variations of the UV background with redshift and/or density. Systematic uncertainties in the measurement of the He II/H I ratio may arise from the assumption of turbulent line broadening. Therefore, taking into account a temperature when analyzing the He II Lyα forest could improve the estimation of scales and amplitudes of UV background fluctuations. Furthermore, an independent estimate of the temperature of the absorbing material could provide, at least in principle, constraints on the thermal state of the IGM.

1 Idea

According to [3] there is a well-defined power law relation between temperature and density, $T = T_0\,(1+\delta)^{\gamma-1}$, for the low density IGM (gas overdensities $\delta < 10$. i.e. $N_{\mathrm{H\,I}} < 10^{15}\,\mathrm{cm}^{-2}$), where T_0 is the temperature at the mean density. Assuming that the gas is almost fully ionized and in photoionization equilibrium, [4] derived a relation between the H I column density and the overdensity δ. Inserting the temperature-density relation yields

$$\log N_{\mathrm{H\,I}} \approx 11.833 + 4.5\,\log(1+z) - \frac{1.5}{\gamma-1}\log T_0 + \left(\frac{1.5}{\gamma-1} - 0.255\right)\cdot \log T\,, \quad (1)$$

where $N_{\mathrm{H\,I}}$ is measured in cm^{-2} and the temperatures are given in K.

The parameters of the temperature-density relation can be estimated from an analysis of the H I and He II Lyα forest based on Doppler profile fits since the temperature T of the absorbers is related to the Doppler parameters

$$b_{\mathrm{He\,II}} = \sqrt{b_{\mathrm{H\,I}}^2 - 2\,k\,T\left(\frac{1}{m_{\mathrm{H}}} - \frac{1}{m_{\mathrm{He}}}\right)}\,. \quad (2)$$

Measuring the H I column density and the temperature T, Eq. (1) can be used to estimate T_0 and γ. Furthermore the He II/H I ratio is given by $\eta = N_{\mathrm{He\,II}}/N_{\mathrm{H\,I}}$.

2 Application to Data

We have developed a procedure which fits a spectrum based on the H I line list to the observed He II data optimizing $\log \eta$ as well as $\log T_0$ and γ, simultaneously. The procedure is tested with artificial data which are created on the basis of the statistical properties of the Lyα forest assuming $\log T_0 = 4.3$, $\gamma = 1.3$, and $\eta = 80$. Different noise levels are chosen for the artificial He II data ($S/N = 5 \ldots 100$), while $S/N = 100$ for the H I spectrum.

For the He II data a signal-to-noise ratio of $S/N > 20$ is required to obtain reasonable constraints of $\log T_0$ and γ (Fig. 1). The η values are recovered well if a temperature-density relation is taken into account. Assuming pure turbulently broadened absoption features ($b_{\mathrm{He\,II}} = b_{\mathrm{H\,I}}$) would underestimate the η value by roughly 0.05 dex.

The observed He II data towards the quasars HE 2347-4342 [6] and HS 1700+6416 [1] are fitted in the range $2.58 < z < 2.74$ by a modified version of the spectrum fit procedure introduced by [2]. Based on the H I line list the He II spectrum is modeled optimizing the η value, the temperature of each absorber, and the spatial scales of the η variation. A temperature can be estimated for 55% of the lines, however, with extremely large error bars. The mean temperature is $T = 10^{4.41 \pm 0.46}$ K. Pure turbulent broadening is favored by 45% of the lines. The distribution of η values changes when the thermal component of the line width is neglected and $b_{\mathrm{He\,II}} = b_{\mathrm{H\,I}}$ is assumed. The average η value is lower by $0.1 - 0.2$ dex and the distribution gets broader. The correlation between η and the strength of the H I absorption found in former studies (e.g. [5]) diminishes if a temperature is taken into account.

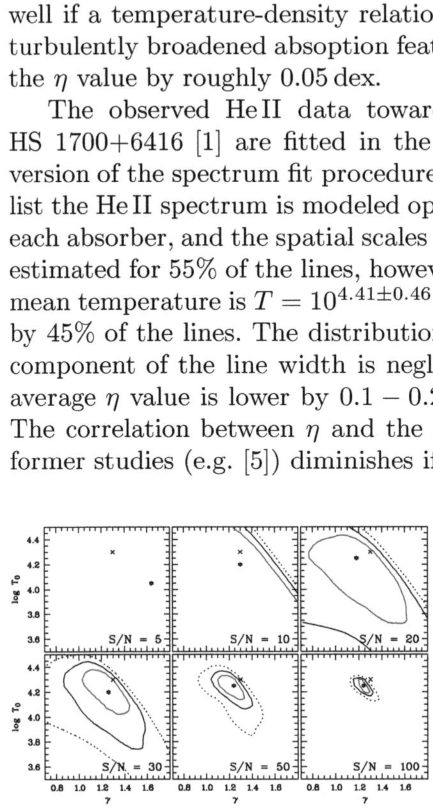

Fig. 1. Contour plots for the parameters of the T-δ relation derived from simulated data with different S/N. The stars (crosses) represent the position of the inferred (true) values. The outer (inner) contours give the joint (individual) $1\,\sigma$ (solid) and $2\,\sigma$ (dotted) confidence levels. The confidence levels in case of $S/N = 5$ are outside the presented parameter range.

References

1. C. Fechner, D. Reimers, G.A. Kriss et al: A&A **455**, 91 (2006)
2. C. Fechner, D. Reimers: A&A accepted (2006), astro-ph/0610366
3. L. Hui, N.Y. Gnedin: MNRAS **292**, 27 (1997)
4. J. Schaye: ApJ **559**, 507 (2001)
5. J.M. Shull, J. Tumlinson, M.L. Giroux et al: ApJ **600**, 570 (2004)
6. W. Zheng, G.A. Kriss, J.-M. Deharveng et al: ApJ **605**, 631 (2004)

Production of H_3^+ and D_3^+ from $(CH_3)_2CO$ and $(CD_3)_2CO$ in PDR'S

A.M. Ferreira-Rodrigues[1*], S. Pilling[2], A.C.F. Santos[1], G.G.B de Souza[1] and H.M. Boechat-Roberty[1]

[1] Universidade Federal do Rio de Janeiro – UFRJ, CEP 20080-090 Rio de Janeiro-RJ, Brazil *anarodrigues@iq.ufrj.br
[2] Laboratório Nacional de Luz Síncrotron, Caixa Postal 6192, CEP 13084-971 Campinas, SP, Brazil

1 Introduction

Acetone ($(CH_3)_2CO$) was the first 10 atom molecule reported in the interstellar medium. The confirmation of acetone has been recently reported in the giant molecular cloud Sgr B2, which undergoes extensive star formation [1]. Chemistry in Sgr B2 has been strongly modified by the photons coming from the hot stars in the nebulae. The formation of H_3^+ through photoionization and photodissociation processes of abundant interstellar organic molecules like $(CH_3)_2CO$ plays an important role in diverse fields from chemistry to astronomy [2]. As an example we mention the chains of reaction that lead to the production of many complex molecular species observed in the interstellar medium [3]. H_3^+ was recently discovered both in molecular clouds [4] and in the diffuse interstellar medium [5].

2 Experimental Setup

Time-Of-Flight Mass Spectrometry (TOF MS) was employed in the study of the fragmentation of acetone and d_6-acetone molecules (purity greater than 99.5 %) at room temperature. The studies were performed at the Brazilian Synchrotron Light Laboratory (LNLS) near the C 1s ionization edge ($\approx$280 eV). Soft X-rays photons ($\sim 10^{12}$ photons/s) from a toroidal grating monochromator (TGM) beamline (100-310 eV), perpendicularly intersect the gas sample inside a high vacuum chamber. The complete description of the experimental setup can be found elsewhere [6, 7].

Conventional time-of-flight mass spectra (TOF-MS) were obtained using the correlation between one Photoelectron and a Photoion Coincidence (PEPICO). The ionized recoil fragments produced by the interaction with the photon beam were accelerated by a two-stage electric field and detected by two micro-channel plate detectors in a chevron configuration, after mass-to-charge (m/q) analysis by a time-of-flight mass spectrometer (297 mm long). They produced up to three stop signals to a time-to-digital converter (TDC) started by the signal from one of the electrons accelerated in the opposite direction and recorded without energy analysis by two micro-channel plate detectors.

3 Results and Discussion

The excellent signal to noise ratio has allowed for the observation of reactive ions of astrochemical interest, H_2^+, H_3^+, CO^+, CH_3^+, CH_3CO^+ and D_3^+ as shown in Figure 1. The presence of H_3^+ and D_3^+ is clearly shown in the 0-7 a.m.u. enlargement (Figure 2).

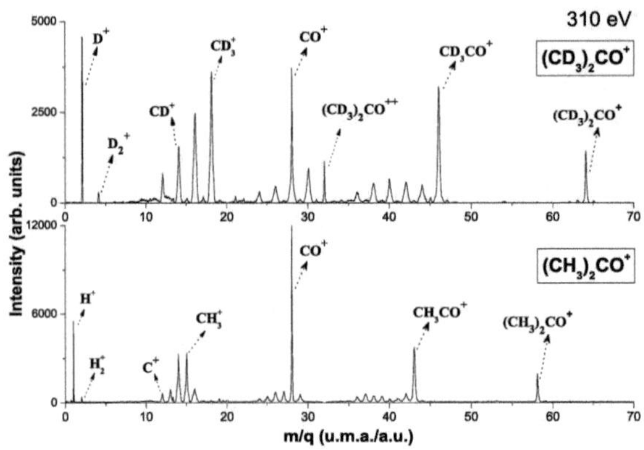

Fig. 1. Time of flight mass spectra (PEPICO) recorded at 310 eV

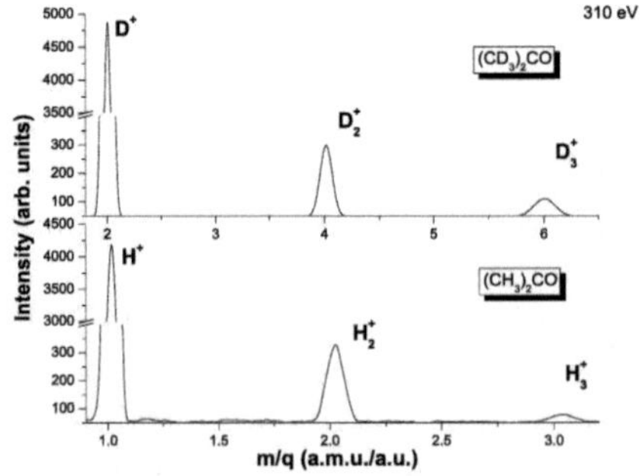

Fig. 2. PEPICO obtained by the dissociation of single photoionized molecules recorded at 310 eV recorded the production of H_3^+ and D_3^+ ions

New branching ratios have been obtained for the observed H^+ and D^+ ions, and are the subject of future publications. Based on the present experimental results, we propose the photodissociation of methylated molecules as an additional source of H_3^+ and D_3^+ in PDR's.

References

1. L.E. Snyder , F.J. Lovas, D.M. Mehringer, N.Y. Miao and Y-J. Kuan: ApJ **578**, 245 (2002)
2. S. Pilling, R. Neves, A.M. Ferreira-Rodrigues, A.C.F. Santos and H.M. Boechat-Roberty: MNRAS, submited (2006)
3. E. Herbst and W. Klemperer: ApJ **185**, 505 (1973)
4. T.R. Geballe and T. Oka: Nature **384**, 334 (1996)
5. McCall B.J., Geballe T.R., Hinkle K.H. and Oka T.: Science **279**, 1910 (1998)
6. H.M. Boechat-Roberty, S. Pilling and A.C.F. Santos: A&A **438**, 915 (2005)
7. S. Pilling, A.C.F Santos and H.M. Boechat-Roberty: A&A **449**, 1289 (2006a)

The decorative 'azulejos', characteristic of Aveiro, on the walls of the dining room, depict the boats on the lagoon surrounding the town.

Bisectors as Distance Estimators for Microquasars?

C. Foellmi[1,2]

[1] European Southern Observatory, 3107 Alonso de Cordova, Vitacura, Casilla 19., Santiago, Chile cfoellmi@eso.org
[2] Laboratoire d'Astrophysique, Observatoire de Grenoble, 414 rue de la Piscine, 38400 Saint-Martin d'Hères, France cfoellmi@obs.ujf-grenoble.fr

1 Scientific Context

Microquasars are a small-scaled version of their extragalactic parents, the quasars with which they share the same physics [5]. They have a central black-hole (here of stellar mass), an accretion disk and (sometime persistent) jets. These jets are sometimes relativistic, and can reach apparent superluminal motion. However, the superluminal "property" depends on the distance between the object and the observer.

Recently, the distance of the second known superluminal microquasar in our Galaxy, namely GRO J1655-40 [4], has been challenged by [2]. At a new distance smaller than 1.7 kpc (instead of 3.2 kpc), the jets are not superluminal anymore. It is thus of central importance to obtain reliable values of the distance in order to achieve a consistent understanding of these relativistic objects. At an inferred distance of 1.0 kpc, GRO J1655-40 competes with the microquasar called 1A 0620-00 to be the closest black-hole to the sun.

2 New Exploratory Method to Obtain Distance Estimate

We propose here a new and exploratory method to obtain an *estimate* of the distance of microquasars. A distance estimate, say within 10-20%, is already a great progress for microquasars, where distances are sometimes uncertain by a factor of 2. This new method is made of two ingredients.

First, it has been shown recently by [3] that the bluemost point of the spectroscopic bisector of a star is linearly related to its absolute magnitude. It is an empirical relation working (at least) for late-type stars. Unfortunately, to obtain a good quality *single-line* bisector, it is necessary to achieve a Signal-to-Noise ratio of at least 300 (often even higher) in the considered line, and at the same time a high spectral resolution (say, above 40 000). For faint objects in the optical such as GRO J1655-40 (V $\sim$ 18), a bisector to be made with UVES (R=45 000) would require more than two continuous weeks of 24-hours observations...

Second, it has been shown recently by [1] using the HARPS instrument installed in the La Silla Observatory (Chile)), that the bisector of the cross-correlation function (CCF) can be used as much the same way as single-line bisectors. This is rather natural, since well-chosen atmospheric lines should behave mostly the same way.

The combination of the two ingredients opens the door to the observation of faint targets. As a matter of fact, it is much easier to obtain a good quality bisector of the CCF, since the information of numerous lines can be combined together. The Signal-to-Noise ratio required now depends also on the number of lines used to compute the CCF.

3 Known Caveats and Prospects

The idea behind this method is to obtain a good-quality spectrum of the secondary star of microquasars in quiescence, compute their CCF and measure the position of the bluemost point of it. This point being then compared to the linear relation as described in [1]. Of course, there are obvious caveats in this method that we list here.

- The correlation between (CCF) bisector and absolute magnitude is not fully understood yet.
- The correlation is known to work only for late-type stars, in addition to have been tested on a small sample only.
- Microquasars have often Roche-filling secondary stars that could influence in an unclear manner the bisector and the absolute magnitude.
- The spectrum of microquasar can be polluted by black-hole accretion disk

Despite these caveats, we are currently investigating this relation using the newly commissionned near-infrared echelle spectrograph at the VLT: CRIRES. With such instrument, it is possible to achieve the necessary high resolution, and obtain a good quality spectrum of microquasars in the near-infrared, where microquasars are much brighter than in the optical (for GRO J1655-40, K$\sim$12.5).

References

1. T. H. Dall, N. C. Santos, T. Arentoft et al.: A&A **454**, 341 (2006)
2. C. Foellmi, E. Depagne, T. H. Dall, I. F. Mirabel: A&A, in press (2006) *ArXiv Astrophysics e-prints (astro-ph/0606269)*
3. D. F. Gray: PASP, **117**, 711 (2005)
4. R. M. Hjellming and M. P. Rupen: Nature, **375**, 464 (1995)
5. I. F. Mirabel: Microquasar-AGN-GRB Connections. In: *ESA SP-552: 5th INTEGRAL Workshop on the INTEGRAL Universe*, ed by V. Schŝnfelder, G. Lichti & C. Winkler (ESA SP-552 2004) pp 175–183

Metallicity of Pleiades Dwarf

H. Funayama[1] Y. Itoh[1] Y. Oasa[1] E. Toyota[1] and T. Mukai[1]

Graduate School of Science and Technology, Kobe University, 1-1 Rokkodai, Nada, Kobe, Hyogo, 657-8501, Japan funayama@kobe-u.ac.jp

Summary. The goal of our study is to discuss how planet-hosting stars (PHSs) are born by analyzing the chemical homogeneity in open clusters. We obtained the spectra of 18 A,F,G stars in the Pleiades by the high-resolution spectroscopic observations in last January. Out of the 18 stars, the metallcities of 10 stars were derived, and their mean metallicity was obtained to be [Fe/H]mean = +0.02±0.14dex. This high dispersion may imply the chemical diversity in Pleiades. We report the results of our analysis.

1 Introduction

1.1 Extrasolar Planet vs. Stellar Metallicity

The study of extrasolar planets suggest that about 5% of stars in our Galaxy possess planets (3) and most of the stars have higher metallicity ([Fe/H]~0.25dex) than stars which no planet has been detected (5). It is suggested that the most of stars (~90%) are formed in clusters (2), hence it is natural to consider PHSs used to be members of clusters. From these evidences, we propose two hypotheses:
1) About 5% of stars in each open cluster have high metallicity.
2) About 5% of open clusters in our Galaxy are star aggregates of stars with high metallicities.
To verify these hypotheses requires analyzing the chemical homogeneities in many open clusters.

1.2 Metallicity of Open Cluster

An open cluster is a star aggregate born in a molecular cloud. Although the cluster members have different masses, their ages and initial chemical compositions are thought to be the same. Paulson et al. (2003) supported this idea by showing the uniform metallicities of 53 F,G,K stars in the Hyades ([Fe/H]mean=+0.13±0.01dex). Besides the Hyades, however, there is no observation discussing the chemical homogeneity of any cluster by determining the metallicities of the large number of the members. Although there are many observations to measure the metallicities of the Pleiades members, the largest number of the samples were only 20 stars (1).

2 Observation and Results

In last January, we obtained the high-resolution spectra, with S/N ratios of 130-200, of 13 late A and F stars with Okayama 188cm/HIDES and 5 F, G stars with Gunma 150cm/GAOES, respectively. This is the first time to determine the metallicities of the stars except one object. The resolutions of both observations were $R\sim30000$ and the wavelength range was 5311∼6590Å for Okayama and 4853∼6690Å for Gunma, respectively. We used the analysis program SPTOOL(6). To date, we determined the atmospheric parameters such as the effective temperature (Teff) and metallicities ([Fe/H]) for 10 stars with the precision of σTeff∼80K, and σ[Fe/H]∼0.1dex. It is found that the mean value of their metallicities (Fig. 1) implies there is a dispersion in metallicity of Pleiades (i.e. [Fe/H]mean=+0.02±0.14dex). However, the number of samples is too small to discuss the chemical homogeneity in the Pleiades quantitatively. We confirmed the other 7 stars to be rapid rotators and consider the possibilities of the adoption of spectral synthesis method to determine their metallicities. In this winter, the planned observations will hopefully increase the number of our samples to 50.

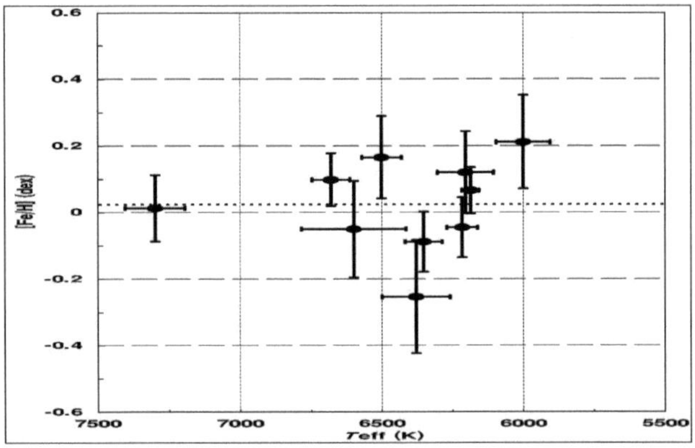

Fig. 1. The metallicities of Pleiades members

References

1. A. M. Boesgaard: ASPC. **336**, 39B (2005)
2. C. J. Lada & E. A. Lada: ARA&A. **41**, 115L (2003)
3. G. Marcy et al: PThPS. **158**, 24M (2005)
4. D. B. Paulson et al: AJ. **125**, 3185 (2003)
5. N. C. Santos et al: A&A. **415**, 1153 (2004)
6. Y. Takeda et al: PASJ. **54**, 451 (2002)

Precision of Radial Velocity Surveys using Multiobject Spectrographs – Experiences with Hectochelle

Gábor Fűrész[1,2], Andrew H. Szentgyorgyi[1], and Søren Meibom[1]

[1] Harvard-Smithsonian CfA, 60 Garden St, MA, 02138, USA
gfuresz@cfa.harvard.edu
[2] University of Szeged, 6720, Szeged, Dóm tér 9, Hungary

1 The Hectochelle Instrument

There are two multiobject spectrographs at the 6.5m MMT telescope (Arizona, USA), sharing the same fiber feed and fiber positioner. A robotic positioner places all 300 fibers over the 600 mm (1 degree wide field [1]) focal surface in five minutes with an accuracy of 25μm or better. The fiber diameter is 250μm, or 1.5 arcsec at the MMT f/5 plate scale. The 30m long fiber run ends in a room accommodating two bench mounted spectrographs, the moderate resolution Hectospec and the high resolution (R=34,000) Hectochelle [2]. As there is no cross disperser, the orders of Hectochelle are separated by interference filters in front of the pseudo-slit, one order being $\sim$ 150 Å wide. Efficiency peaks at $\sim$ 7%, at a wavelength $\sim$ 5200 Å .

2 The Multiobject Advantage — and Drawback

Radial velocity (RV) surveys on large number of stars can be very time consuming using single object instruments. Sacrificing the multi-order advantage of echelle spectrographs for multi-object capabilities clearly have the time-efficiency advantage, but the limited wavelength coverage means less precise measurements. And there are several other factors decreasing the RV precision of multiobject spectrographs. In order to pack a few hundred spectra on the detector format, Hectochelle uses a very fast camera optics which delivers sharp images, but affected by pincushion and strong color. With the anamorphic magnification these render the round fiber inputs into elongated, tilted line profiles, which show different behavior over the focal plane (Fig. 1). In addition, mounting errors of the fibers in the slit (e.g. tiny tilts respect to each other) results further changes in the line profiles not just between apertures, but between wavelength ranges within one spectrum produced by the same fiber.

Further problem is calibration. Illuminating a screen in front of the telescope makes calibration light passing through the entire optical train just like starlight. But this ideal lighting of fiber inputs does not work with faint

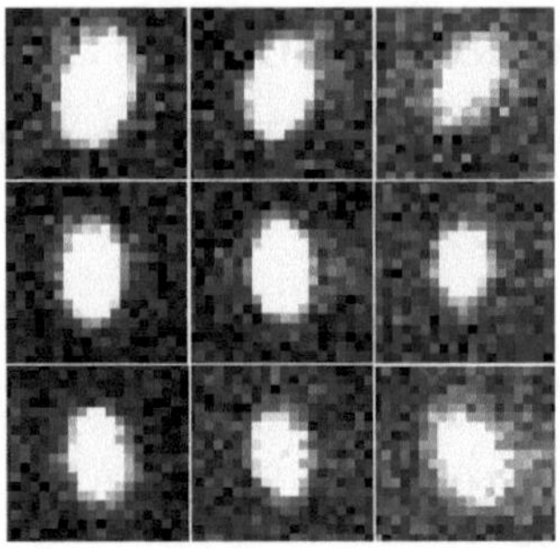

Fig. 1. Line profiles from a Hectochelle ThAr image. Spots extracted from aperture 1, 120 and 240 (top, middle and bottom, respectively – just like the position of these apertures on the detector), at three different wavelengths in each aperture (red end, middle and blue end of the 31d echelle order, from left to right, respectively)

ThAr lamps used to provide the wavelength solution. With Hectochelle, in order to decrease calibration time under 5 min, we use a direct illumination source mounted right in front of the secondary. (Being within the shadow cone of the secondary it takes away no starlight.) As the focal surface is curved, the fiber axes point to the calibration system this way, independently of their focal position. Illumination of fibers at the same angle was found to be very important in order to reduce RV errors.
Unfortunately due to space limitations the spectrograph room of Hectochelle could not be well isolated in thermal means from its environment. This seems to be the a main source of time dependency in the observed RV values. Although the instrument was designed with thermal stability in mind, the 6m long bench and 1m class optics are very sensitive for a few degree variations in temperature.

3 Key: Continuous Calibration and Optimal Extraction

Taking an exposure of the dusk/dawn sky offers a high signal-to-noise calibration method to map the net result of the above listed effects. Cross-correlating the same solar spectra, recorded through the different fibers, with a synthetic template results a fiber-to-fiber RV variation (Fig. 2). Taking such calibration frames over a year we can state that a given fiber always exhibit the same RV deviation, so a correction value can be determined for each aperture on the CCD. Even this can be as large as 1.4 kms^{-1} between two apertures, for a given fiber this correction factor is stable at a 80 m/s RMS level.
We believe that optimal extraction of the spectra could further improve the RV precision of multiobject spectrographs, like Hectochelle. In our current extraction, performed by standard IRAF routines, the curvature in the trace of a given order is taken into account but collapsing the 12 pixel wide trace

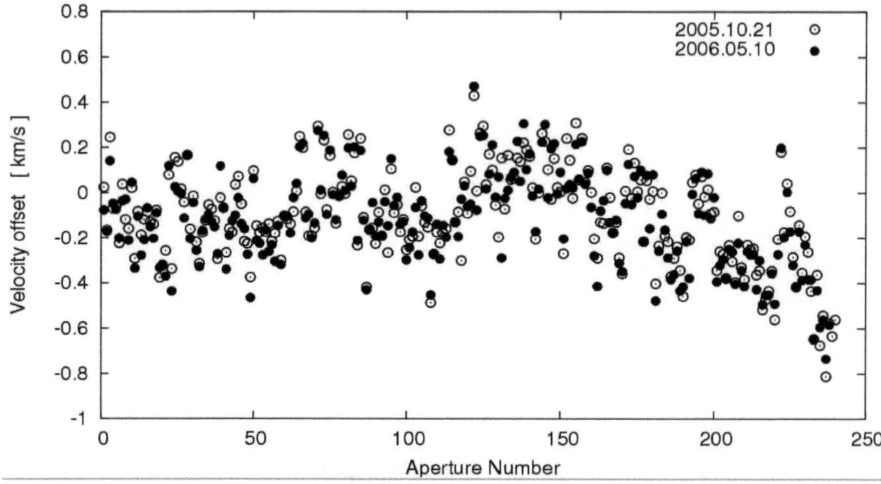

Fig. 2. Fiber-to-fiber RV offsets measured for Hectochelle, at two epochs. For a given aperture the offset value varies only with a 80m/s RMS, based on 11 measurements taken over a year.

into the one dimensional spectra still takes place strictly along columns/rows of the CCD. In presence of line profiles reported on Fig. 1 it is obvious that extracting spectra this way broadens/modifies line profiles/centroids. As the tilt of the lines are changing not just from one aperture to the other, but even within a given aperture, spectra within one image can not be compared to each other accurately, and each spectrum would differ from an ideal, synthetic spectra. To reach higher RV precision an optimal extraction seems to be the key, locally mapping the line profiles at each wavelengths and therefore correcting for the line tilts.

References

1. Fabricant, D.G., et al. Proceedings of the SPIE, **5492**, 767 (2004)
2. Szentgyorgyi, A. H., Cheimets, P., Eng, R., Fabricant, D. G., Geary, J. C., Hartmann, L., Pieri, M. R., & Roll, J. B. Proceedings of the SPIE, **3355**, 242 (1998)

Many thanks to our LOC, Britt Sjöberg and Susana Fernandes.

High Resolution Study of the Young Quadruple System AO Vel with an Eclipsing BpSi Primary

J. F. González[1], N. Nesvacil[2], and S. Hubrig[3]

[1] CASLEO and University of San Juan, Argentina fgonzalez@casleo.gov.ar
[2] University of Vienna, Austria nicole@jan.astro.univie.ac.at
[3] European Southern Observatory, Chile shubrig@eso.org

1 Introduction

During last year's analysis of FEROS spectra [1], we discovered that the triple system AO Vel with an eclipsing BpSi primary is in fact a remarkable quadruple system with ZAMS and PMS companions. The four stars form two spectroscopic binaries orbiting around each other in a wide orbit with a 41 yr period. For the first time, direct determination of the radius and the mass have been obtained for a BpSi star. We present here our recent study of chemical abundances of all four components using recent high-resolution, high signal-to-noise UVES spectra.

2 Observations and Spectral Analysis

Three high resolution (R=80,000–100,000) spectra were obtained in October 2005 with UVES. In order to obtain separate spectra for all four components of the system and to measure their radial velocities, we applied the iterative method described by [2], adapted here for multiple systems. We obtained abundances via equivalent width measurements and using ATLAS9 atmospheres [3] and the Kurucz WIDTH9 code. Stellar temperatures and microturbulent velocities were determined from the correlation of abundances with excitation potentials and equivalent widths of Fe lines. The temperature finally adopted for components A, B, C, and D were 13 700, 12 200, 12 000, and 11 500 K, respectively.

3 Results

The subsystem A+B is an eclipsing binary, with a BpSi primary. The star A exhibits an overabundance of Si by 1.0 dex and an underabundance of He by 0.5 dex. The star B is a normal late B-type star with abundances close to solar. Fig.1 presents the spectra of all four companions, A, B, C, D around He I 5016 and Si II 5041, 5056.

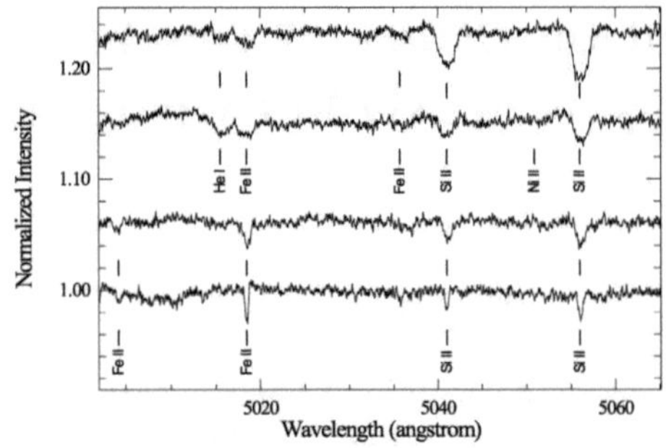

Fig. 1. Spectra of all four companions, A, B, C, D (from top to bottom) around He I 5016 and Si II 5041, 5056.

In our spectra of the C and D components the most interesting, but rather unexpected fact is the presence of spectral lines of Y, Hg and Pt, which are typical for HgMn stars. However, very few Mn II lines have been identified in the spectrum of the C component and none in D component. In Fig.2 we present some spectral regions showing Hg, Pt, Y, and Sr lines.

Star C shows a strong overabundance of Hg by 6.7 dex, while other metals (Fe, Ti, Mg and Si) show solar abundances. Star D exhibits, besides Hg, overabundance of Y II (3.1 dex), Pt I (5 dex), and Sr II (2.2 dex). The measured wavelength of Hg II 3984 indicates that the heavy Hg isotopes 202 and 204 are predominant in components C and D. A more detailed study, including spectral synthesis is planned to provide limits on the Mn abundance.

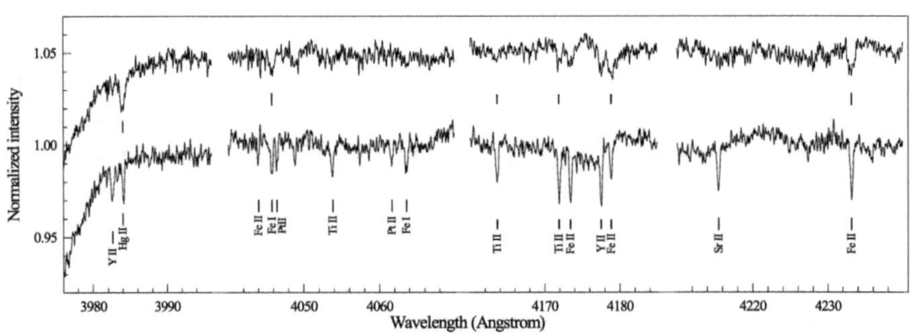

Fig. 2. Spectra of components C (top) and D (bottom) in regions around Hg II 3984, Pt II 4061, Y II 4178 and Sr II 4215.

References

1. J. F. González, S. Hubrig, N. Nesvacil, P.North: A&A **449**, 327 (2006)
2. J. F. González, H. Levato: A&A **448**, 283 (2006)
3. R. L. Kurucz: CDROM13, SAO Cambridge (1993)

A Survey for Extrasolar Planets Around A–F Type Stars

M. Hartmann[1], A.P. Hatzes[1], E.W. Guenther[1], and M. Esposito[1,2]

[1] Thüringer Landessternwarte Tautenburg, Sternwarte 5, D-07778 Tautenburg, Germany `michael@tls-tautenburg.de`

[2] Dipartimento di Fisica "E.R. Caianiello", Università di Salerno, via S. Allende, 84081 Baronissi (Salerno), Italy

1 Introduction

Since the discovery of the first extrasolar planet 51 Peg b [1] more than 190 extrasolar planets have been detected in orbit around other stars by means of precise stellar radial velocity (RV) measurements. But all RV surveys are concentrating almost entirely on solar-type stars covering only a narrow spectral and stellar mass range (spectral type F8–K1, 0.7–1.3 $M_\odot$) and up to now only six substellar companions around the intermediate mass stars HD 47536 [2], HD 104985 [3], γ Cep [4], HD 11977 [5], HD 13189 [6], and β Gem [7] have been identified. Interestingly, all these planets have masses of several Jupiter masses.

Thus, we have little knowledge about how planet formation depends on the most fundamental parameter of a star: its mass. In order to understand the influence of stellar mass on the process of planet formation and to test formation theories, it is essential to probe the mass dependence of planet formation by extending RV surveys over a wide range of stellar masses. As part of the Tautenburg Observatory Planet Search (TOPS) program we started an extensive survey for extrasolar planets around A–F type stars.

2 Monitored Sample and Observations

In this study we selected a large sample of stars containing about 200 F-type and roughly 100 A-type stars lying in the mass range of 1–3 $M_\odot$. This sample size will allow us to draw statistically meaningful conclusions about planets around more massive stars. The observations are carried out with the 2.0 m Alfred Jensch telescope of the Thüringer Landessternwarte Tautenburg (TLS), which is equipped with a high resolution coudé echelle spectrograph (resolving power $R = 67\,000$) and an iodine absorption cell placed in the optical light path in front of the spectrographs slit. This grism crossed-dispersed echelle spectrograph has a wavelength coverage of 4700–7400 Å.

3 The New Planet

We have discovered a new massive planet around a F4V star in a 329-day low-eccentric orbit with an $m \sin i = 11\, M_{\rm Jup}$. The phase-folded RV measurements are shown in Fig. 1, as well as the best Keplerian orbital solution. The corresponding periodogram is given in Fig. 2.

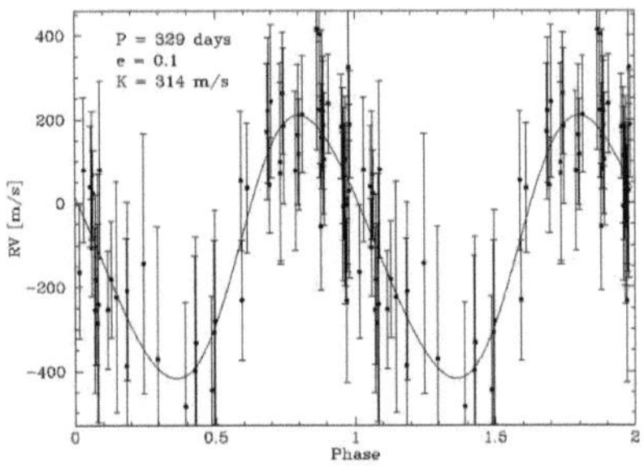

Fig. 1. Phase-folded radial velocity measurements of the planet with 329 day period and $m \sin i = 11\, M_{\rm Jup}$. The solid line represents the best-fit Keplerian orbital solution.

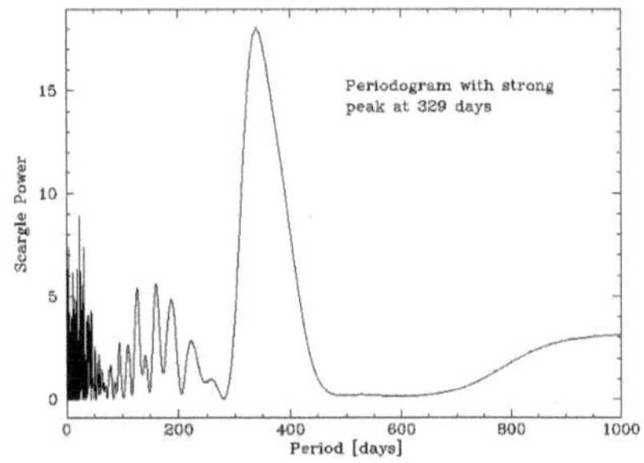

Fig. 2. Periodogram of this planet with strong peak at 329 days.

4 Conclusions

Although the project is still in its initial phase, this example shows RV variations which are caused by a substellar companion having a mass in the range of a super planet or brown dwarf. In addition to the discovered six planets around intermediate mass stars this could be a first hint that more massive stars harbor more massive planets. The future aim is to determine the frequency of planets around A and F stars as well as to compare their properties with those of planets around solar-like stars.

References

1. M. Mayor, D. Queloz: Nature **378**, 355 (1995)
2. J. Setiawan, A.P. Hatzes, O. von der Lühe, et al.: A&A **398**, L19 (2003)
3. B. Sato, H. Ando, E. Kambe, et al.: ApJ **597**, L157 (2003)
4. A.P. Hatzes, W.D. Cochran, M. Endl, et al.: ApJ **599**, 1383 (2003)
5. J. Setiawan, J. Rodmann, L. da Silva, et al.: A&A **437**, L31 (2005)
6. A.P. Hatzes, E.W. Guenther, M. Endl, et al.: A&A **437**, 743 (2005)
7. A.P. Hatzes, W.D. Cochran, M. Endl, et al.: A&A **457**, 335 (2006)

A Study of the Magnetic Helium Variable Emission-line Star HD 125823.

S. Hubrig[1], N. Nesvacil[2], F. González[3], B. Wolff[4], I. Savanov[5]

[1] European Southern Observatory, Casilla 19001, Santiago, Chile shubrig@eso.org,
[2] Institut für Astronomie, Universität Wien, Türkenschanzstr. 17, 1180 Vienna, Austria nicole@jan.astro.univie.ac.at,
[3] Complejo Astronómico El Leoncito, Casilla 467, 5400 San Juan, Argentina fgonzalez@casleo.gov.ar,
[4] European Southern Observatory, Karl-Schwarzschild-Str. 2, 85748 Garching, Germany bwolff@eso.org,
[5] Astrophysikalisches Institut Potsdam, An der Sternwarte 16, 14482 Potsdam, Germany isavanov@aip.de

The 5.9 $M_\odot$ star HD 125823 is a striking helium variable with a period of 8.82 d, ranging in helium spectral type from He-strong B2 to B8 (e.g., Norris 1968 [3]). In fact, HD 125823 seems to be a transition object between the He-weak and He-strong stars with unreddened colors just at the boundary between the two groups in the UBV color-color diagram. Although high-dispersion studies have been carried out in the past by several authors (e.g. Wolff & Morrison 1974 [6]), no abundance analysis is available.

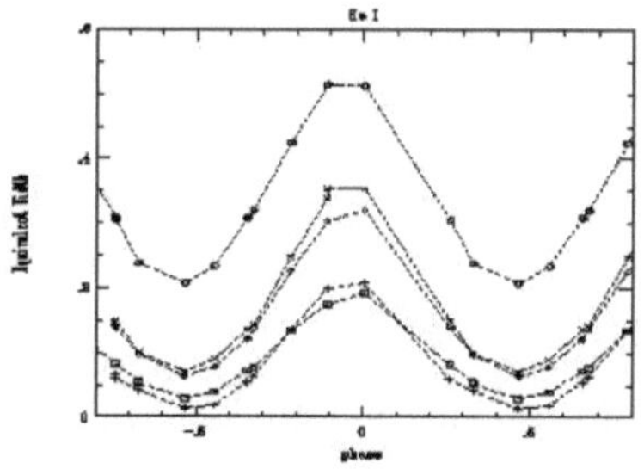

Fig. 1. Variations of the equivalent widths of the He I lines λλ4121, 4438, 4713, 5047, and 6678 over the rotation period.

Recently, we obtained high spectral resolution high signal-to-noise UVES spectra of HD 125823 over the rotation period of 8.82 d in order to study the surface distribution of various chemical elements. In Fig. 1 we show the behaviour of the He I lines λλ4121, 4438, 4713, 5047, and 6678 over the rotation period. The magnetic field observations reported in the literature (Wolff & Morrison 1974 [6]; Borra et al. 1983 [1]) indicate that the negative extremum coincides closely in phase with the maximum helium strength.

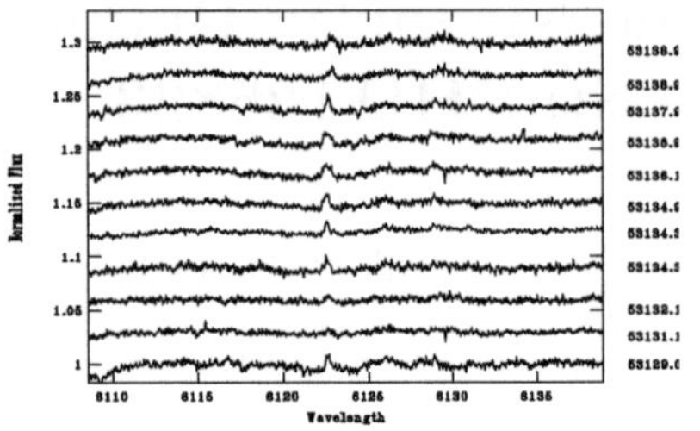

Fig. 2. Mn II emission lines of multiplet 13 in the spectra of HD 125823 obtained on nine consecutive nights with UVES.

However, the positive magnetic pole shows a He-deficient cap, so that the two magnetic poles are associated with very different helium abundances.

An additional interest to study HD 125823 comes from the recent detection of weak emission lines of various ions (Mn II, Fe II, Cr II, Ti II, etc.) in optical spectra of B-type stars (e.g., Sigut et al. 2000 [4]; Wahlgren & Hubrig 2000 [5]; Castelli & Hubrig 2004 [2]). For the first time we present here observational evidence for the appearance of emission lines of Mn and Fe in a variable magnetic star. In Fig. 2 we show the behaviour of the emission line profiles of the Mn II multiplet 13. It is remarkable that the emission lines of Mn II are definitely variable, changing their appearance from the first night to the last. To date, explanations of this phenomenon have been put forward in the context of non-LTE line formation and possible fluorescence mechanisms. The qualitative and quantitative assessments of the spectra of B type stars suggest a possible correlation of the appearance of diverse emission lines with the non-magnetic P-Ga and HgMn groups as well as with spectral type.

From the behaviour of the line profile variations of various elements we can conclude that helium is enhanced in regions of the stellar surface where silicon and other metals are depleted, and helium is depleted in regions where the metals are enhanced. A future goal is to derive surface abundance maps for various elements. Such a study is an important step towards understanding the effect of magnetic fields on the development of surface chemical peculiarities in the photospheres of hot stars.

References

1. Borra E. F., Landstreet J. D., Thompson I., 1983, ApJS, 53, 151
2. Castelli F., Hubrig S., 2004, A&A, 425, 263
3. Norris J., 1968, Natur, 219, 1342
4. Sigut T. A. A., Landstreet J. D., Shorlin S. L. S., 2000, ApJ, 530, L89
5. Wahlgren G. M., Hubrig S., 2000, A&A, 362, L13
6. Wolff S. C., Morrison N. D., 1974, PASP, 86, 935

bHROS: The High-Resolution Optical Spectrograph at Gemini South

Steven J Margheim[1]

Gemini Observatory†, Southern Operations Center c/o AURA, Casilla 603, La Serena, Chile `smargheim@gemini.edu`

The Gemini bench-mounted High-Resolution Spectrograph (bHROS) is the highest resolution (R=150,000, 3 pixel sampling) optical echelle spectrograph optimized for use on an 8-meter telescope. Located within the thermally stable pier of the Gemini South telescope, bHROS is fiber-fed via GMOS-S from the focal plane of the telescope.

bHROS has 2 modes of operation; an "object-sky" mode which uses two 0.7" diameter fibers separated by 20 arcsecs, one each for object and sky and an "object-only" mode where only a single fiber with a projected diameter of 0.9 arcsec is used. Each fiber feeds its own dedicated image slicer to produce a 'slit' 0.14 arcsec wide in the spectral direction. A slicer rotation mechanism is used to ensure a 'vertical' slit at the central wavelength.

Cross-dispersion is provided by a set of fused silica prisms. Prism cross-dispersion is used because of the increased efficiency and lower scattered light as compared to grating cross-dispersion. The echelle unit has a a ruled length of 408mm and a blaze angle of 63 and a ruling of 87g/mm. The echelle has 2 degrees of motion to scan the echellogram on the detector in order to configure the spectrograph for a specific wavelength range.

The detector is a single 2048x4608 E2V CCD with 13.5μm pixels. The CCD mosaic is located at the prime focus of the camera, immediately behind a field flattener lens. bHROS covers the wavelength range of 400nm to 1000nm at a fixed resolution of R=150,000 with 3 pixel sampling. The simultaneous wavelength coverage varies with wavelength and ranges from 3.5nm at 4000nm to nearly 10nm at 1000nm. The number of orders falling on the detector also varies with wavelength ranging from 5 in the blue to 9 in the red.

A java applet is available on the bHROS instrument web page (http://www.gemini.edu/sciops/instruments/hros/) to interactively examine the coverage available at a given grating setting. User-defined spectral fea-

†The Gemini Observatory is operated by the Association of Universities for Research in Astronomy, Inc., under a cooperative agreement with the NSF on behalf of the Gemini partnership: the National Science Foundation (United States), the Particle Physics and Astronomy Research Council (United Kingdom), the National Research Council (Canada), CONICYT (Chile), the Australian Research Council (Australia), CNPq (Brazil), and CONICET (Argentina)

tures can be plotted within the echellogram applet to aid in the determination of the appropriate grating settings.

bHROS underwent commissioning in July 2005 and became available for scientific use in February 2006 and is currently available. Data taken as Demonstration Science is available through the bHROS instrument web pages and the Gemini Science Archive. Additionally, a sample extracted spectra (a portion of which is shown in Figure 1) is also available for download. Further information on the instrument including information about performance and use can be found on the bHROS instrument web pages located at http://www.gemini.edu/sciops/instruments/hros/.

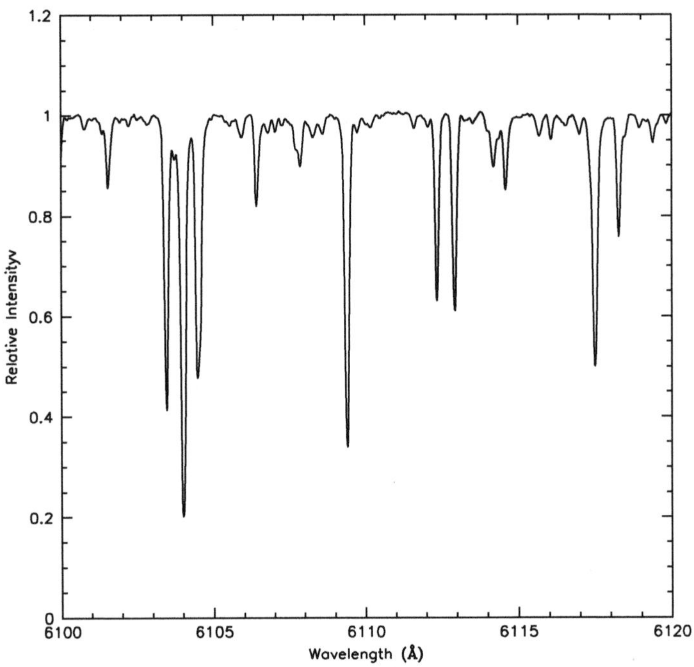

Fig. 1. Example processed and extracted bHROS spectrum of HD 192879 with a signal-to-noise ratio of 800 per spectral pixel. The complete extracted spectrum is available for download from the bHROS instrument web page.

Stellar Wobble Caused by a Binary System: Investigation in the Framework of the General Three Body Problem

M.H.M. Morais[1] and A.C.M. Correia[2]

[1] Grupo de Astrofísica da Universidade de Coimbra, Departamento de Matemática FCTUC, Largo D. Dinis, 3000 Coimbra, Portugal
hmorais@mat.uc.pt

[2] Departamento de Física da Universidade de Aveiro, Campus Universitário de Santiago, 3810-193 Aveiro, Portugal acorreia@fis.ua.pt

1 The Problem

Consider the general three body consisting of a star of mass m_1 (whose spectrum we can analyse) at a distance a_A from the center of mass of a binary star system with masses $m_2 = m_3$ separated by $2a_B$. Furthermore, suppose that we do not know about the presence of the binary system (although we may know about the presence of one of the stars). It has been shown by [1] that we may be lead to think that the star m_1 has a planet companion while it is in fact being perturbed by the binary star system.

The authors in [1] studied the case with $m_1 = 0$, $a_A >> a_B$, $m_2 = m_3$ moving in a circular orbit around each other. They showed that it has the following solution: the star moves slowly around the center of mass of the binary at angular velocity $\Omega \approx \sqrt{G(m_2 + m_3)/{a_A}^3}$; the motion of the binary induces a superimposed oscillation with amplitude[3] $\sim \delta = a_B(a_B/a_A)^4$ and angular velocity $\sim 2\omega$, where $\omega = \sqrt{G(m_2 + m_3)/(2a_B)^3}$.

Under these assumptions, the fourier spectrum of the star's barycentric velocity will exhibit one frequency $\sim 2\omega$ with associated amplitude $\sim 2\omega\delta$ which can be mistaken by a planet with semi-major axis $a = (Gm_1/(2\omega)^2)^{1/3}$ and mass $m \sim m_1\delta/a$.

2 Our Work - Some Results

The model presented in [1] is the restricted 3 body problem (R3BP). We performed some numerical integrations of the general 3 body problem choosing $a_A = 10$ AU, $2a_B = 1$ AU and masses (in units of the solar mass) $m_2 = m_3 = 0.5$, $m_1 = 0$, $m_1 = 0.001$, $m1 = 0.01$ and $m_1 = 0.02$.

[3] According to [1] this superimposed oscillation is, in fact, an ellipse with semi-major axis $\delta_x = 4.5\delta$, along the line that joins m_1 to the binary's center of mass, and semi-minor axis $\delta_y = 3\delta$.

We then applied the Fast Fourier Transform (FFT) algorithm [2] to obtain the main frequencies and amplitudes present in the time series of the barycentric radius and radial velocity of the star m_1. These results are shown in Table 1 together with the theoretical predicted values for the R3BP ($m_1 = 0$).

As we increase m_1 it becomes more difficult to choose initial conditions that are associated with stable orbits. However, our preliminary results seem to indicate that the main characteristics of the FFT spectra are preserved. As noted previously, if we did not know about the presence of the binary system then the analysis of the FFT spectrum of the barycentric velocity of m_1 would lead us to conclude that this star had a planetary companion (with orbital frequency approximately 2ω). In general we can still detect the frequency Ω if our data covers a long enough timespan but we could think that this was caused by the presence of a single star of one solar mass located at 10 AU.

m_1	f0 (cicles/y)	f (cicles/y)	ampl. radius (AU)	ampl. rad. vel. (AU/y)
0.0	0.0316	1.937	1.42×10^{-5}	1.19×10^{-4}
0.001	0.0316	1.937	1.37×10^{-5}	1.11×10^{-4}
0.01	0.0310	1.938	1.45×10^{-5}	1.15×10^{-4}
0.02	0.0301	1.940	1.71×10^{-5}	1.05×10^{-4}
R3BP	0.0316	1.937	1.41×10^{-5}	1.18×10^{-4}
	$\Omega/2\pi$	$2(\omega - \Omega)/2\pi$	δ_x	$2\omega\delta_y$

Table 1.

3 Conclusion

We confirmed that the results of [1] are approximately valid in the framework of the general 3 body problem. However, we saw that even the effect of a close binary star system like the one studied here will lead to a very small perturbation of the observed star (of the order of 10^{-4} AU/years). Such small perturbations are at the detection limit of the current spectroscopy techniques (the current resolution is around 1 meter/second $\approx 2 \times 10^{-4}$ AU/years)

This work has been financed by grant SFRH/BPD/19155/2004 from the Fundação para a Ciência e a Tecnologia (Portugal).

References

1. J. Schneider, J. Cabrera: Astronomy and Astrophysics **445**, 1159-1163 (2006).
2. W. Press, B. Flannery, S. Teukolsky, W. Vetterling: *Numerical Recipes in Fortran*, 2nd edn (Cambridge University Press 1992).

The Chemical Composition of B-type Pulsators: Some Unexpected Results

T. Morel[1,2], M. Briquet[1], and C. Aerts[1,3]

[1] Katholieke Universiteit Leuven, Departement Natuurkunde en Sterrenkunde, Instituut voor Sterrenkunde, Celestijnenlaan 200D, B-3001 Leuven, Belgium
[2] European Space Agency (ESA) postdoctoral external fellow
[3] Department of Astrophysics, University of Nijmegen, PO Box 9010, 6500 GL Nijmegen, The Netherlands
thierry,maryline,conny@ster.kuleuven.be

Summary. We present a project aimed at self-consistently deriving the physical parameters and chemical composition of massive pulsators (β Cephei stars, Slowly Pulsating B stars) based on high-resolution optical spectra. Such data will be essential for a proper theoretical interpretation of their oscillation spectrum, but may also contribute in a broader context to our understanding of mixing and diffusion processes in B-type stars. As an illustration, our first results reveal the existence of core-processed material at the surface of some slowly-rotating β Cephei stars which is not predicted by current evolutionary models including rotation.

1 B-type Pulsators and the Rationale of the Project

The β Cephei and slowly pulsating B stars (SPBs) are two well-established classes of pulsating variables found in the upper part of the main sequence. The β Cephei stars (B0–B3) pulsate through low radial order p/g modes with periods in the range 2–12 hours, while SPBs (B2–B9) exhibit high radial order g modes and longer periods in the range 0.5–5 days.

Dramatic advances in our knowledge of the pulsational properties of these objects are presently being made from multisite, ground-based campaigns or from space (e.g. *MOST*). In turn, theoretical modelling of these data is expected to give unique insights into some fundamental, yet poorly-known parameters of B-type stars, such as the amount of convective core overshooting or the rotation profile in the interior (e.g. [1]). Accurate estimates of the physical parameters and metallicity of these targets is needed to constrain the theoretical models and, ultimately, for correct inferences regarding their internal structure.

This has motivated us to launch an NLTE abundance study of a large sample of prime B-type targets for asteroseismology using fully line-blanketed Kurucz atmospheric models, the line formation codes DETAIL/SURFACE and high-resolution optical spectra obtained with a variety of echelle spectrographs offering a resolving power $R\sim 50\,000$ (mainly CORALIE, ELODIE and FEROS). Our analysis is based on a large number of time-resolved spectra covering in most cases the entire oscillation cycle of the stars.

2 Nitrogen Excess in Slowly-rotating β Cephei Stars: Deep Mixing or Diffusion?

Our initial study of a number of prototypical β Cephei stars [6] reveals that their photospheric metal content is indistinguishable from the values previously reported for early B dwarfs in the solar neighbourhood. However, we clearly detect in some objects an unexpected nitrogen excess accompanied by a strong boron depletion, as expected if core-processed material were dredged up to the surface by rotationally-induced mixing. Although appealing, one serious difficulty with this interpretation lies in the fact that the equatorial velocities of most N-enriched stars appear well below what is required by the most recent evolutionary models including rotation (e.g. [4]). Alternatively, recent theoretical work suggests that diffusion effects might be invoked [2]. In any case, any attempt to explain this phenomenon should account for the fact that a magnetic field of up to a few hundreds Gauss has been detected in most N-rich stars (see, e.g. [5]).

3 Future Developments

We are currently carrying out an abundance study of about 3/4 of all confirmed SPBs (∼30 stars) using the same techniques (Briquet et al., in prep.). Most stars in this sample have already been the subject of sensitive magnetic field measurements [5], a fact which will allow us to further investigate the incidence of chemical peculiarities as a function of the field strength. Preliminary results [3] also suggest the existence of N-rich, B-depleted stars in this class of objects, e.g. the magnetic star ζ Cas with a polar field strength B_p=340±90 G and a rotational velocity of only about 55 km s^{-1} [8].

A vast number of B stars will be observed during the course of the asteroseismology programme of the *CoRoT* mission. High-resolution optical spectra have so far been obtained for ∼200 stars in the zones of the sky which will be surveyed by the satellite as part of the ground-based preparatory campaign. We plan to analyze the full sample using (semi-)automated, spectral synthesis techniques (see [7] for the very first results).

References

1. C. Aerts, A. Thoul, J. Daszyńska, et al.: Science **300**, 1926 (2003)
2. P.-O. Bourge, S. Théado, A. Thoul: MNRAS, submitted (2006)
3. M. Briquet, T. Morel: CoAst, **148**, in press (2007)
4. A. Heger, N. Langer: ApJ, **544**, 1016 (2000)
5. S. Hubrig, M. Briquet, M. Schöller, et al.: MNRAS **369**, L61 (2006)
6. T. Morel, K. Butler, C. Aerts, et al.: A&A **457**, 651 (2006)
7. T. Morel, C. Aerts: CoAst, **148**, in press (2007)
8. C. Neiner, V. C. Geers, H. F. Henrichs, et al.: A&A **406**, 1019 (2003)

Radial Velocity Precision in the Near-Infrared with T-EDI

Philip S. Muirhead[1], David J. Erskine[2], Jerry Edelstein[3], Travis S. Barman[4], James P. Lloyd[5]

[1] Cornell University, Ithaca, NY, USA, muirhead@astro.cornell.edu
[2] Lawrence Livermoore National Laboratory, Livermoore, CA, USA, erskine1@llnl.gov
[3] University of California, Berkeley, CA, USA, jerrye@ssl.berkeley.edu
[4] Lowell Observatory, Flagstaff, AZ, USA, barman@lowell.edu
[5] Cornell University, Ithaca, NY, USA, jpl@astro.cornell.edu

Summary. Presented are calculations of the photon-limited radial velocity sensitivity of the TripleSpec Externally Dispersed Interferometer for a series of M-type stellar models. This instrument uses an interferometer before the dispersing element of a moderate-resolution spectrograph to boost radial velocity precision and reject systematic noise.

1 The T-EDI Instrument

TripleSpec is a moderate resolution near-IR spectrograph (R=2700, simultaneous coverage of J, H and K bands) to be commissioned as a facility instrument on the Cassegrain mount of the Palomar 200-inch Hale telescope in April of 2007 [1]. By placing a Michelson interferometer before the dispersing element of the spectrograph, the narrow features in a stellar spectrum are heterodyned with a sinusoidal interferometer comb [2]. Together, the TripleSpec Externally Dispersed Interferometer (T-EDI) will survey nearby M dwarfs for planetary companions using the radial velocity technique (see Figure 1 for T-EDI layout).

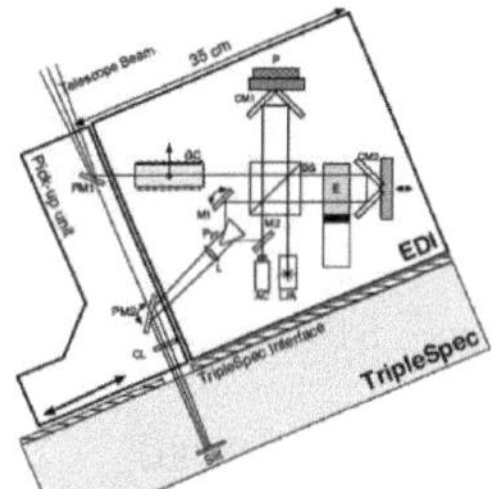

Fig. 1. T-EDI layout showing the Michelson interferometer before the spectrograph

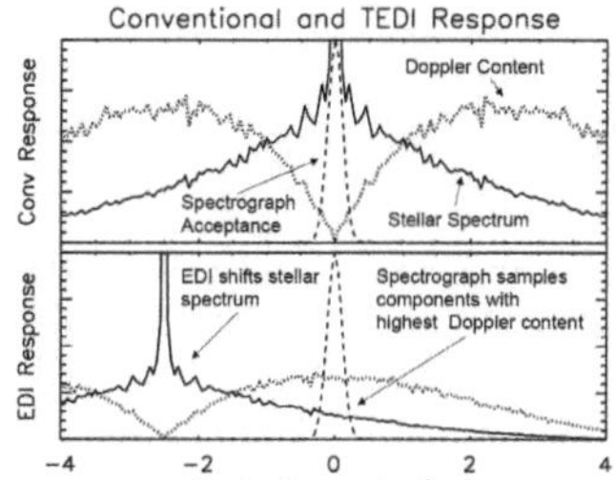

Fig. 2. J-band power spectra showing the effect of the interferometer in Fourier space.

The Michelson interferometer shifts the Fourier components of the stellar spectrum by the optical path difference between mirrors, such that the Fourier components with the highest Doppler content are selected by the spectrograph [3] (see Figure 2). This method has recently met with success at optical wavelengths where a new planet was discovered around HD 102195 by measuring the stellar reflex motion induced on the host star [4].

2 Photon-limited Radial Velocity Precision

The photon-limited RV precision of T-EDI is calculated by analyzing the change in the signal due to a stellar Doppler shift, relative to the photon-noise per pixel [5]. Using high-resolution models of late-type stellar spectra provided by Didier Saumon and Mark Marley of the Los Alamos National Laboratory and Travis Barman of Lowell Observatory , T-EDI has a calculated photon-limited precision of $\sim$ 5m/s for a typical M-dwarf with modest throughput assumptions (see Table 1).

Table 1. Calulated photon-limited radial velocity precision of T-EDI

Spectral Type	**Eff. Temp**	**Rotation**	**Int. Time**	**J-Mag**	**Throughput**	**δ RV**
M3-M4	3200 K	3 km/s	10 min	8	15%	4.0 m/s
M5	2800 K	3 km/s	10 min	9	15%	5.1 m/s
M8	2400 K	3 km/s	10 min	10	15%	4.7 m/s

With m/s radial velocity precision on M-type stars, T-EDI is capable of measuring the stellar reflex motion induced by a 5 Earth-mass planet orbiting within the habitable zone of an mid-to-late M dwarf.

We would like to thank Didier Saumon and Mark Marley of Los Alamos National Laboratory for generously providing high resolution models. This work is partially funded by the National Science Foundation under grant AST-0504874.

References

1. J. C. Wilson, C. P. Henderson, T. L. Herter et al: Proc. SPIE **5492**, 1295 (2004)
2. J. Edelstein, D. J. Erskine, J. P. Lloyd et al: Proc. SPIE **6269**, 46 (2006)
3. D. J. Erskine: PASP **115**, 255 (2003)
4. J. Ge, J. van Eyken, S. Mahadevan et al: ApJ **648**, 683 (2006)
5. R. P. Butler, G. W. Marcy, E. Williams et al: PASP **108**, 500 (1996)

HD154708 - The Challenging Abundance Analysis of an Extremely Magnetic Star

N. Nesvacil[1], S. Hubrig[2] and S. Khan[3]

[1] Department of Astronomy, University of Vienna, Türkenschanzstrasse 17, 1180 Vienna, Austria nnesvaci@eso.org
[2] European Southern Observatory, Casilla 19001, Santiago, Chile shubrig@eso.org
[3] Physics and Astronomy Department, University of Western Ontario, London, ON, N6A 3K7, Canada shkan@astro.uwo.ca

1 Observations

During a systematic study of magnetic fields in chemically peculiar stars with FORS1 at the ESO VLT, we discorvered an extremely large field in HD 154708 [1]. Using the high resolution echelle spectrograph UVES ($R \approx 110000$), we determined the mean magnetic field modulus of 24.5 kG, which is the second largest modulus ever measured in a cool Ap star. 8.0-min radial velocity variations have recently been discovered by Kurtz et al. [2]. The fact that rapidly oscillating Ap (roAp) stars normally do not display large magnetic fields makes HD 154708 rather unique.

2 Abundance analysis

We used Stroemgren and Geneva photometry provided by the SIMBAD database in order to calculate atmospheric parameters. Spectra computed with parameters derived from Geneva photometry ($T_{\rm eff} = 6800$K, $\log g = 4.1$) provide a better fit to the Hβ line profile in our FORS1 observations, therefore we used these parameters as a starting point for the analysis. From interpolation of the evolutionary tracks of Schaller et al. [3] we derived a stellar mass of $1.5 \pm 0.1 M_{\odot}$. According to the above assumptions, in the HR diagram HD 154708 lies close to the coolest and lowest mass roAp stars. From Fe I lines with low Landé factors we measured a projected rotational velocity $v\sin i = 1.5$km/s.

As a first step an ATLAS9 [4] model atmosphere was calculated for the given atmospheric parameters. Due to the large splitting of lines due to the Zeeman effect, line blends are difficult to model. We therefore selected a sample of unblended lines with well determined atomic parameters and especially well known Zeeman patterns. Synthetic spectra were calculated using an evolved version of synthmag [5], taking into account magnetic radiative transfer. Atomic line data were taken from VALD [6]. Iteratively, the derived abundances were then used to calculate a model with individual abundances using the LLmodels code [7]. With this model, line abundances were recalculated for each element. Fig.1 shows the abundance pattern of HD 154708, which is typical for a roAp star, in comparison with solar abundances [8]. Light elements as well as Ti, Fe and Ni are underabundant whereas Sc and

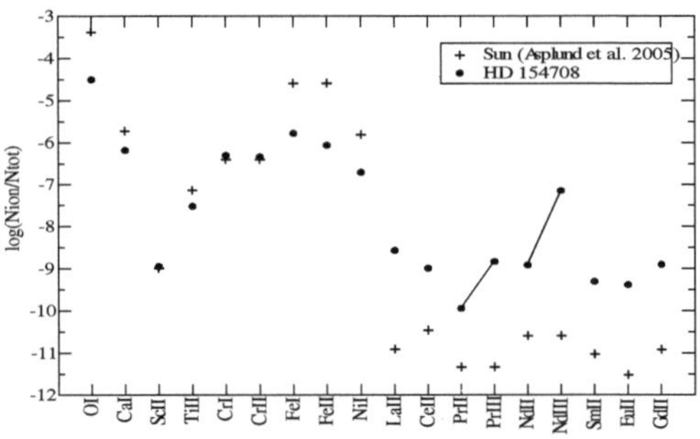

Fig. 1. Abundance pattern of HD 154708 (circles) and the sun (crosses). Lines highlight REE-anomaly.

Cr are solar. Heavy elements and rare earth elements are strongly overabundant and the rare earth anomaly which is typical for roAp stars is clearly present, i.e., Nd III is 0.86 dex overabundant in respect to Nd II and Pr III is overabundant by 1.11 dex compared to Pr II.

Next we calculated a magnetic model atmosphere using the newly developed software by Khan & Shulyak [9], taking into account the influence of the Zeeman effect on the model structure. Abundances derived with classical and magnetic model atmospheres differ only by 0.05 dex, which is much smaller than the typical abundance error (±0.2 dex). Hence, elemental abundances derived with LLModels can be considered reliable assuming the correctness of the atmospheric parameters.

References

1. S. Hubrig et al., A&A 440, 2005
2. D. Kurtz et al., MNRAS 372, 2006
3. G. Schaller et al., A&AS 96, 1992
4. R.L. Kurucz, CD-ROM No. 13, SAO, 1993
5. N.E. Piskunov, Contr. of the Astr. Obs. Skalnate Pleso, 27, 1998
6. F. Kupka et al. A&AS 138, 1999
7. D. Shulyak et al. A&A 428,2004
8. M. Asplund et al. ASPC 336, 2005
9. S. Khan & D.Shulyak A&A 454, 2006

A Search for Disk-Locking in the Chamaeleon I Star Forming Region

Duy Cuong Nguyen, Ray Jayawardhana, Marten van Kerkwijk, Alexis Brandeker and Aleks Scholz

University of Toronto, Department of Astronomy and Astrophysics, 50 St. George Street, Toronto, Ontario, Canada M5S 3H4 nguyen@astro.utoronto.ca, rayjay@astro.utoronto.ca, mhvk@astro.utoronto.ca, brandeker@astro.utoronto.ca, scholz@astro.utoronto.ca

Summary. We investigated the connection between disks and stellar rotation. Disk-locking theory predicts that accreting stars are preferentially slow rotators compared to their peers. If true, classical T Tauri stars (CTTS), which are accreting based on strong Hα emission, should have observably lower $v \sin i$ values compared to weak-lined T Tauri stars (WTTS), which are not accreting. We present our findings from high-resolution optical spectra taken with the Magellan Inamori Kyocera Echelle (MIKE) spectrograph on the Magellan Clay 6.5-m telescope located at the Las Campanas Observatory of 63 T Tauri stars in the Cha I SFR.

1 Introduction

From magnetospheric accretion theory, the stellar magnetic field is thought to channel gas from the inner disk. One would expect the star to be connected to the disk via magnetic field lines, which may retard the spin-up of the star as it contracts during the pre-main sequence phase. This process is often referred to as disk-locking [1, 5, 6].

Several photometric studies have found a correlation between rotation properties and near-infrared colour excess, suggestive of disks, while other groups fail to detect such a correlation [2, 3, 7]. However, IR signatures suggest the presence of a dusty disk and not the coupling between star and disk as required by the disk-locking scenario. A more sensible diagnostic to correlate with stellar rotation rates is on-going accretion which signifies a direct link between the inner disk and the central star.

2 Observations and Analysis

We took a total of 256 high-resolution optical spectra of 27 CTTS and 36 WTTS in the Chamaeleon I star forming region using the MIKE echelle spectrograph on the Magellan Clay 6.5-m telescope at Las Campanas, Chile. The data were collected during two observing runs in Feb 2006 and Apr 2006.

The projected rotation rates of our sample were estimated by correlating data from 11 echelle orders (6 800Å – 8 900Å) for the observed spectra with

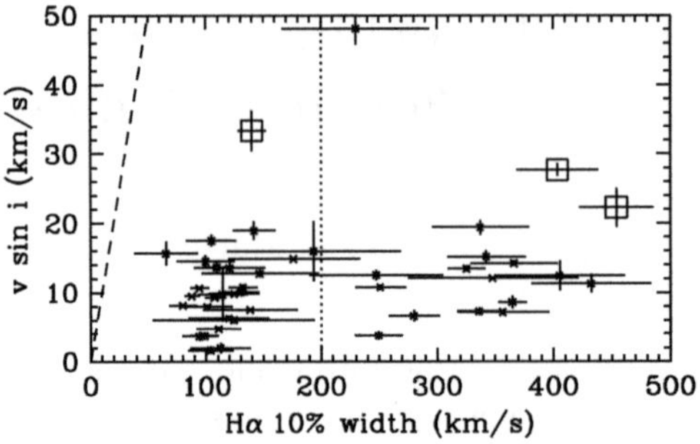

Fig. 1. Projected rotational velocity vs. Hα 10% width for a sample of T Tauri stars in the Cha I SFR. Weakly suspected binaries are plotted as squares. The error bars do not correspond to the measurement uncertainty, but to the scatter in our multi-epoch data. The adopted boundary between accretors and non-accretors and the intrinsic contribution of rotation to line width is shown by the dotted and dashed lines, respectively. Most stars are observed to be slow rotators.

artificially broadened template spectra of slowly rotating stars of similar spectral type and discarding the two highest and lowest estimates. We used the full width of Hα at 10% of the peak (hereafter, 10% width) as a diagnostic for accretion [8] and adopted a width of 200 km s^{-1} or greater as an indicator of on-going accretion for our analysis [4].

From our sample, 17 CTTS and 27 WTTS had consistent estimates in both projected rotational velocity and 10% width, and showed little sign of close-in companions which would spuriously inflate rotational velocity estimates. We plot the $v \sin i$ of these 44 stars against 10% width in Fig. 1 and the trend is roughly consistent with disk-locking theory. All but one of the presumed accreting and non-accreting single stars in this sample are slow rotators. The fast rotator is a non-accretor if we account for the intrinsic contribution of rotation to the line width. The large number of slow rotators is surprising and we have recently taken spectra of 17 additional T Tauri stars in the Chamaeleon I star forming region to augment our sample.

References

1. M. Camenzind: RvMA **3**, 234 (1990)
2. S. Edwards et al: AJ **106**, 372 (1993)
3. W. Herbst et al: A&A **396**, 513 (2002)
4. R. Jayawardhana, S. Mohanty, G. Basri: ApJ **592**, 282 (2003)
5. A. Königl: ApJ **370**, L39 (1991)
6. F. Shu et al: ApJ **429**, 781 (1994)
7. K.G. Stassun et al: AJ **117**, 2941 (1999)
8. R.J. White, G. Basri: ApJ **582**, 1109 (2003)

A Precision Radial Velocity Survey of Red Giants

Andrzej Niedzielski[1,2] and Alex Wolszczan[2,1]

[1] Centrum Astronomii, Uniwersytet Mikolaja Kopernika w Toruniu
aniedzi@astri.uni.torun.pl
[2] Pennsylvania State University alex@astro.psu.edu

Summary. We present our large, long-term, precision radial velocity (RV) survey of red giants (σ_{RV} $\leq$10 m/s), currently underway with the Hobby-Eberly Telescope.

1 Motivation and Observations

The extrasolar planet searches of the last decade have been mainly concerned with the low-mass, solar-type stars. Consequently, there have been very few detections of planets around the evolved descendants of higher-mass stars (e.g. [1] and references therein) and little has been known about planet formation and their frequency of occurence around stars of masses $\geq$1 $M_{\odot}$. As precision Doppler velocity surveys of post-MS stars with progenitor masses as high as ~5 $M_{\odot}$ are entirely feasible, detections of planets around GK-giants will almost certainly lead to a much more complete understanding of a relationship between stellar mass and the planet formation process. In addition, evolution of a star out of the MS can dramatically affect its planetary system. Planets within ~1 AU will be consumed by an expanding stellar envelope. An evolving giant may cause a dramatic evolution of planetary orbits. Our survey should provide enough planet detections around giants at different stages of evolution to enable statistical studies of the evolution of planetary systems from the time of formation almost all the way to the final white dwarf stage of their parent stars.

Our target list contains about 900 northern hemisphere stars with declinations between -10° and +75° and V magnitudes brighter than 12. On the HRD, they are located above the MS and between the instability strip and the coronal dividing line, either within the red giant "clump", or in the process of evolving toward it. Observations are conducted with the Hobby-Eberly Telescope (HET), a 9.2-meter telescope located at the McDonald Observatory,[3] equipped with the High Resolution Spectrograph (HRS), a single channel adaptation of the ESO UVES spectrometer [2]. Spectra are acquired with R=60,000 and a typical SNR of 150-250.

2 The Survey

Over the first 3 years of the survey we have collected multi-epoch observations for more than 300 targets. About 50 % of stars have been found to be relatively stable with RV variation amplitudes below 50 m/s, after 2-3 exposures. The other half of stars in the sample does show RV variations in excess of 50 m/s (10% of these are previously unknown binaries). The remaining 40% of the objects are stars showing intrinsic variability and, possibly, RV-signatures of orbiting planets. For most promising objects, we have made multi–epoch RV measurements. Some of them show correlated RV and line bisector changes indicating intrinsic (stellar activity) nature of the observed variability. For many others, however, no such correlation exists, which suggests a Keplerian origin of these variations. More observations of the planet candidate stars are needed to obtain unambiguous orbital solutions. Analysis of intrinsic RV rms scatter for stars, for which enough observations were gathered reveals that ~60% of stars show RV scatter ≤50 m/s and the scatter reaches a maximum at about 20 m/s in agreement with [4].

Acknowledgements

Kind cooperation of the Hobby-Eberly Telescope staff is acknowledged with thanks.

The Hobby-Eberly Telescope (HET) is a joint project of the University of Texas at Austin, the Pennsylvania State University, Stanford University, Ludwig-Maximilians-Universität München, and Georg-August-Universität Göttingen. The HET is named in honor of its principal benefactors, William P. Hobby and Robert E. Eberly.

This project is supported in part by Polish Ministry of Science and Education grant 1P03D 007 30.

References

1. A.P. Hatzes, E.W. Guenther, M. Endl, W.D. Cochran, M.P. Döllinger, A. Bedalov: A&A **437**, 743 (2005)
2. R.G. Tull: SPIE **3355** 387 (1998)
3. L.W. Ramsey, M.T. Adams, T.G. Barnes et al.: SPIE **3352**, 34 (1998)
4. D.S., Mitchell, S. Frink, A. Quirrenbach, D.A. Fischer: BAAS,**33**, 1440 (2001)

Chromospheric Lines as Diagnostics of Stellar Oscillations

Diane B. Paulson[1], W. Dean Pesnell[1], L. Drake Deming[1] and Martin Snow[2], Travis S. Metcalfe[3], Tom Woods[2], Brigette Hesman[1]

[1] NASA Goddard Space Flight Center, Greenbelt MD USA
[2] LASP, Boulder CO USA
[3] HAO/NCAR, Boulder CO USA

1 Background

Gravitational waves in the chromosphere, theorized as early as 1963 [10], are thoroughly explored in the more recent papers by [7, 8]. Theory predicts that the convective overshoot in the upper photosphere and low chromosphere will readily excite gravity waves. [9] note that these waves are not easily detected because of the long periods, short wavelengths required and slanted propagation angles of the waves themselves (causing small velocity shifts and short duration on individual detector pixels). Recently, [9] find evidence for gravity waves manifested in the 1700Å chromospheric line with frequencies <1 mHz.

Other types of waves which may be present globally are the result of a conversion of acoustic waves into longitudinal or transverse waves. These waves could be generated within the chromosphere by collisions of flux tubes with convective cells [1, 3], generated solely within convective regions [4], or be upwardly propagating p-modes interacting with the magnetic canopy [2]. The theoretical waves have frequencies which correspond almost precisely to observed frequencies in spatially resolved studies [5, 6].

The present study incorporates data from ground-based and space-based spectra of chromospherically sensitive lines in integrated sunlight. By using integrated light, we may be able to observe wave phenomena which occur on global scales.

2 Observations

Irradiance spectra of the Mg II h&k lines were obtained with the SOLSTICE spectrometer aboard the SORCE spacecraft. The irradiance was summed in 0.1 nm "filters" centered on the h & k lines as well as 0.1 and 0.2 nm filters centered on the inter-line (continuum) region. The resulting filter irradiances were analyzed using a Lomb-Scargle periodogram. While we initially presented a signal detection at 2.4 mHz, further inspection revealed an error in our software and ultimately the observations yielded no significant peaks above the 50% FAP level. There may be some indication of a weak signal around 1.5 mHz but this will require further observations for confirmation.

Ground-based observations of scattered-light (sky) in the Ca II H & K bands were taken using the McDonald Observatory cross dispersed coude spectrograph with R=60,000. For this analysis, we measured the line-depth-ratios (LDRs) of Ca II H & K cores relative to nearby Al I lines (with similar depth, but formed under different temperature and pressure conditions). Additionally, LDRs were measured against the portion of the H & K wings just outside of the chromospheric K_{2V} and H_{2V} grains which are formed slightly lower in the chromosphere than the core. LDRs are required as the scattering of daytime sky causes a filling-in of lines. Equivalent width (and therefore flux) measurements are adversely affected by this filling-in process, but LDR measurements remain mostly unaffected. Upon completion of these measurements, we found no significant signal in the peridogram analysis indicating oscillatory behavior. We presume that the lack of signal in these measurements is a result of inadequate comparison lines. LDRs work well for weaker lines or for H/K ratios, but it is likely that the precision of such measurements breaks down for these particular lines, where the cores are formed in vastly different environments than other nearby lines.

3 Conclusions

The data we have collected to-date has not resulted in the detection of global oscillations in the chromosphere. There is convincing evidence of the existence of chromospheric gravity waves [9] and these should be manifested in variable irradiance levels in chromospheric lines. Instruments such as SOLSTICE are ideal for this type of observing program, while similar measurements of the more accessible Ca II H&K lines will be equally as interesting.

References

1. Hasan, S. S. & Kalkofen, W. *ApJ*, **512**, 899 (1999)
2. Judge, P. G., Tarbell, T. D., & Wilhelm, K. *ApJ*, **554**, 424 (2001)
3. Kalkofen, W. *ApJ*, **486**, L145 (1997)
4. Lou, Y.-Q. *MNRAS*, **274**, L1 (1995)
5. McAteer, R. T. J. PhDT (2003)
6. McAteer, R. T. J., Gallagher, P. T., Bloomfield, D. S., Williams, D. R., Mathioudakis, M., & Keenan, F. P. *ApJ*, **602**, 436 (2004)
7. Mihalas, B. W. & Toomre, J. *ApJ*, **249**, 349 (1981)
8. Mihalas, B. W. & Toomre, J. *ApJ*, **263**, 386 (1982)
9. Rutten, R. J. & Krijger, J. M. *A&A*, **407**, 735 (2003)
10. Whitaker, W. *ApJ*, **137**, 914 (1963)

Comparing 3D Solar Model Atmospheres with Observations: Hydrogen Lines and Centre-to-limb Variations

Tiago M. D. Pereira[1], Martin Asplund[1], and Regner Trampedach[1]

Research School of Astronomy & Astrophysics, The Australian National University, Mount Stromlo Observatory, Cotter Road, Weston ACT 2611, Australia tiago@mso.anu.edu.au

Three dimensional hydrodynamical stellar model atmospheres represent a major step forward in stellar spectroscopy. Making use of radiative-hydrodynamical convection simulations that contain no adjustable free parameters, the model atmospheres provide a robust and realistic treatment of convection. These models have been applied to several lines in the Sun and other stars, yielding an excellent agreement with observations (*e.g.*, Asplund et al. (2000) [1]).

The present work aims to provide additional observational tests to the 3D solar model atmospheres: hydrogen lines and continuum centre-to-limb variations. The Hα and Hβ lines are good probes of the temperature of a model (in their wings, as they are mainly sensitive to temperature). The continuum limb-darkening profile (center-to-limb variations) is a useful tool to probe the atmospheric structure as a function of depth. By changing the viewpoint, one effectively probes for different depths in the atmosphere and continuum intensity yields information about the model's temperature structure.

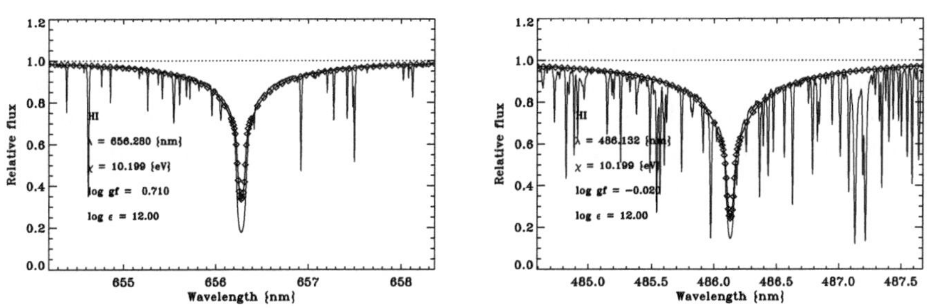

Fig. 1. The predicted spatially and temporally averaged flux profiles of the Hα and Hβ lines (left and right panel, respectively). Diamonds and thin line: 3D model; thick line: solar flux atlas of Kurucz et al. (1984) [2].

We present preliminary results for the normalized flux profiles of Hα, Hβ and for the centre-to-limb variations of the continuum disk-centre intensity. For these results, we used a 3D solar model atmosphere taken from a sequence

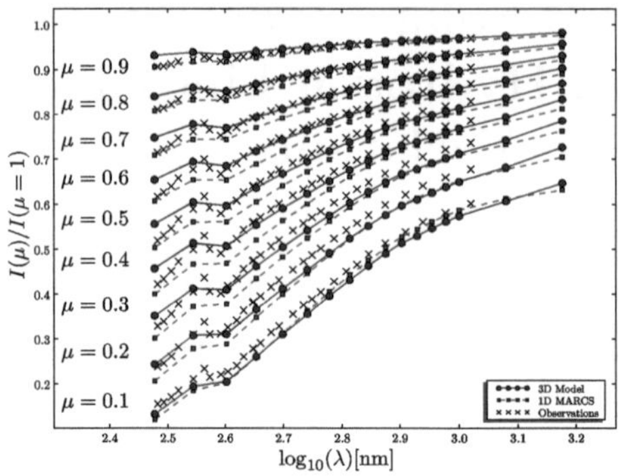

Fig. 2. The limb-darkening profile for the 3D solar simulation in the range of 3000-20000 Åcomputed for the continuum intensity. The μ parameter defines the angle of the viewpoint ($\mu = \cos\theta$). Solar observations data taken from Neckel & Labs (1994) [3].

of snapshots (covering 50 min of solar time) from the full solar convection simulation.

The results for Hα and Hβ are plotted in Fig. 1. A reasonable agreement can be found in the wings but not at the core (which is believed to be formed at chromospheric layers under non-LTE conditions). In the wings, the line profiles from the 3D model seem to be a bit stronger than the observed line profiles.

For the centre-to-limb variations, results are plotted in Fig. 2. While generally the agreement with observations seems to be better than for the 1D MARCS model, it is clear that there is a systematic difference between the 3D model and the observations, in the sense that the 3D has a slightly too steep temperature gradient.

Both these two preliminary results give us some hint that the temperatures and temperature structure of the 3D model might not be exactly the same that one observes in the Sun. Despite the excellent agreement of the 3D solar model atmospheres with many spectral lines, it is clear that there is still room for improvement. We are currently working on improving the radiative transfer in the 3D models.

References

1. Asplund, M., Nordlund, Å., Trampedach, R., Allende Prieto, C., & Stein, R. F. 2000, A&A, 359, 729
2. Kurucz, R. L., Furenlid, I., Brault, J., & Testerman, L. 1984, Solar flux atlas from 296 to 1300 nm (National Solar Observatory Atlas, Sunspot, New Mexico: National Solar Observatory, 1984)
3. Neckel, H. & Labs, D. 1994, Solar Physics, 153, 91

Towards the Detection of Reflected Light from Exo-planets: a Comparison of Two Methods

Florian Rodler[1,2] and Martin Kürster[1]

[1]Max-Planck-Institut für Astonomie, Königstuhl 17, 69117 Heidelberg, Germany
[2]Institut für Astronomie, Univ. Wien, Türkenschanzstr. 17, 1180 Wien, Austria
rodler@mpia.de

1 Introduction

For exo-planets the huge brightness contrast between the star and the planet constitutes an enormous challenge when attempting to observe some kind of direct signal from the planet. With high resolution spectroscopy in the visual one can exploit the fact that the spectrum reflected from the planet is essentially a copy of the rich stellar absorption line spectrum. This spectrum is shifted in wavelength according to the orbital RV of the planet and strongly scaled down in brightness by a factor of a few times 10^{-5}, and therefore deeply buried in the noise. The S/N of the plantetary signal can be increased by applying one of the following methods. The *Least Squares Deconvolution Method* (LSDM, eg. Collier Cameron et al. 2002) combines the observed spectral lines into a high S/N mean line profile (star + planet), determined by least-squares deconvolution of the observed spectrum with a template spectrum (from VALD, Kupka et al. 1999). Another approach is the *Data Synthesis Method* (DSM, eg. Charbonneau et al. 1999), a forward data modelling technique in which the planetary signal is modelled as a scaled-down and RV-shifted version of the stellar spectrum.

2 Simulation and Conclusion

We performed simulations for the F7V star τ Boo creating 900 stellar spectra with an average S/N of 350. We added the phase-shifted planetary spectrum, representing a planet fainter than the star by a factor of 40000. This factor is about twice as high as the lower limits of exo-planets found by other authors (eg. Leigh et al. 2003). A Venus-like phase function was used to calculate the illumination of the planet with respect to the orbital inclination i and phase. We chose $i = 45°$ resulting in an RV semi-amplitude of K=105 km/s in the case of τ Boo. We selected the simulated observations to cover the high illumination phases 0.3-0.45 and 0.55-0.7, but avoiding phases around 0.5 (upper conjunction) where the stellar and the planetary spectra blend.

We evaluated models for the data with the two mentioned approaches and searched for those pairs of K and contrast which yield the smallest χ^2. As shown in Fig. 1, the DSM recovers the planet with the correct parameters, but the LSDM fails to retrieve the planetary signal. We determined the significance of the detection via bootstrap simulation (eg. Kürster et al. 1997). In each of 1000 trial runs, we retained the phase values, but randomly redistributed the spectra. This bootstrap analysis reveiled no other detection with a χ^2 smaller or equal than the χ^2 for the original data set. So we conclude that the detection is significant at a level of $\geq$99.9% confidence.

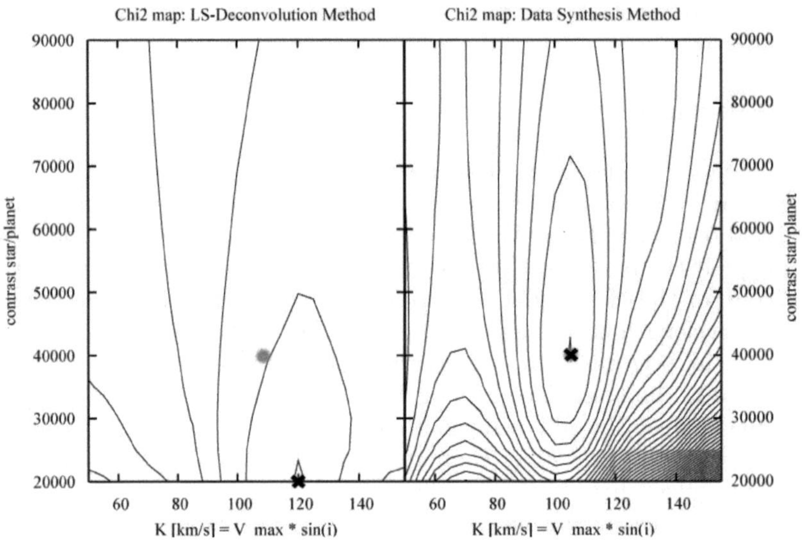

Fig. 1. Contour maps showing χ^2 as a function of the RV semi-amplitude K and the star-to-planet contrast for the two different methods. These χ^2 maps clearly reveal that the DSM recovers the planetary signal from the simulated spectra, but not so the LSDM. The correct position of the planet in the maps is: K=105, contrast = 40000. The input value is depicted by the grey circles, and the values reconstructed are represented by the x's.

References

1. Charbonneau D., Noyes R.W., Korzennik P.N., et al., 1999, ApJ, 522, L145
2. Collier Cameron A., Horne K., Penny A., Leigh C., 2002, MNRAS, 330, 160
3. Kupka F., Ryabchikova T.A., Piskunov N.R., et al., 1999, A&A, 138, 119
4. Kürster M., Schmitt J.H.M.M., Cutispoto G., et al., 1997, A&A, 320, 831
5. Leigh C., Collier Cameron A., Horne K., et al., 2003, MNRAS, 344, 1271

A Correlation Between the Activity Level and the Radial-velocity for Solar-type Stars?*

N.C. Santos[1,2,3], C. Melo[4], C. Lovis[3], and M. Billéres[4]

[1] Centro de Astronomia e Astrofísica da Universidade de Lisboa, Portugal
`nuno.santos@oal.ul.pt`
[2] Centro de Geofísica de Évora, Portugal
[3] Observatoire de Genève, Switzerland
[4] European Southern Observatory, Chile

Summary. In this paper we present preliminary results suggesting the existence of a correlation between the systemic radial velocity of solar-type stars and their activity level.

1 Introduction

Phenomena such as stellar pulsations and oscillations [2], inhomogeneous convection or spots [5] can prevent us from finding planets or even give us false candidates if they produce a stable periodic signal in the radial-velocity measurements (e.g. a rotating spot) [4]. These effects are important when dealing with short period signals, with timescales of the order of the stellar rotational period. For long period planets, however, an important obstacle to their detection may be the existence of stellar magnetic cycles with similar timescales to the 11 year solar magnetic cycle [1]. Whether such cycles can induce radial-velocity variations has never been tested in solar-type FGK dwarfs. The only studies up to now were done for one M-dwarf [3].

Here we present the first results of a long term project to measure simultaneously radial-velocities, stellar activity indicators, and cross-correlation function (CCF) bisectors for a set of FGK stars. The goals are to understand if stellar magnetic cycles may induce periodic radial-velocity variations, and if it is possible to distinguish these putative signals from the ones induced by real long-period planetary candidates.

2 Observations and Preliminary Results

In order to address the above issue, we took a set of 7 FGK stars known to have well defined magnetic activity solar-type cycles. These stars were chosen from the study of Baliunas et al. [1], and are being followed using the HARPS spectrograph, at the 3.6-m ESO telescope (ESO, La Silla).

*Based on observations collected using the HARPS spectrograph (ESO 3.6-m telescope), runs ID 072.C-0096, 073.D-0038, 074.D-0131, 075.D-0194, 076.D-0130.

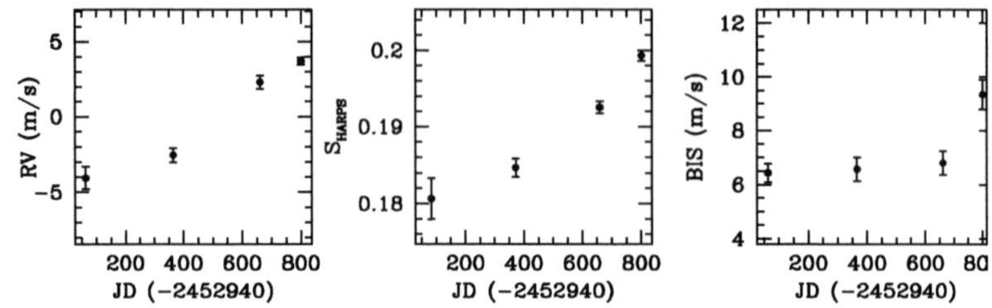

Fig. 1. Radial-velocities, S-index and bisector measurements for one star in our sample. See text for more details.

The spectra are used to derive i) accurate radial-velocities (at the $1\,\mathrm{m\,s^{-1}}$ level), ii) measurements of the bisector of the CCF [4], and iii) a chromospheric activity index similar to the Mount Wilson "S", based on the flux in the center of the CaII H and K lines [5].

These measurements will give the chance to determine the level to which the solar-type magnetic cycles in other stars might influence the radial-velocity measurements down to a level of a few m/s. This will also enable us to verify if we are able to distinguish magnetic-cycle induced signals from real long period planetary candidates.

If Fig.1 we show the radial velocities (left panel), chromospheric activity index (middle panel), and bisector measurements (right panel) for one of the stars in our sample. Each point in the figure represents the average of one season of measurements. The error bars represent the rms of the N measurements divided by $\sqrt{N}$. The results show a clear correlation between the radial-velocity and the activity level of the star. A weaker correlation may also be present between the radial-velocity and the bisector measurements. More data is needed to confirm whether the observed trends in RV are due to the presence of unknown companions or to the effects of the varying chromospheric activity level of the star. The results of this program may have a strong impact on current (e.g. HARPS) and future radial-velocity planet search projects (e.g. CODEX and ESPRESSO on the ESO ELT telescope).

N.C.S. would like to thank the support from FCT (Portugal), in the form of a fellowship (SFRH/BPD/8116/2002) and a grant (POCI/CTE-AST/56453/2004).

References

1. S. Baliunas, R.A. Donahue, W. Soon et al.: ApJ **438**, 269 (1995)
2. F. Bouchy, M. Bazot, N.C. Santos et al.: A&A **440**, 609 (2005)
3. M. Kürster, M. Endl, F. Rouesnel et al.: A&A **403**, 1077 (2003)
4. D. Queloz, G.W. Henry, J.-P. Sivan et al.: A&A **379**, 279 (2001)
5. N.C. Santos, M. Mayor, D. Naef et al.:A&A **361**, 265 (2000)

Spectroscopic Parameters for a Sample of Metal-rich Solar-type Stars

S.G. Sousa[1,2,6], N.C. Santos[1,3,4], G. Israelian[5], M. Mayor[3], and M.J.P.F.G. Monteiro[2,6]

[1] CAAUL, Observatório Astronómico de Lisboa, Tapada da Ajuda, 1349-018 Lisboa sousasag@oal.ul.pt
[2] CAUP, Rua das Estrelas, 4150-762 Porto Portugal
[3] Observatoire de Genève, 51 Ch. des Mailletes, 1290 Sauverny, Switzerland
[4] Centro de Geofísica de Évora, Colégio Luis Antonio Verney, 7002-554 Évora, Portugal
[5] Instituto de Astrofísica de Canarias, 38200 La Laguna, Tenerife, Spain
[6] DMA, Faculdade de Ciências da Universidade do Porto, Portugal.

Summary. We present stellar parameters and metallicities for a sample of 64 high metal content stars not known to harbor any planet. This sample provides the reference for investigating new correlations between stars and the existence of an orbiting planet.

1 Motivation

There are around 170 known planet host stars at the moment. One well established characteristic of these stars is that they are very metal-rich when compared with "single" stars, i.e. stars not hosting any planet [5, 6].

The main results of a uniform study, concerning the metallicity of planet host stars, have been presented in Santos et al. 2004, 2005 [7, 8]. Left panel of Fig. 1 shows the metallicity distribution of the planet host sample versus the comparision sample presented in Santos et al. 2005 [8]. The difference between the average metallicity of the two samples is 0.24 dex. As we can see from the figure, there is a reduced number of comparison stars, without known planets, with $[Fe/H] \geq 0.2$.

This sample will then be used for element abundance analysis to improve previous works in this field [2, 1, 3, 4]. With this high metal content sample we can search for new possible correlations between planet host and "single" stars by comparing abundances relative to iron for several other elements in the high [Fe/H] regime.

2 A New Sample of Metal-rich Stars

The stars in this new sample belong to the CORALIE southern planet search program [10], have spectral types between F and K, and were not announced

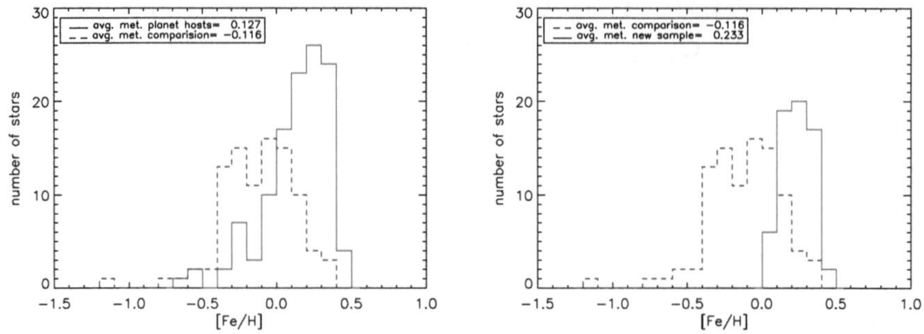

Fig. 1. Left panel: Planet host stars (full line) vs. Comparison Sample (dashed line). Data from Santos et al. 2005 [8]; Right panel: Previous Comparison Sample (dashed line) vs. New Comparison Sample (full line). Data from Sousa et al. 2006 [9]

to harbor any planetary companion. Moreover we chose the ones that had [Fe/H] ≥ 0.1 determined using the CORALIE cross-correlation function (see e.g. Santos et al. 2004 [7]).

Stellar parameters and metallicities were derived as in Santos et al. 2004 [7], based on the EWs of Fe I and Fe II weak lines, and by imposing excitation and ionization equilibrium. The list of the stellar parameters for this sample is presented in Sousa et al. 2006 [9].

In the right panel of Fig. 1 we show the metallicity distribution of the previous comparison sample versus the new comparison sample presented here.

References

1. Beirão, P., Santos, N.C., Israelian, G. and Mayor, M.: A&A 438, 251, 2005
2. Bodaghee, A., Santos, N.C., Israelian, G. and Mayor, M.: A&A 404, 715, 2003
3. Ecuvillon, A., Israelian, G., Santos, N.C., Shchukina, N.G., Mayor, M. and Rebolo, R.: A&A 445, 633, 2006
4. Gilli, G., Israelian, G., Ecuvillon, A., Santos, N.C. and Mayor, M.: A&A 449, 723, 2006
5. Gonzalez, G.: MNRAS 285, 403, 1997
6. Santos, N.C., Israelian, G. and Mayor, M.: A&A 373, 1019, 2001
7. Santos, N.C., Israelian, G. and Mayor, M.: A&A 415, 1153, 2004
8. Santos, N.C., Israelian, G., Mayor, M., Bento, J.P., Almeida, P.C., Sousa, S.G. and Ecuvillon, A.: A&A 437, 1127, 2005
9. Sousa, S.G., Santos, N.C., Israelian, G., Mayor, M. and Monteiro, M.J.P.F.G.: A&A 458, 873, 2006
10. Udry, S., Mayor, M., Naef, D., Pepe, F., Queloz, D., Santos, N.C., Burnet, M., Confino, B. and Melo, C.: A&A 356, 590, 2000

Radial Velocity Search for Extrasolar Planets in Binary Systems

E. Toyota[1] and Y. Itoh[1], S. Ishiguma[1], D. Murata[1], Y. Oasa[1], B. Sato[2], T. Mukai[1]

[1] Graduate School of Science and Technology, Kobe University, 1-1 Rokkodai-cho, Nada, Kobe, Hyogo, 657-8501, Japan toyota@kobe-u.ac.jp
[2] Okayama Astrophysical Observatory, National Astronomical Observatory of Japan, Honjo, Kamogata, Asakuchi, Okayama, 719-0232, Japan

Summary. We have started a search for extrasolar planets in visual binary systems from 2003. We monitor them by precise radial velocity measurements, using HIgh Dispersion Echelle Spectrograph(HIDES) equipped on the Okayama Astrophysical Observatory's(OAO) 188 cm telescope. Radial velocity precision of better than 10 m/s has achieved with an iodine absorption cell during our observational span. We here report the current status of the survey.

1 Introduction

Multiple star systems are common in the solar neighborhood(1). Nevertheless, among more than 200 extrasolar planets discovered to date, it is only about 40 planets that have been discovered in binary or multiple star systems(2)(3)(4)(5). It has been considered that a planet cannot be dynamically stable in a binary system. However, recent theoretical study predicts that in the case of a binary system with large semi-major axis, a planet survives for a long time(6).
From recent obsrvational results, there are marginal differences in the mass–period or eccentricity–period distributions between planets of single stars and planets of binaries(2)(4)(7). These differences may suggest that planets found around single stars and in binary systems have different formation and evolution mechanisms.

2 Observation and Our Target Stars

We have started search for extrasolar planets in binary systems by precise Doppler shift measurements from 2003, using HIDES equipped on the OAO's 188cm reflector telescope. We obtain the spectra with a wavelength resolution of R=70000 for the slit width of 0.76 arcsec. We have monitored stadard star known to be stable in radial velocity(RV)(e.g.(8)). We achieved a RV precision of better than 10 m/s during our observational span(Fig.1).

Our target stars are late F – early K type visual binary systems with semi-major axis larger than 100 AU and brighter than 7 mag at the V-band.

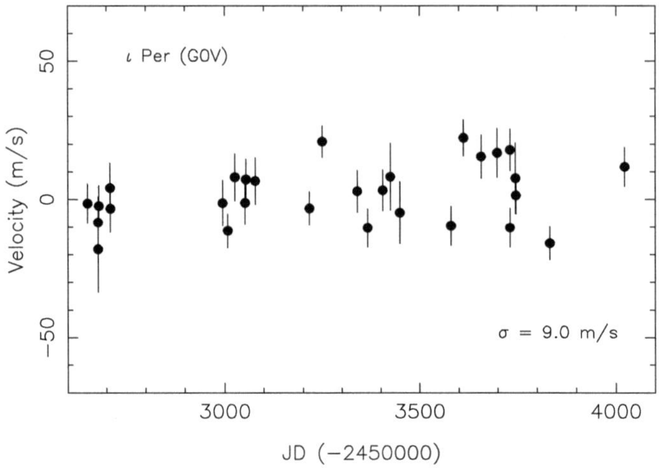

Fig. 1. Radial velocity of ιPer.

We have currently monitored RV variations 8 pairs of binary systems and 3 primary stars every 1 or 2 months. The RV dispersions of 7 objects are smaller than 40 m/s, and have not been detected any periodic variations. 9 of 19 stars are turned out to be SB1s or SB2s, namely they are triple star systems. 3 objects show RV dispersions larger than 90 m/s. One out of three has a possibility that it has a long term RV variations. For these stars, we will investigate the cause of the RV variations in detail such as CaIIH and K lines to check the stellar activity, and spectral line bisectors to measure the variation of absorption line shapes.

References

1. A. Duquennoy, & M. Mayor:A&A, **248**, 485 (1991)
2. A. Eggenberger, S. Udry, M. Mayor:A&A, **417**, 353 (2004)
3. D. Raghavan, & T. J. Henry, B. D. Mason et al.:ApJ, **646**,523 (2006)
4. S. Desidera, M. Barbieri:A&A accepted, (2006)
5. G. Chauvin, A. M. Lagrange, S. Udry et al.:A&A, **456** 1165 (2006)
6. M. J. Holman, and P. A. Wiegert:AJ, **17**, 621 (1999)
7. S. Zucker, and T. Mazeh:ApJ, **568**, 113 (2002)
8. D. A. Fischer, G. W. Marcy, R. P. Butler et al.:ApJ, **551**, 1107 (2001)

Inferring Photospheric Velocities from P Cygni Lines in Type IIP Supernova Atmospheres

József Vinkó[1]

Department of Optics & Quantum Electronics, University of Szeged, Hungary
vinko@physx.u-szeged.hu

1 Introduction

The precise measurement of the expansion velocities at the photosphere of supernova atmospheres is very important, because *i*) they can be used to constrain theoretical models, and *ii*) distance measurement methods (e.g. the Expanding Photosphere Method) rely heavily on observed expansion velocities.

However, the line formation in a SN ejecta is different from that in normal stellar atmospheres. When the inner part of the ejecta is optically thick (i.e. in the photospheric phase), the lines are mostly formed above the photosphere via resonant scattering. If the expansion is homologous ($v \sim r$) and the velocity gradient is large, the observed flux can be derived via the Sobolev-approximation (e.g. [1]).

The emergent spectrum contains classical P Cygni-lines consisting of an emission component centered on rest wavelength and a blueshifted absorption component. It can be shown that the absorption minimum of optically thin lines corresponds to the Doppler-shift of the photospheric expansion velocity.

2 Model Computations

We have implemented this approximation in the computer code *SOBOLEV*, similar to the widely-used supernova code *SYNOW* ([2]). As in *SYNOW*, the line strengths are computed from Boltzmann-excitation relative to a reference line of the particular ion. The optical depths of the reference lines at the photosphere are chosen to match the observed spectra. *SOBOLEV* is written in C and the user has the capability of scaling the source function of the given line. This is particularly useful for fitting the observed hydrogen line profiles ([3]).

We have applied *SOBOLEV* for modeling spectra of Type IIP (plateau) SNe and compared the model expansion velocities with those derived from spectral lines. Four sets of models have been defined corresponding to +5, +15, +50 and +100 days after explosion. The model parameters have been adjusted to fit real SNe spectra observed in similar phases. For each epoch, the expansion velocity is varied by ±50 % around its mean value.

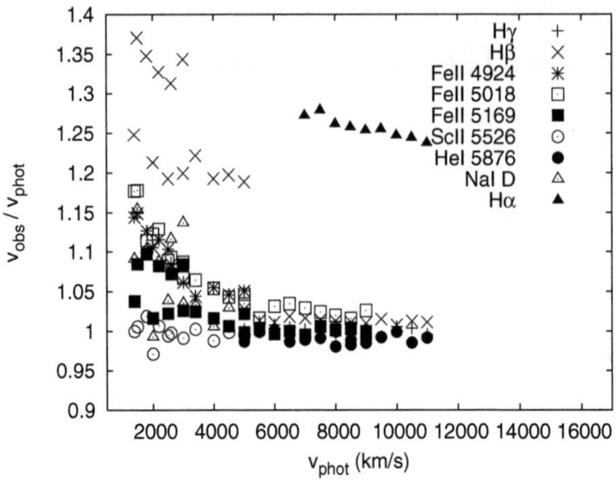

Fig. 1. Ratio of observed and model velocities as function of the photospheric velocity.

3 Results

The expansion velocities have been derived for a selected linelist (see legend in Fig.1) from the absorption minimum of the P Cygni profiles. Fig.1 shows the ratio of the measured and the model velocities. In ideal case, these should be identical. However, Fig.1 shows that strong lines ($H\alpha$, $H\beta$) give excess velocities of $\sim 20 - 30$ % ([4]). Ionized metals (FeII, ScII) are usually better velocity indicators, but these are visible only at later phases.

In the early ($\sim +5$ days) phases the best velocity indicator is $H\gamma$, but the observed spectra often do not sample this spectral region. Alternatively, HeI λ5875 also gives good velocity estimates. Later, the FeII λ5169 line (which is usually applied in SN studies for computing photospheric velocities) results in slightly higher velocities when $v_{ph} < 4000$ km s^{-1}. In the middle- and late plateau phases the ScII λ5527 feature gives a better estimate, at least in our models. However, this line may be blended by FeII and other unidentified lines in real SN spectra.

This research was supported by Hungarian OTKA Grants No. TS 049872 and T042509.

References

1. D. Kasen, D. Branch, E. Baron, D. Jeffery: ApJ **565**, 380 (2002)
2. A. Fisher, A.: PhD Thesis, University of Oklahoma (2000)
3. J. Vinkó et al.: MNRAS **369**. 1780 (2006)
4. L. Dessart & D. J. Hillier: A&A **439**, 617 (2005)

High-resolution Spectroscopic Characterization of Young Stars

Patrick Weise[1], Johny Setiawan[1], Thomas Henning[1] and André Müller[1,2]

[1] Max-Planck-Institut für Astronomie Heidelberg, Germany
weise@mpia.de
[2] Department of Physics, University of Jena, Germany

Summary. We present the first results of our spectroscopic measurements of nearby young stars with ages between few Myrs until few hundreds Myrs. Our sample consists of 98 stars in nearby young associations and 54 field young stars. We measured the Lithium 6708 Å equivalent width, which is a good age indicator for our targets. Furthermore, we measured stellar radial velocities by using a cross-correlation technique. From the equivalent widths of Fe and other metallic lines, we determined the spectral type, effective temperature, surface gravity ($\log g$) and metallicity [Fe/H]. Based on our radial velocity measurements and the shape of the line profiles we detected stars in binary/multiple systems. At least a binarity fraction of 45% of the young stars in our sample has been found. From the cross-correlation functions we computed the projected rotational velocity. We found that around 57% of young stars among our sample have $v \sin i$ lower than 20 km/s.

– **Spectral type:**

Following Stock & Stock (1999), we used CaI, CrI, NaI, FeI, Hγ and Hγ as neutral elements as well as FeII as an ionized element to determine the spectral type. We avoided the use of Hα and CaII H&K because of strong emission features in these lines due to the stellar activity. The equivalent width (EW) has been measured and calibrated on a set of spectral standard stars with near solar metallicity (Cayrel de Strobel 2001). This method is reliable for spectral classes F until K5. For the spectral type M the TiO index can be applied for the determination (Allen 1996).
The mean error of this method is 0.25 in subclasses within the spectral type. Our measured spectral types are in good agreement with known literature values.

– **$\mathbf{T_{eff}}$, $\mathbf{\log g}$, [Fe/H]:**

We used the EW of 40 FeI and 6 FeII lines and computed the T_{eff}, $\log g$ and [Fe/H] values for the measured EW by using the iteration procedure *TGV* by Takeda et al. (2002). The lines are chosen by their excitation potential χ and known $\log gf$ values. In addition, they should be unblended.
With this method we derived the solar T_{eff} as 5765±48 K and the $\log g$ as 4.623±0.135 [cm/s^2]. Therefore, this method seems to be reliable for the stars of our sample.

– **Projected rotational velocity** $v \sin i$**:**
Rotational velocity can be used as an indicator of stellar age, with young stars being fast rotators compared to older stars (Zuckerman & Song 2004). But even the projected rotational velocity is also a good indicator for stellar activity and can be used to estimate other stellar parameters, like the maximum rotational period (P/sin i) and starspots coverage. We used the cross-correlation technique to determine $v \sin i$ (Queloz et al. 1998). To our surprise, we found that 57% of our targets show $v \sin i$ lower than 20 km/s. Thus, these particular stars are appropriate for planet search programs via radial velocity technique.

– **LiI as an age indicator:**
The Lithium absorption line at $\lambda = 6707.8$ Å can be used as an indicator for the youth of low-mass stars (e.g., Wichmann et al. 2003). Lithium is depleted in low-mass stars after about 100 Myrs (for F-, G-type) and much later for M-type. By comparing our measurements to the EW(LiI) of other stars of known age within the same spectral type, we found that the majority of our targets are younger or at the age of the Pleiades.

– **Stellar activity index:**
The CaII K&H emission lines are also good stellar activity indicators, because they are sensitive to magnetic activity in the stellar chromosphere (Baliunas et al. 1995). We found correlation between the activity and ages within our sample. However, we found no correlation between $v \sin i$ and the activity index, which may be caused by the uncertainty of the inclination angle.

– **Stellar multiplicity:**
Spectroscopic binaries were identified by the line profile (double lined SB) or by RV measurements. We marked a target as single-lined SB, when the long-term RV scatter is higher than 400 ms^{-1} (below this, it can also be induced by a substellar companion or rotational modulation). We found that ≈45% of our sample are binaries. For field stars, we obtain a binarity fraction of 26%, whereas for the stars in young associations we have 55% binarity fraction.

References

1. Allen, Ph.D. Thesis, University of Massachusetts, Massachusetts (1996)
2. Baliunas et al.: ApJ **438**, 269 (1995)
3. Cayrel de Strobel et al.: A&A **373**, 159 (2001)
4. Queloz et al.: A&A **335**, 183 (1998)
5. Stock & Stock: RMxAA **35**, 143 (1999)
6. Takeda et al.: PASJ **54**, 451 (2002)
7. Wichmann et al.: A&A **399**, 983 (2003)
8. Zuckerman & Song: ARAA **42**, 685 (2004)

TIRAVEL – Template Independent RAdial VELocity Measurement

Shay Zucker[1] and Tsevi Mazeh[2]

[1] Dept. of Geophysics & Planetary Sciences, Raymond and Beverly Sackler Faculty of Exact Sciences, Tel Aviv University, Tel Aviv 69978, Israel shayz@post.tau.ac.il
[2] School of Physics & Astronomy, Raymond and Beverly Sackler Faculty of Exact Sciences, Tel Aviv University, Tel Aviv 69978, Israel mazeh@post.tau.ac.il

1 Introduction

The popular way to measure astronomical radial velocities is by cross-correlating the spectrum of the studied object with a template spectrum (e.g., [1]). Some studies use an extensive grid of theoretical spectra and cross-correlate the observed spectra against the templates in the grid, looking for the best correlation [2]. However, imperfections in the correspondence between the template and the true nature of the object may lead to sub-optimal precision and systematic errors. In this work we present a new approach – TIRAVEL – which does not use a predetermined template. Instead, it finds the set of relative radial velocities that maximizes the overall alignment of all observed spectra. A more detailed presentation of TIRAVEL is presented in [3].

2 TIRAVEL

TIRAVEL starts by calculating the cross-correlation functions of all pairs of observed spectra:

$$R_{ij}(s_i - s_j) = \frac{1}{N} \sum_n f_i(n - s_i) f_j(n - s_j) \ .$$

Thus, for each trial set of K Doppler shifts $(s_1, s_2, \cdots, s_K)$, we get a $K \times K$ correlation matrix. TIRAVEL uses the largest eigenvalue of this matrix, λ_M, as a measure of the overall alignment among the K spectra, and searches for the set of shifts which maximizes this measure. The theory of Principal Component Analysis provides the rationale: when the alignment is perfect, there is a single principal component with which we can generate all the observed spectra.

3 Real Test Cases

In [3] we have tested TIRAVEL on single-lined spectroscopic binaries which were already solved and published by Latham et al. [2]. The original published solutions used velocities obtained by cross-correlating the observed spectra of each binary against a template chosen from an extensive grid of synthetic spectra. We ran TIRAVEL on each set of spectra and fit orbital models to the TIRAVEL velocities. We present here the results for three cases that demonstrate the potential of TIRAVEL. Table 1 below shows the RMS of the residuals from the orbital fits. The first two stars show a marginal difference, while the third one (G48–54) seems to have benefited considerably from the use of TIRAVEL, and its residual RMS has reduced by 26%. Therefore, the first two cases show that the performance of TIRAVEL is at least as good as cross-correlating against a perfect template. The third case was the latest-type star in the survey by Latham et al., with a template temperature of 3750K. It is well known that atmospheres of such low temperatures are not modeled as well as those of the earlier-type stars. This is probably the reason TIRAVEL obtains a better solution than the synthetic template.

Table 1. Comparing TIRAVEL with previously published solutions

Star	Published RMS [$\mathrm{km\,s^{-1}}$]	TIRAVEL RMS [$\mathrm{km\,s^{-1}}$]
G72–59	0.94	0.96
G178–27	0.61	0.56
G48–54	1.20	0.89

4 Summary

We have shown that TIRAVEL performs at least as good as the conventional cross-correlation technique. Its advantages are most pronounced when a satisfactory template is unavailable. By improving the precision of radial velocity measurements, TIRAVEL will hopefully contribute to the search for extrasolar planets, in particular around late-type stars.

References

1. J. Tonry, M. Davis: AJ **84**, 1511 (1979)
2. D.W. Latham et al: AJ **124**, 1144 (2002)
3. S. Zucker, T. Mazeh: MNRAS **371**, 1513 (2006)

ESO ASTROPHYSICS SYMPOSIA
European Southern Observatory

Series Editor: Bruno Leibundgut

L. Kaper, E.P.J. van den Heuvel, P.A. Woudt (Eds.),
Black Holes in Binaries and Galactic Nuclei: Diagnostics, Demography and Formation
Proceedings, 1999. XXIII, 378 pages. 2001.

G. Setti, J.-P. Swings (Eds.)
Quasars, AGNs and Related Research Across 2000
Proceedings, 2000. XVII, 220 pages. 2001.

A.J. Banday, S. Zaroubi, M. Bartelmann (Eds.),
Mining the Sky
Proceedings, 2000. XV, 705 pages. 2001.

E. Costa, F. Frontera, J. Hjorth (Eds.),
Gamma-Ray Bursts in the Afterglow Era
Proceedings, 2000. XIX, 459 pages. 2001.

S. Cristiani, A. Renzini, R.E. Williams (Eds.),
Deep Fields
Proceedings, 2000. XXVI, 379 pages. 2001.

J.F. Alves, M.J. McCaughrean (Eds.),
The Origins of Stars and Planets: The VLT View
Proceedings, 2001. XXVII, 515 pages. 2002.

J. Bergeron, G. Monnet (Eds.),
Scientific Drivers for ESO Future VLT/VLTI Instrumentation
Proceedings, 2001. XVII, 356 pages. 2002.

M. Gilfanov, R. Sunyaev, E. Churazov (Eds.),
Lighthouses of the Universe: The Most Luminous Celestial Objects and Their Use for Cosmology
Proceedings, 2001. XIV, 618 pages. 2002.

R. Bender, A. Renzini (Eds.),
The Mass of Galaxies at Low and High Redshift
Proceedings, 2001. XXII, 363 pages. 2003.

W. Hillebrandt, B. Leibundgut (Eds.),
From Twilight to Highlight: The Physics of Supernovae
Proceedings, 2002. XVII, 414 pages. 2003.

P.A. Shaver, L. DiLella, A. Giménez (Eds.),
Astronomy, Cosmology and Fundamental Physics
Proceedings, 2002. XXI, 501 pages. 2003.

M. Kissler-Patig (Ed.),
Extragalactic Globular Cluster Systems
Proceedings, 2002. XVI, 356 pages. 2003.

P.J. Quinn, K.M. Górski (Eds.),
Toward an International Virtual Observatory
Proceedings, 2002. XXVII, 341 pages. 2004.

W. Brander, M. Kasper (Eds.),
Science with Adaptive Optics
Proceedings, 2003. XX, 387 pages. 2005.

A. Merloni, S. Nayakshin, R.A. Sunyaev (Eds.),
Growing Black Holes: Accretion in a Cosmological Context
Proceedings, 2004. XIV, 506 pages. 2005.

L. Stanghellini, J.R. Walsh, N.G. Douglas (Eds.)
Planetary Nebulae Beyond the Milky Way
Proceedings, 2004. XVI, 372 pages. 2006.

S. Randich, L. Pasquini (Eds.)
Chemical Abundances and Mixing in Stars in the Milky Way and its Satellites
Proceedings, 2004. XXIV, 411 pages. 2006.

A.P. Lobanov, J.A. Zensus, C. Cesarsky, P.J. Diamond (Eds.)
Exploring the Cosmic Frontier
Astrophysical Instruments for the 21st Century.
XXVI, 270 pages. 2006.

I. Saviane, V.D. Ivanov, J. Borissova (Eds.)
Groups of Galaxies in the Nearby Universe
Proceedings, 2005. XX, 390 pages. 2007.

G.W. Pratt, P. Schuecker, H. Boehringer (Eds.)
Heating versus Cooling in Galaxies and Clusters of Galaxies
Proceedings, 2006. XIV, 468 pages. 2007.

M. Kissler-Patig, R.J. Walsh, M.M. Roth (Eds.)
Science Perspectives for 3D Spectroscopy
Proceedings of the ESO Workshop held in Garching, Germany, 10-14 October 2005
Proceedings, 2006. XVIII, 402 pages. 2007.

S. Hubrig, M. Petr-Gotzens, A. Tokovinin (Eds.)
Multiple Stars Across the H-R Diagram
Proceedings of the ESO Workshop held in Garching, Germany, 12-15 July 2005
Proceedings, 2006. XIV, 304 pages. 2007.

N.C. Santos, L. Pasquini, A.C.M. Correia, M. Romaniello (Eds.)
Precision Spectroscopy in Astrophysics
Proceedings of the ESO/Lisbon/Aveiro Conference held in Aveiro, Portugal, 11-15 September 2006
Proceedings, 2006. XXII, 330 pages. 2008.

MIX
Papier aus verantwortungsvollen Quellen
Paper from responsible sources
FSC® C105338

If you have any concerns about our products,
you can contact us on
ProductSafety@springernature.com

In case Publisher is established outside the EU,
the EU authorized representative is:
Springer Nature Customer Service Center GmbH
Europaplatz 3, 69115 Heidelberg, Germany

Printed by Libri Plureos GmbH
in Hamburg, Germany